Craftsman

2002 NATIONAL BUILDING COST MANUAL

Twenty-Sixth Edition

Edited by

Dave Ogershok

Craftsman Book Company

6058 Corte del Cedro / P.O. Box 6500 / Carlsbad, CA 92018

Illustrations by Laura Knight

ISBN 1-57218-107-9

Contents of This Manual

Explanation of the Cost Tables

This manual provides construction or replacement costs for a wide variety of residential, commercial, industrial, agricultural and military buildings. For your convenience and to avoid possible errors, all the cost and reference information you need for each building type is listed with the primary cost figures for that building. After reading this and the following two pages you should be able to turn directly to any building type and make an error-free estimate or appraisal.

The costs are per square foot of floor area for the basic building and additional costs for optional or extra components that differ from building to building. Building shape, floor area, design elements, materials used, and overall quality influence the basic structure cost. These and other cost variables are isolated for the building types. Components included in the basic square foot cost are listed with each building type. Instructions for using the basic building costs are included above the cost tables. These instructions include a list of components that may have to be added to the basic cost to find the total cost for your structure.

The figures in this manual are intended to reflect the amount that would be paid by the end user of a building as of mid 2002.

They show the total construction cost including all design fees, permits, and the builder's supervision, overhead, and profit. These figures do not include land value, site development costs, or the cost of modifying unusual soil conditions or grades.

Building Quality

Structures vary widely in quality and the quality of construction is the most significant variable in the finished cost. For estimating purposes the structure should be placed in one or more quality classes. These classes are numbered from 1 which is the highest quality generally encountered. Each section of this manual has a page describing typical specifications which define the quality class. Each number class has been assigned a word description (such as best, good, average or low) for convenience and to help avoid possible errors.

The quality specifications do not reflect some design features and construction details that can make a building both more desirable and more costly. When substantially more than basic design elements are present, and when these elements add significantly to the cost, it is appropriate to classify the quality of the building as higher than would be warranted by the materials used in construction.

Many structures do not fall into a single class and have features of two quality classes. The tables have "half classes" which apply to structures which have some features of one class and some features of a higher or lower class. Classify a building into a "half class" when the quality elements are fairly evenly divided between two classes. Generally quality elements do not vary widely in a single building. For example, it would be unusual to find a top quality single family residence with minimum quality roof cover. The most weight should be given to quality elements that have the greatest cost. For example, the type of wall and roof framing or the quality of interior finish are more significant than the roof cover or bathroom wall finish. Careful evaluation may determine that certain structures fall into two distinct classes. In this case the cost of each part of the building should be evaluated separately.

Building Shapes

Shape classification considers any cost differences that arise from variations in building outline. Shape classification considerations vary somewhat with different building types. Where the building shape often varies widely between buildings and shape has a significant effect on the building cost, basic building costs are given for several shapes. Use the table that most closely matches the shape of the building you are evaluating. If the shape falls near the division between two basic building cost tables, it is appropriate to average the square foot cost from those two tables.

Area of Buildings

The basic building cost tables reflect the fact that larger buildings generally cost less per square foot than smaller buildings. The cost tables are based on square foot areas which include the following:

1. All floor area within and including the exterior walls of the main building.
2. Inset areas such as vestibules, entrances or porches outside of the exterior wall but under the main roof.
3. Any enclosed additions, annexes or lean-tos with a square foot cost greater than three-fourths of the square foot cost of the main building.

Select the basic building cost listed below the area which falls closest to the actual area of your building. If the area of your building falls nearly mid-way between two listed building areas, it is appropriate to average the square foot costs for the listed areas.

Wall Heights

Building costs are based on the wall heights given in the instructions for each building cost table. Wall height for the various floors of a building are computed as follows: The basement is measured from the bottom of floor slab to the bottom of the first floor slab or joist. The main or first floor extends from the bottom of the first floor slab or joist to the top of the roof slab or ceiling joist. Upper floors are measured from the top of the floor slab or floor joist to the top of the roof slab or ceiling joist. These measurements may be illustrated as follows:

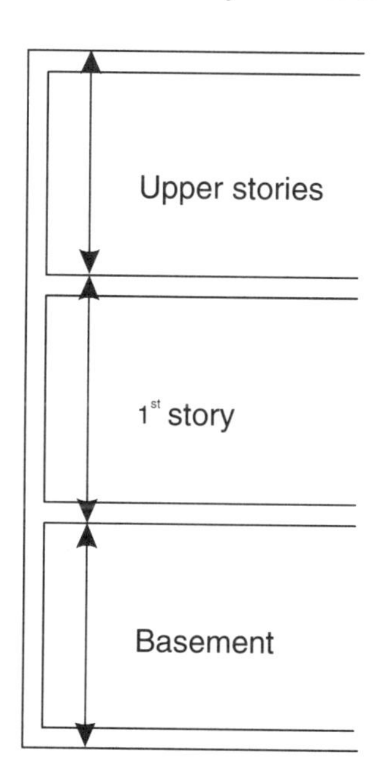

Square foot costs of most building design types must be adjusted if the actual wall height differs from the listed wall height. Wall height adjustment tables are included for buildings requiring this adjustment. Wall height adjustment tables list square foot costs for a foot of difference in perimeter wall height of buildings of various areas. The amount applicable to the actual building area is added or deducted for each foot of difference from the basic wall height.

Buildings such as residences, medical-dental buildings, funeral homes and convalescent hospitals usually have a standard 8-foot ceiling height except in chapels or day room areas. If a significant cost difference exists due to a wall height variation, this factor should be considered in establishing the quality class.

Other Adjustments

A common wall exists when two buildings share one wall. Common wall adjustments are made by deducting the in-place cost of the exterior wall finish plus one-half of the in-place cost of the structural portion of the common wall area.

If an owner has no ownership in a wall, the in-place cost of the exterior wall finish plus the in-place cost of the structural portion of the wall should be deducted from the total building costs. Suggested common wall and no wall ownership costs are included for many of the building types.

Some square foot costs include the cost of expensive veneer finishes on the entire perimeter wall. When these buildings butt against other buildings, adjustments should be made for the lack of this finish. Where applicable, linear foot cost deductions are provided.

The square foot costs in this manual are based on composite costs of total buildings including usual work room or storage areas. They are intended to be applied on a 100% basis to the total building area even though certain areas may or may not have interior finish. Only in rare instances will it be necessary to modify the square foot cost of a portion of a building.

Multiple story buildings usually share a common roof structure and cover, a common foundation and common floor or ceiling structures. The costs of these components are included in the various floor levels as follows:

The first or main floor includes the cost of a floor structure built at ground level, foundation costs for a one-story building, a complete ceiling and roof structure, and a roof cover. The basement includes the basement floor structure and the difference between the cost of the first floor structure built at ground level and its cost built over a basement. The second floor includes the difference between the cost of a foundation for a one-story building and the cost of a foundation for a two-story building and the cost of the second story floor structure.

Location Adjustments

The figures in this manual are intended as national averages for metropolitan areas of the United States. Use the information on page 7 to adapt the basic building costs to any area listed. Frequently building costs outside metropolitan areas are 2% to 6% lower if skilled, productive, lower cost labor is available in the area. The factors on page 7 can be applied to nearly all the square foot costs and some of the "additional" costs in this book.

Depreciation

Depreciation is the loss in value of a structure from all causes and is caused primarily by three forms of obsolescence: (1) physical (2) functional, and (3) economic.

Physical obsolescence is the deterioration of building components such as paint, carpets or roofing. This deterioration may be partially curable. The tables on pages 40, 200 and 233 consider only typical physical obsolescence. Individual judgments will have to be made of functional and economic obsolescence.

Functional obsolescence is due to a deficiency or inadequacy in some characteristic of the building, such as too few bathrooms for the number of bedrooms, or some excess, such as a 10 foot ceiling in a residence. This obsolescence may be curable. The tables do not include functional obsolescence considerations.

Economic obsolescence is caused by factors not directly concerning the structure, but rather, by adverse environmental factors, resulting in loss of desirability. Examples include the obsolescence of a store in an area of declining economic activity, or obsolescence resulting from governmental regulation changing the zone of an area. Because this kind of obsolescence is particularly difficult to measure, it is not included in the tables.

"Effective age" considers all forms of depreciation. It may be less than chronological age, if recently remodeled or improved, or more than the actual age, if deterioration is particularly bad. Though effective age is not considered in the physical life tables, it may yield a better picture of a structure's life than the actual physical age. Once the effective age is determined, considering physical, functional and economic deterioration, use the percent good tables on pages 40 or 200 to determine the present value of a depreciated building. Present value is the result of multiplying the replacement cost (found by using the cost tables) by the appropriate percent good.

Limitations

This manual will be a useful reference for anyone who has to develop budget estimates or replacement costs for buildings. Anyone familiar with construction estimating understands that even very competent estimators with complete working drawings, full specifications and precise labor and material costs can disagree on the cost of a building. Frequently exhaustive estimates for even relatively simple structures can vary 5% or more. The range of competitive bids on some building projects is as much as 10%. Estimating costs is not an exact science and there is room for legitimate disagreement on what the "right" cost is. This manual can not help you do in a few minutes what skilled estimators may not be able to do in several hours. This manual will help you determine a reasonable replacement or construction cost for most buildings. It is not intended as a substitute for judgment or as a replacement for sound professional practice, but should prove a valuable aid to developing an informed opinion of value.

Area Modification Factors

Construction costs are higher in some cities than in other cities. Add or deduct the percentage shown on this page or page 8 to adapt the costs in this book to your job site. Adjust your estimated total project cost by the percentage shown for the appropriate city in this table to find your total estimated cost. Where 0% is shown it means no modification is required. Factors for Canada adjust to Canadian dollars.

These percentages were compiled by comparing the construction cost of buildings in nearly 600 communities throughout North America. Because these percentages are based on completed projects, they consider all construction cost variables, including labor, equipment and material cost, labor productivity, climate, job conditions and markup.

Modification factors are listed alphabetically by state. Areas within each state are listed by the first three digits of the postal zip code. For convenience, one representative city is identified in each zip code or range of zip codes.

These percentages are composites of many costs and will not necessarily be accurate when estimating the cost of any particular part of a building. But when used to modify costs for an entire structure, they should improve the accuracy of your estimates.

Alabama Average		**-14**
350-352	Birmingham	-13
354	Tuscaloosa	-13
355	Jasper	-13
356	Sheffield	-16
357	Scottsboro	-16
358	Huntsville	-13
359	Gadsden	-13
360-361	Montgomery	-13
362	Anniston	-13
363	Dothan	-16
364	Evergreen	-16
365-366	Mobile	-11
367	Selma	-13
368	Auburn	-13
369	Bellamy	-13

Alaska Average		**43**
995	Anchorage	30
996	King Salmon	51
997	Fairbanks	45
998	Juneau	44
999	Ketchikan	46

Arizona Average		**1**
850	Phoenix	-1
852-853	Mesa	-1
855	Douglas	-1
856-857	Tucson	-2
859	Show Low	4
860	Flagstaff	-1
863	Prescott	3
864	Kingman	3
865	Chambers	3

Arkansas Average		**-12**
716	Pine Bluff	-11
717	Camden	-12
718	Hope	-12
719	Hot Springs	-12
720-722	Little Rock	-12
723	West Memphis	-12
724	Jonesboro	-12
725	Batesville	-12
726	Harrison	-12
727	Fayetteville	11
728	Russellville	-9
729	Fort Smith	-9

California Average		**14**
900-901	Los Angeles	14
902-905	Inglewood	13
906	Whittier	13
907-908	Long Beach	14
910-912	Pasadena	13
913-916	Van Nuys	13
917-918	Alhambra	13
919-921	San Diego	10
922	El Centro	10
923-925	San Bernardino	13
926-928	Anaheim	13
930	Oxnard	11
931	Santa Barbara	11
932-933	Bakersfield	13
934	Lompoc	11
935	Mojave	10
936-938	Fresno	13
939	Salinas	14
940	Sunnyvale	17
941	San Francisco	24
942	Sacramento	10
943-944	San Mateo	16
945-947	Oakland	16
948	Richmond	22
949	Novato	22
950-951	San Jose	17
952-953	Stockton	11
954	Santa Rosa	22
955	Eureka	16
956-958	Rancho Cordova	11
959	Marysville	11
960	Redding	12
961	Herlong	18

Colorado Average		**-1**
800-801	Aurora	-3
802-804	Denver	0
805	Longmont	-3
806	Greeley	-3
807	Fort Morgan	-3
808-809	Colorado Springs	1
810	Pueblo	-4
811	Pagosa Springs	-3
812	Salida	-3
813	Durango	-3
814-815	Grand Junction	-3
816	Glenwood Spng	-3
	Ski resorts	13

Connecticut Average		**9**
060-061	Hartford	9
062	West Hartford	9
063	Norwich	8
064	Fairfield	10
065	New Haven	10
066	Bridgeport	8
067	Waterbury	11
068-069	Stamford	11

Delaware Average		**3**
197	Newark	3
198	Wilmington	4
199	Dover	3

District of Columbia		**2**
200-201	Washington	2
202-205	Washington	2

Florida Average		**-8**
320	Saint Augustine	-10
321	Daytona Beach	-10
322	Jacksonville	-3
323	Tallahassee	-14
324	Panama City	-13
325	Pensacola	-13
326	Gainesville	-10
327	Altamonte Spng	-1
328	Orlando	-1
329	Melbourne	-1
330-332	Miami	-6
333	Fort Lauderdale	-1
334	W. Palm Beach	-1
335-336	Tampa	-11
337	St Petersburg	-11
338	Lakeland	-10
339	Fort Myers	-10
342	Bradenton	-10
344	Ocala	-10
346	Brooksville	-10
347	Saint Cloud	-10
349	Fort Pierce	-10

Georgia Average		**-11**
300-302	Marietta	-1
303	Atlanta	-1
304	Statesboro	-11
305	Buford	-11
306	Athens	-11
307	Calhoun	-11
308-309	Augusta	-17
310	Dublin	-11
311	Fort Valley	-11
312	Macon	-11
313	Hinesville	-13
314	Savannah	-13
315	Kings Bay	-12
316	Valdosta	-11
317	Albany	-13
318-319	Columbus	-16

Hawaii Average		**34**
967	Ewa	34
967	Halawa Heights	34
967	Hilo	34
967	Lualualei	34
967	Mililani Town	34
967	Pearl City	34
967	Wahiawa	34
967	Waianae	34
967	Wailuku (Maui)	34
968	Alimanu	34
968	Honolulu	34
968	Kailua	35

Idaho Average		**-6**
832	Pocatello	-6
833	Sun Valley	-5
834	Idaho Falls	-6
835	Lewiston	-8
836	Meridian	-8
837	Boise	-6
838	Coeur d'Alene	-2

Illinois Average		**6**
600	Arlington Hts.	6
601	Carol Stream	10
602	Quincy	6
603	Oak Park	8
604	Joliet	6
605	Aurora	6
606-608	Chicago	10
609	Kankakee	5
610-611	Rockford	2
612	Green River	5
613	Peru	6
614	Galesburg	5
615-616	Peoria	6
617	Bloomington	5
618-619	Urbana	6
620	Granite City	5
622	Belleville	3
623	Decatur	5
624	Lawrenceville	6
625-627	Springfield	6
628	Centralia	5
629	Carbondale	5

Indiana Average		**3**
460-462	Indianapolis	3
463-464	Gary	4
465-466	South Bend	4
467-468	Fort Wayne	4
469	Kokomo	3
470	Aurora	3
471	Jeffersonville	3
472	Columbus	4
473	Muncie	4
474	Bloomington	4
475	Jasper	1
476-477	Evansville	1
478	Terre Haute	4
479	Lafayette	4

Iowa Average		**-4**
500-503	Des Moines	1
504	Mason City	-4
505	Fort Dodge	-4
506-507	Waterloo	-4
508	Creston	-4
510	Cherokee	-4
511	Sioux City	-4
512	Sheldon	-4
513	Spencer	-4
514	Carroll	-5
515	Council Bluffs	-4
516	Shenandoah	-5
520	Dubuque	-4
522-524	Cedar Rapids	-4
521	Decorah	-4
525	Ottumwa	-4
526	Burlington	-5
527-528	Davenport	-2

Kansas Average		**-5**
660-662	Kansas City	-5
664-666	Topeka	-5
667	Fort Scott	-5
668	Emporia	-6
669	Concordia	-5
670-672	Wichita	-4
673	Independence	-5
674	Salina	-5
675	Hutchinson	-6
676	Hays	-5
677	Colby	-5
678	Dodge City	-5
679	Liberal	-5

Kentucky Average		**4**
400-402	Louisville	-5
403-406	Lexington	-5
407-409	London	-7
410	Covington	-5
411-412	Ashland	-5
413-414	Campton	-5
415-416	Pikeville	-5
417-418	Hazard	-5
420	Paducah	-5
421	Bowling Green	-5
422	Hopkinsville	-4
423	Owensboro	-5
424	White Plains	-3
425-426	Somerset	-3
427	Elizabethtown	-3

Louisiana Average		**-4**
700-701	New Orleans	1
703	Houma	1
704	Mandeville	-6
705	Lafayette	-5
706	Lake Charles	-5
707-708	Baton Rouge	-5
710	Minden	-8
711	Shreveport	-6
712	Monroe	-6
713-714	Alexandria	-6

Maine Average		**-4**
039-040	Brunswick	-2
041	Portland	-2
042	Auburn	-4
043	Augusta	-4
044	Bangor	-6
045	Bath	-4
046	Cutler	-4
047	Northern Area	-4
048	Camden	-5
049	Dexter	-4

Maryland Average		**0**
206-207	Laurel	2
208-209	Bethesda	0
210-212	Baltimore	0
214	Annapolis	1
215	Cumberland	0
216	Church Hill	0
217	Frederick	0
218	Salisbury	0
219	Elkton	0

Massachusetts Average		**15**
015-016	Ayer	15
010	Chicopee	15
011	Springfield	15
012	Pittsfield	15
013	Northfield	15
014	Fitchburg	14
017	Bedford	15
018	Lawrence	15
019	Dedham	15
020	Hingham	16
021-022	Boston	16
023-024	Brockton	15
025	Nantucket	15
026	Centerville	15
027	New Bedford	15

Michigan Average		**6**
480	Royal Oak	7
481-482	Detroit	6
483	Pontiac	6
484-485	Flint	6
486-487	Saginaw	1
488-489	Lansing	6
490-491	Battle Creek	7
492	Jackson	6
493-495	Grand Rapids	7
496	Traverse City	6
497	Grayling	6
498-499	Marquette	1

Minnesota Average		**5**
550-551	St Paul	9
553-555	Minneapolis	9
556-558	Duluth	3
559	Rochester	5
560	Mankato	5
561	Magnolia	5
562	Willmar	5
563	St Cloud	5
564	Brainerd	5
565	Fergus Falls	5
566	Bemidji	5
567	Thief River Falls	5

Area Modification Factors

Mississippi Average		-13
386	Clarksdale	-14
387	Greenville	-5
388	Tupelo	-14
389	Greenwood	-14
390-392	Jackson	-14
393	Meridian	-14
394	Laurel	-14
395	Gulfport	-12
396	McComb	-12
397	Columbus	-16

Missouri Average		1
630-631	St Louis	5
633	Saint Charles	5
634	Hannibal	1
635	Kirksville	1
636	Farmington	1
637	Cape Girardeau	1
638	Caruthersville	1
639	Poplar Bluff	0
640-641	Independence	1
644-645	Saint Joseph	1
646	Chillicothe	1
647	East Lynne	1
648	Joplin	1
650-651	Jefferson City	-1
652	Columbia	0
653	Knob Noster	-1
654-655	Lebanon	2
656-658	Springfield	1

Montana Average		1
590-591	Billings	1
592	Fairview	1
593	Miles City	1
594	Great Falls	2
595	Havre	1
596	Helena	1
597	Butte	1
598	Missoula	1
599	Kalispell	1

Nebraska Average		-11
680-681	Omaha	-8
683-685	Lincoln	-11
686	Columbus	-11
687	Norfolk	-11
688	Grand Island	-11
689	Hastings	-11
690	McCook	-11
691	North Platte	-11
692	Valentine	-11
693	Alliance	-11

Nevada Average		5
889-891	Las Vegas	4
893	Ely	4
894	Fallon	7
895	Reno	4
897	Carson City	6
898	Elko	6

New Hampshire Average		2
030-031	New Boston	3
032-033	Manchester	2
034	Concord	3
035	Littleton	2
036	Charlestown	2
037	Lebanon	2
038	Dover	2

New Jersey Average		9
070-073	Newark	9
074-075	Paterson	9
076	Hackensack	9
077	Monmouth	8
078	Dover	9
079	Summit	9
080-084	Atlantic City	9
085	Princeton	9
086	Trenton	8
087	Brick	9
088-089	Edison	9

New Mexico Average		-5
870-871	Albuquerque	-4
873	Gallup	-4
874	Farmington	-4
875	Santa Fe	-4
877	Holman	-4
878	Socorro	-5
879	Truth or Con.	-5
880	Las Cruces	-5
881	Clovis	-6
882	Fort Sumner	-4
883	Alamogordo	-5
884	Tucumcari	-4

New York Average		9
100	NYC (Manhattan)	26
100	New York City	13
103	Staten Island	9
104	Bronx	9
105-108	White Plains	9
109	West Point	9
110	Queens	16
111	Long Island	16
112	Brooklyn	9
113	Flushing	9
114	Jamaica	9
115	Garden City	9
116	Rockaway	9
117	Amityville	9
118	Hicksville	9
119	Montauk	9
120-123	Albany	6
124	Kingston	7
125-126	Poughkeepsie	7
127	Stewart	7
128	Newcomb	6
129	Plattsburgh	7
130-132	Syracuse	8
133-135	Utica	8
136	Watertown	6
137-139	Binghampton	8
140	Batavia	8
141	Tonawanda	8
142	Buffalo	8
143	Niagara Falls	8
144-146	Rochester	7
147	Jamestown	5
148-149	Ithaca	5

North Carolina Avg.		-16
270-274	Winston-Salem	-17
275-277	Raleigh	-17
278	Rocky Mount	-17
279	Elizabeth City	-17
280-282	Charlotte	-17
283	Fayetteville	-11
284	Wilmington	-11
285	Kinston	-17
286	Hickory	-17
287-289	Asheville	-17

North Dakota Average		-4
580-581	Fargo	-4
582	Grand Forks	-4
583	Nekoma	-4
584	Jamestown	-4
585	Bismarck	-4
586	Dickinson	-4
587	Minot	-3
588	Williston	-4

Ohio Average		1
430-431	Newark	1
432	Columbus	1
433	Marion	0
434-436	Toledo	2
437-438	Zanesville	0
439	Stubenville	1
440-441	Cleveland	3
442-443	Akron	1
444-445	Youngstown	1
446-447	Canton	1
448-449	Sandusky	0
450-452	Cincinnati	0
453-455	Dayton	0
456	Chillicothe	0
457	Marietta	1
458	Lima	2

Oklahoma Average		-8
730-731	Oklahoma City	-7
734	Ardmore	-8
735	Lawton	-8
736	Clinton	-8
737	Enid	-8
738	Woodward	-8
739	Adams	-8
740-741	Tulsa	-8
743	Pryor	-8
744	Muskogee	-6
745	McAlester	-12
746	Ponca City	-8
747	Durant	-8
748	Shawnee	-8
749	Poteau	-8

Oregon Average		6
970-972	Portland	6
973	Salem	7
974	Eugene	7
975	Grants Pass	7
976	Klamath Falls	7
977	Bend	7
978	Pendleton	6
979	Adrian	6

Pennsylvania Average		5
150-151	Warrendale	2
152	Pittsburgh	2
153	Washington	2
154	Uniontown	2
155	Somerset	2
156	Greensburg	2
157	Punxsutawney	5
158	DuBois	5
159	Johnstown	5
160	Butler	5
161	New Castle	5
162	Kittanning	5
163	Meadville	5
164-165	Erie	5
166	Altoona	5
167	Bradford	5
168	Clearfield	5
169	Genesee	5
170-171	Harrisburg	5
172	Chambersburg	2
173-174	York	5
175-176	Lancaster	5
177	Williamsport	5
178	Beaver Springs	5
179	Pottsville	5
180-181	Allentown	5
182	Hazleton	5
183	East Stroudsburg	5
184-185	Scranton	4
186-187	Wilkes Barre	5
188	Montrose	5
189	Warminster	8
190-191	Philadelphia	9
195-196	Reading	5
193	Southeastern	5
194	Valley Forge	5

Rhode Island Average		10
028	Bristol	9
028	Coventry	9
028	Davisville	9
028	Narragansett	9
028	Newport	10
028	Warwick	10
029	Cranston	10
029	Providence	10

South Carolina Avg.		-13
290-292	Columbia	-15
293	Spartanburg	-15
294	Charleston	-13
295	Myrtle Beach	-13
296	Greenville	-15
297	Rock Hill	-15
298	Aiken	-9
299	Beaufort	-9

South Dakota Average		-8
570-571	Sioux Falls	-10
572	Watertown	-8
573	Mitchell	-8
574	Aberdeen	-7
575	Pierre	-7
576	Mobridge	-8
577	Rapid City	-8

Tennessee Average		-8
370-372	Nashville	-9
373	Cleveland	-9
374	Chattanooga	-8
376	Kingsport	-8
377-379	Knoxville	-8
380-381	Memphis	-7
382	McKenzie	-8
383	Jackson	-8
384	Columbia	-8
385	Cookeville	-8

Texas Average		-9
750	Plano	-5
751-753	Dallas	-5
754	Greenville	-11
755	Texarkana	-8
756	Longview	-11
757	Tyler	-11
758	Palestine	-11
759	Lufkin	-11
760	Arlington	-11
761-762	Ft Worth	-5
763	Wichita Falls	-11
764	Woodson	-11
765-767	Waco	-11
768	Brownwood	-12
769	San Angelo	-12
770-772	Houston	-10
773	Huntsville	-9
774	Bay City	-10
775	Galveston	-11
776-777	Beaumont	-11
778	Bryan	-11
779	Victoria	-10
780-782	San Antonio	-8
783-784	Corpus Chrsti	-11
785	McAllen	-11
786-787	Austin	-11
788	Del Rio	-2
789	Giddings	-2
790-791	Amarillo	-11
792	Childress	-9
793-794	Lubbock	-8
795-796	Abilene	-11
797	Midland	-10
798-799	El Paso	-11

Utah Average		-7
840	Clearfield	-7
841	Salt Lake City	-8
843-844	Ogden	-6
845	Green River	-7
846-847	Provo	-8

Vermont Average		-4
050	White Rv. Junct.	-4
051	Spingfield	-4
052	Bennington	-4
053	Battleboro	-4
054	Burlington	-3
056	Montpelier	-5
057	Rutland	-4
058	Albany	-4
059	Beecher Falls	-5

Virginia Average		-10
220-223	Alexandria	-11
224-225	Fredericksburg	-15
226	Winchester	-10
227	Culpeper	-10
228	Harrisonburg	-10
229	Charlottesville	-10
230-231	Williamsburg	-7
232	Richmond	-10
233-237	Norfolk	-7
238	Petersburg	-9
239	Farmville	-8
240-241	Radford	-10
242	Abingdon	-10
243	Galax	-10
244	Staunton	-10
245	Lynchburg	-10
246	Tazewell	-10

Washington Average		8
980-981	Seattle	8
982	Everett	8
983-984	Tacoma	8
985	Olympia	8
986	Vancouver	8
988	Wenatchee	7
989	Yakima	6
990-992	Spokane	10
993	Pasco	8
994	Clarkston	8

West Virginia Average		-1
247-248	Bluefield	-3
249	Lewisburg	-1
250-253	Charleston	-1
254	Martinsburg	-1
255-257	Huntington	-1
258-259	Beckley	-1
260	Wheeling	-1
261	Parkersburg	-1
262	New Martinsville	-1
263-264	Clarksburg	-1
265	Morgantown	-1
266	Fairmont	-1
267	Romney	-1
268	Sugar Grove	-3

Wisconsin Average		-2
530-534	Milwaukee	4
535	Beloit	-2
537	Madison	-6
538	Prairie du Chien	-6
539	Portage	-7
540	Amery	-2
541-543	Green Bay	-1
544	Wausau	-2
545	Clam Lake	-2
546	La Crosse	-2
547	Eau Claire	-2
548	Ladysmith	-2
549	Oshkosh	-2

Wyoming Average		-11
820	Cheyenne	-10
821	Laramie	-11
822	Wheatland	-11
823	Rawlins	-10
824	Powell	-11
825	Riverton	-11
826	Casper	-11
827	Gillette	-11
828	Sheridan	-11
829-831	Rock Springs	-10

Canada

Alberta Average	53.0
Calgary	49.3
Edmonton	54.1
Ft. McMurray	55.7

British Columbia Avg.	51.6
Fraser Valley	50.9
Okanagan HRCC	45.6
Vancouver	62.5

Manitoba Average	57.3
North Manitoba	67.7
South Manitoba	57.0
Selkirk	49.7
Winnipeg	54.6

New Brunswick Avg.	49.1
Moncton	49.1

Nova Scotia Average	57.8
Amherst	56.9
Nova Scotia	52.7
Sydney	63.9

Newfoundland Avg.	49.8
Newfoundland & Laborador	49.8

Ontario Average	51.1
London	59.1
Thunder Bay	48.3
Toronto	59.1

Quebec Average	59.5
Montreal	63.4
Quebec	55.4

Saskatchewan Avg.	58.6
La Ronge	62.3
Prince Albert	57.7
Saskatoon	55.7

Building Cost Historical Index

Use this table to find the approximate current dollar building cost when the actual cost is known for any year since 1935. Multiply the figure listed below for the building type and year of construction by the known cost. The result is the estimated 2002 construction cost.

Year	Masonry Buildings	Concrete Buildings	Steel Buildings	Wood-Frame Buildings	Agricultural Buildings	Year of Construction
1935	24.41	24.70	24.86	22.02	22.35	1935
1936	23.95	24.49	24.86	21.26	21.00	1936
1937	21.41	22.49	21.55	19.28	19.11	1937
1938	20.76	22.42	21.42	19.28	18.91	1938
1939	20.55	22.25	21.15	18.72	18.88	1939
1940	20.31	22.03	20.62	18.48	18.34	1940
1941	19.00	20.39	19.72	18.16	17.01	1941
1942	17.33	18.98	18.55	15.56	16.01	1942
1943	17.10	18.17	17.73	14.83	14.35	1943
1944	16.18	17.62	17.06	13.57	13.70	1944
1945	15.29	16.94	16.53	12.86	12.89	1945
1946	13.12	14.87	14.70	11.51	11.15	1946
1947	11.20	12.28	12.85	9.25	9.03	1947
1948	9.81	10.52	11.47	8.47	8.26	1948
1949	9.88	10.41	11.40	8.58	8.51	1949
1950	9.39	9.94	11.19	8.19	7.92	1950
1951	8.78	9.37	10.16	7.66	7.35	1951
1952	8.48	9.16	9.93	7.53	7.28	1952
1953	8.36	8.85	9.49	7.35	7.12	1953
1954	8.21	8.54	9.49	7.35	7.12	1954
1955	7.86	8.14	8.99	6.96	6.80	1955
1956	7.47	7.77	8.28	6.67	6.53	1956
1957	7.25	7.49	7.95	6.61	6.38	1957
1958	7.05	7.20	7.57	6.60	7.60	1958
1959	6.83	6.97	7.38	6.32	6.09	1959
1960	6.67	6.85	7.25	6.22	5.97	1960
1961	6.52	6.82	7.14	6.11	5.95	1961
1962	6.38	6.62	6.98	6.04	5.85	1962
1963	6.28	6.45	6.89	5.92	5.32	1963
1964	6.11	6.36	6.79	5.71	5.59	1964
1965	5.91	6.21	6.56	5.60	5.43	1965
1966	5.64	6.04	6.31	5.36	5.28	1966
1967	5.51	5.74	5.90	5.10	5.07	1967
1968	5.28	5.43	5.63	4.81	4.85	1968
1969	4.99	5.19	5.44	4.63	4.57	1969
1970	4.80	4.95	5.17	4.41	4.35	1970
1971	4.50	4.55	4.80	3.80	4.05	1971
1972	4.18	4.20	4.47	3.81	3.76	1972
1973	3.81	3.99	3.97	3.51	3.53	1973
1974	3.39	3.65	3.72	3.28	3.29	1974
1975	3.10	3.23	3.37	3.09	2.92	1975
1976	2.89	3.08	3.19	2.97	2.77	1976
1977	2.69	2.88	3.02	2.76	2.60	1977
1978	2.62	2.60	2.79	2.53	2.37	1978
1979	2.31	2.40	2.49	2.33	2.25	1979
1980	2.09	2.18	2.22	2.09	2.03	1980
1981	1.06	2.05	2.04	2.00	1.89	1981
1982	1.89	1.96	1.98	1.93	1.81	1982
1983	1.82	1.89	1.93	1.83	1.72	1983
1984	1.71	1.79	1.85	1.70	1.66	1984
1985	1.66	1.71	1.79	1.64	1.63	1985
1986	1.62	1.69	1.77	1.61	1.61	1986
1987	1.61	1.66	1.75	1.59	1.59	1987
1988	1.57	1.59	1.70	1.57	1.56	1988
1989	1.53	1.56	1.64	1.53	1.52	1989
1990	1.44	1.51	1.54	1.44	1.45	1990
1991	1.56	1.48	1.48	1.36	1.37	1991
1992	1.39	1.46	1.45	1.35	1.36	1992
1993	1.35	1.44	1.40	1.33	1.33	1993
1994	1.32	1.34	1.35	1.29	1.25	1994
1995	1.25	1.22	1.26	1.21	1.18	1995
1996	1.21	1.20	1.21	1.19	1.16	1996
1997	1.17	1.17	1.16	1.16	1.13	1997
1998	1.12	1.12	1.12	1.11	1.11	1998
1999	1.09	1.09	1.09	1.09	1.09	1999
2000	1.06	1.06	1.06	1.05	1.05	2000
2001	1.03	1.03	1.03	1.02	1.02	2001
2002 (estimated)	1.00	1.00	1.00	1.00	1.00	2002

Residential Structures Section

Section Contents

The figures in this section include all costs associated with normal construction:

Foundations as required for normal soil conditions. Excavation for foundations, piers, and other foundation components given a fairly level construction site. Floor, wall, and roof structures. Interior floor, wall, and ceiling finishes. Exterior wall finish and roof cover. Interior partitions as described in the quality class. Finish carpentry, doors, windows, trim, etc. Electric wiring and fixtures. Rough and finish plumbing as described in applicable building specifications. Built-in appliances as described in applicable building specifications. All labor and materials including supervision. All design and engineering fees, if necessary. Permits and fees. Utility hook-ups. Contractors' contingency, overhead and profit.

The square foot costs do not include heating and cooling equipment or the items listed in the section "Additional Costs for Residential Structures" which appear on pages 24 to 28. The costs of the following should be figured separately and added to the basic structure cost: porches, basements, balconies, exterior stairways, built-in equipment beyond that listed in the quality classifications, fireplaces, garages and carports.

Single Family Residences

Single family residential structures vary widely in quality and the quality of construction is a significant factor in the finished cost. For estimating purposes the structure should be placed in one or more classes from 1 to 4. Class 1 construction is top quality work and Class 4 is the minimum required under most building codes. Many structures do not fall into a single class and have features of two quality classes. The tables have 3 "half classes" which apply to structures which have some features of two quality classes. Some structures may fall into two distinct classes and the square foot area of each part should be computed separately.

The shape of the outside perimeter is an important consideration in estimating the total construction cost. Generally, the more complex the shape, the more expensive the structure per square foot of floor area. The shape classification of multiple story or split level homes is based on the outline formed by the outer-most exterior walls, including the garage area, regardless of the varying level. Most structures have 4, 6, 8 or 10 corners, as illustrated in the example. Small insets not requiring a change in the roof shape can be ignored when determining the shape.

These figures can be applied to nearly all single family dwellings built using conventional methods and common materials. There is no simplified method of evaluating the cost of the relatively small number of highly decorative, starkly original or exceptionally well-appointed residences. It is safe to assume, however, that the cost of these highly customized homes will exceed the figures that appear on the following pages.

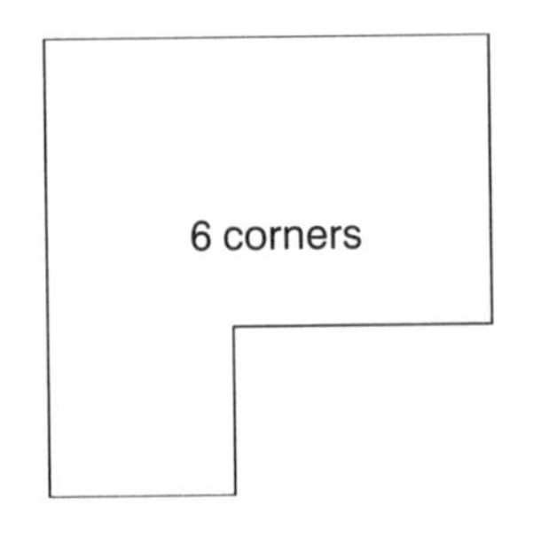

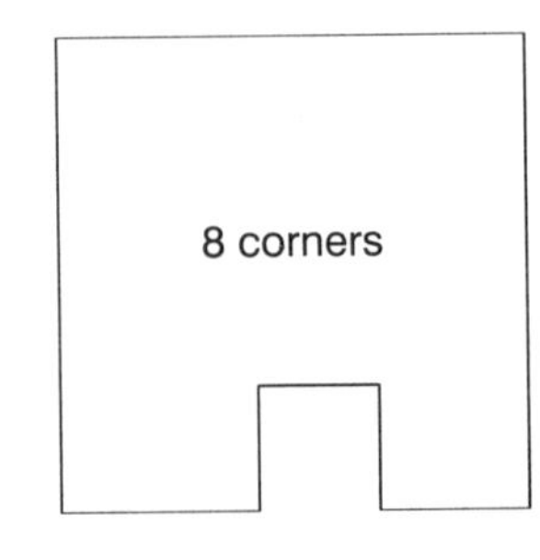

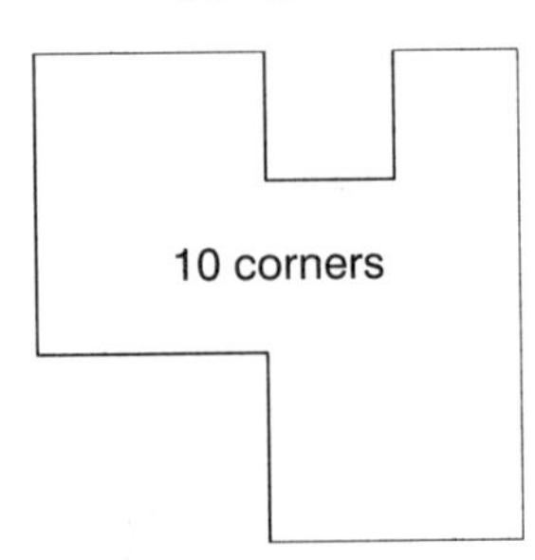

Single Family Residences

Quality Classification

	Class 1 Best Quality	Class 2 Good Quality	Class 3 Average Quality	Class 4 Low Quality
Foundation (9.3%)	Tri-level or built on sloping site	Conventional crawl-space, footing depth over 30"	Conventional crawl-space, footing depth to 30"	Concrete slab
Exterior Wood Frame Walls (25.5%)	Tri-level or complex wood frame floor and roof plans.	Complex wood frame floor and roof plan	Wood frame with complex roof plan.	Simple wood frame.
Exterior Finish (9.2%)	Good quality wood siding, stucco or masonry with good trim.	Good quality hardboard or plywood siding, some masonry veneer or good stucco	Hardwood or shingle or minimum quality stucco	Composition siding or minimum quality stucco.
Windows and Doors(4.8%)	Good quality, large insulated wood or coated aluminum. windows. Most are casement	Average quality insulated aluminum or wood windows.	Minimum quality wood or aluminum windows.	Aluminum sliding. windows
Roofing (7%)	Heavy shake, slate or tile with a 24" to 36" enclosed soffit.	Wood shingle or built-up with rock cover. Less than 24" enclosed soffit.	Minimum shingle or built-up with rock cover or good quality composition shingle roof.	Minimum quality built-up with gravel surface or composition shingle. Simple open cornice.
Interior Finish (18.9%)	Gypsum board with heavy texture or plaster. Some vinyl-covered walls and paneling. Cathedral ceiling at entry and living room. Several ornamental lighting fixtures. Fireplace with brick or natural stone face.	Taped and textured gypsum wallboard, some wallpaper and plywood paneling. Exposed beams and. over 8' ceiling at entry. At least one ornamental lighting fixture. Concrete or masonry fireplace.	Taped and textured gypsum wallboard, some wallpaper. Good quality lighting fixtures. Insulated prefabricated metal fireplace with masonry veneer.	Minimum quality wallboard. Minimum built-in lighting. Freestanding metal fireplace.
Flooring (5.6%)	Terrazzo or tile entry. Carpet or good quality hardwood in most rooms. Sheet vinyl in. kitchen, family and breakfast rooms.	Average quality tile or hardwood entry. Average quality hard-wood or carpet in most rooms. Resilient tile in kitchen, family and breakfast rooms.	3/8" hardwood or medium cost carpet in most rooms, vinyl or resilient tile in remaining rooms.	Composition tile or low cost carpet
Bathrooms (1.5%)	2-1/2 bathrooms, tile floor, ceramic tile shower over tub, glass shower door, 4' to 6' tile or cultured marble pullman in 2 baths.	2 bathrooms, sheet vinyl floor, some average quality tile, 4' to 6' tile or laminated plastic pullman in each bath.	2 bathrooms, usually back to back, resilient tile floor, plastic coated hardboard or fiberglass around shower and over tub. 4' laminated plastic pullman in cach bath.	2 bathrooms back to back, composition tile, plastic shower door, fiberglass enclosures.
Kitchen (8.6%)	20 linear feet of hard-wood veneer wall and base cabinets, 20 linear feet of ceramic tile or simulated stone counter. Built in oven, range, dishwasher, garbage disposer, range hood oven, range, dishwasher, with fan, all good quality.	16 linear feet of average quality hardwood veneer cabinets and ceramic tile or good quality laminated plastic counter. Built- in oven, range, dishwasher, garbage disposer, range hood range, dishwasher, and fan, all average quality.	12 linear feet of low cost hardwood veneer cabinets and laminated plastic counter. Low cost built-in oven, appliances, garbage disposer and range hood.	8 linear feet of low cost cabinets with laminated face. Minimum built-in appliances.
Plumbing (9.6%)	10 to 12 good quality fixtures, copper tubing supply and waste.	8 or 9 good quality fixtures.	8 or 9 average quality fixtures.	8 low cost fixtures, plastic supply.
***For Masonry Walls**	Textured, colored reinforced concrete block or brick with ornamental details.	Colored or textured reinforced concrete block or brick.	Reinforced concrete block or brick.	Concrete block, brick or clay tile.

*When masonry walls are used in lieu of wood frame walls add 5% to the appropriate cost

Note: Use the percent of total cost to help identify the correct quality classification.

Single Family Residences

4 Corners

Estimating Procedure

1. Establish the structure quality class by applying the information on page 11.
2. Multiply the structure floor area (excluding the garage) by the appropriate square foot cost below.
3. Multiply the total from step 2 by the correct location factor listed on page 7.
4. Add, when appropriate, the cost of a porch, garage, heating and cooling equipment, basement, fireplace, carport, appliances and plumbing fixtures beyond that listed in the quality classification. See the cost of these items on pages 24 to 26.

Single Family Residence, Class 3

Single Family Residence, Class 4

Square Foot Area

Quality Class	700	800	900	1,000	1,100	1,200	1,300	1,400	1,500	1,600	1,700
1, Best	98.52	94.39	91.01	88.14	85.69	83.54	81.63	79.96	78.50	77.12	75.85
1 & 2	90.78	86.98	83.83	81.21	78.93	76.97	75.24	73.68	72.30	71.03	69.89
2, Good	85.57	81.98	79.04	76.54	74.40	72.56	70.92	69.46	68.14	66.96	65.88
2 & 3	79.25	75.94	73.21	70.90	68.93	67.19	65.68	64.34	63.11	62.03	61.04
3, Average	73.63	70.57	68.04	65.90	64.07	62.42	61.04	59.79	58.65	57.65	56.72
3 & 4	67.20	64.38	62.07	60.12	58.43	56.96	55.68	54.52	53.52	52.58	51.72
4, Low	60.90	58.37	56.26	54.48	52.97	51.65	50.50	49.42	48.50	47.65	46.90

Square Foot Area

Quality Class	1,800	2,000	2,200	2,400	2,600	2,800	3,000	3,200	3,400	3,600	4,000
1, Best	74.50	74.13	73.59	72.89	72.20	71.44	70.68	69.91	69.16	68.43	67.02
1 & 2	67.61	67.27	66.73	66.13	65.50	64.82	64.11	63.41	62.73	62.07	60.79
2, Good	61.78	61.49	61.05	60.48	59.85	59.25	58.62	58.00	57.35	56.74	55.60
2 & 3	57.03	56.75	56.32	55.80	55.26	54.69	54.10	53.54	52.95	52.39	51.32
3, Average	53.48	53.18	52.79	52.32	51.78	51.26	50.72	50.16	49.64	49.09	48.10
3 & 4	49.28	48.99	48.64	48.20	47.70	47.24	46.74	46.22	45.73	45.24	44.31
4, Low	45.25	44.98	44.68	44.26	43.83	43.38	42.91	42.46	42.00	41.53	40.68

Note: Tract work and highly repetitive jobs may reduce the cost 8 to 12%. Add 4% to the square foot cost of floors above the second floor level. Work outside metropolitan areas may cost 2 to 6% less. When the exterior walls are masonry, add 8 to 9% for class 2 and 1 structures and 6 to 7% for class 4 and 3 structures. Deduct 2% for area built on a concrete slab.

The building area includes all full story (7'6" to 8' high) areas within and including the exterior walls of all floor areas of the building, including small inset areas such as entrances outside the exterior wall but under the main roof. For areas with a ceiling height of less than 80", see the section on half-story areas on page 27.

Single Family Residences

6 Corners

Estimating Procedure

1. Establish the structure quality class by applying the information on page 11.
2. Multiply the structure floor area (excluding the garage) by the appropriate square foot cost below.
3. Multiply the total from step 2 by the correct location factor listed on page 7.
4. Add, when appropriate, the cost of a porch, garage, heating and cooling equipment, basement, fireplace, carport, appliances and plumbing fixtures beyond that listed in the quality classification. See the cost of these items on pages 24 to 26.

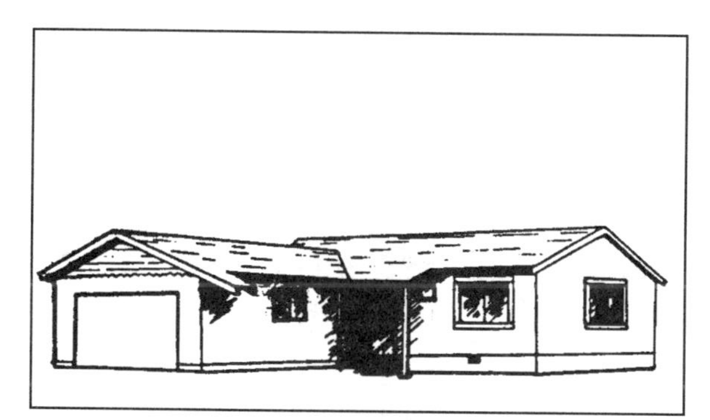

Single Family Residence, Class 4

Single Family Residence, Class 2

Square Foot Area

Quality Class	700	800	900	1,000	1,100	1,200	1,300	1,400	1,500	1,600	1,700
1, Best	101.06	96.25	92.42	89.35	86.81	84.66	82.84	81.25	79.88	78.67	77.59
1 & 2	93.89	89.40	85.86	83.02	80.65	78.67	76.97	75.50	74.21	73.09	72.07
2, Good	87.69	83.53	80.20	77.51	75.34	73.45	71.88	70.52	69.31	68.26	67.31
2 & 3	81.53	77.62	74.56	72.06	70.00	68.28	66.81	65.54	64.42	63.44	62.59
3, Average	75.92	72.27	69.39	67.09	65.19	63.59	62.21	61.00	59.98	59.08	58.26
3 & 4	69.16	65.83	63.23	61.11	59.36	57.91	56.67	55.58	54.67	53.83	53.07
4, Low	62.69	59.71	57.31	55.42	53.85	52.53	51.40	50.41	49.53	48.79	48.11

Square Foot Area

Quality Class	1,800	2,000	2,200	2,400	2,600	2,800	3,000	3,200	3,400	3,600	4,000
1, Best	76.97	76.62	76.07	75.37	74.59	73.82	73.03	72.21	71.41	70.60	69.08
1 & 2	68.83	68.51	68.02	67.39	66.69	65.97	65.26	64.53	63.84	63.14	61.75
2, Good	63.05	62.76	62.28	61.72	61.11	60.47	59.78	59.13	58.47	57.83	56.58
2 & 3	58.68	58.43	57.99	57.47	56.90	56.26	55.64	55.07	54.44	53.84	52.69
3, Average	54.84	54.58	54.20	53.71	53.17	52.63	52.02	51.44	50.87	50.31	49.24
3 & 4	50.54	50.28	49.94	49.47	48.97	48.45	47.92	47.38	46.86	46.34	45.35
4, Low	46.86	46.67	46.31	45.90	45.44	44.93	44.46	43.97	43.46	42.98	42.08

Note: Tract work and highly repetitive jobs may reduce the cost 8 to 12%. Add 4% to the square foot cost of floors above the second floor level. Work outside metropolitan areas may cost 2 to 6% less. When the exterior walls are masonry, add 8 to 9% for class 2 and 1 structures and 6 to 7% for class 4 and 3 structures. Deduct 2% for area built on a concrete slab.

The building area includes all full story (7'6" to 8' high) areas within and including the exterior walls of all floor areas of the building, including small inset areas such as entrances outside the exterior wall but under the main roof. For areas with a ceiling height of less than 80", see the section on half-story areas on page 27.

Single Family Residences

8 Corners

Estimating Procedure

1. Establish the structure quality class by applying the information on page 11.
2. Multiply the structure floor area (excluding the garage) by the appropriate square foot cost below.
3. Multiply the total from step 2 by the correct location factor listed on page 7.
4. Add, when appropriate, the cost of a porch, garage, heating and cooling equipment, basement, fireplace, carport, appliances and plumbing fixtures beyond that listed in the quality classification. See the cost of these items on pages 24 to 26.

Single Family Residence, Class 1

Single Family Residence, Class 3

Square Foot Area

Quality Class	700	800	900	1,000	1,100	1,200	1,300	1,400	1,500	1,600	1,700
1, Best	101.98	97.33	93.64	90.70	88.26	86.21	84.48	82.97	81.63	80.48	79.49
1 & 2	94.83	90.49	87.06	84.34	82.06	80.17	78.56	77.14	75.94	74.86	73.91
2, Good	89.08	85.01	81.82	79.21	77.10	75.32	73.78	72.49	71.33	70.33	69.44
2 & 3	82.51	78.73	75.78	73.38	71.42	69.77	68.33	67.15	66.07	65.13	64.32
3, Average	76.77	73.23	70.50	68.26	66.44	64.91	63.60	62.44	61.48	60.60	59.83
3 & 4	69.94	66.75	64.25	62.24	60.55	59.14	57.97	56.92	56.01	55.23	54.51
4, Low	63.38	60.22	58.21	56.39	54.86	53.59	52.54	51.58	50.78	50.05	49.40

Square Foot Area

Quality Class	1,800	2,000	2,200	2,400	2,600	2,800	3,000	3,200	3,400	3,600	4,000
1, Best	78.81	77.85	76.90	75.99	75.08	74.27	73.46	72.73	72.02	71.35	70.11
1 & 2	71.40	70.51	69.64	68.82	68.02	67.27	66.54	65.87	65.22	64.63	63.52
2, Good	65.51	64.68	63.89	63.12	62.40	61.70	61.07	60.45	59.82	59.28	58.25
2 & 3	63.32	62.54	61.76	61.02	60.33	59.68	59.03	58.44	57.85	57.33	56.33
3, Average	56.96	56.25	55.55	54.89	54.26	53.64	53.08	52.56	52.03	51.55	50.63
3 & 4	52.46	51.78	51.17	50.56	49.99	49.42	48.90	48.39	47.93	47.50	46.67
4, Low	48.23	47.62	47.03	46.49	45.95	45.44	44.94	44.50	44.05	43.65	42.90

Note: Tract work and highly repetitive jobs may reduce the cost 8 to 12%. Add 4% to the square foot cost of floors above the second floor level. Work outside metropolitan areas may cost 2 to 6% less. When the exterior walls are masonry, add 8 to 9% for class 2 and 1 structures and 6 to 7% for class 4 and 3 structures.

The building area includes all full story (7'6" to 8' high) areas within and including the exterior walls of all floor areas of the building, including small inset areas such as entrances outside the exterior wall but under the main roof. For areas with a ceiling height of less than 80", see the section on half-story areas on page 27.

Single Family Residences

10 Corners

Estimating Procedure

1. Establish the structure quality class by applying the information on page 11.
2. Multiply the structure floor area (excluding the garage) by the appropriate square foot cost below.
3. Multiply the total from step 2 by the correct location factor listed on page 7.
4. Add, when appropriate, the cost of a porch, garage, heating and cooling equipment, basement, fireplace, carport, appliances and plumbing fixtures beyond that listed in the quality classification. See the cost of these items on pages 24 to 26.

Single Family Residence, Class 1

Single Family Residence, Class 2

Square Foot Area

Quality Class	700	800	900	1,000	1,100	1,200	1,300	1,400	1,500	1,600	1,700
1, Best	103.59	98.91	95.18	92.14	89.58	87.42	85.57	83.93	82.49	81.20	80.04
1 & 2	97.06	92.69	89.19	86.31	83.92	81.92	80.16	78.64	77.30	76.09	75.01
2, Good	91.18	87.07	83.77	81.09	78.86	76.96	75.30	73.88	72.61	71.50	70.49
2 & 3	84.38	80.56	77.53	75.03	72.95	71.19	69.69	68.33	67.18	66.16	65.22
3, Average	78.39	74.84	72.02	69.71	67.78	66.14	64.73	63.52	62.41	61.47	60.59
3 & 4	71.53	68.27	65.73	63.62	61.85	60.35	59.07	57.97	56.96	56.07	55.29
4, Low	64.82	61.87	59.55	57.64	56.04	54.69	53.53	52.53	51.61	50.84	50.10

Square Foot Area

Quality Class	1,800	2,000	2,200	2,400	2,600	2,800	3,000	3,200	3,400	3,600	4,000
1, Best	79.58	79.13	78.54	77.82	77.03	76.26	75.45	74.65	73.87	73.10	71.60
1 & 2	72.09	71.71	71.18	70.53	69.84	69.12	68.40	67.66	66.94	66.24	64.92
2, Good	66.13	65.77	65.25	64.68	64.06	63.39	62.73	62.05	61.41	60.77	59.53
2 & 3	61.59	61.28	60.79	60.25	59.66	59.03	58.42	57.79	57.17	56.59	55.45
3, Average	57.59	57.27	56.84	56.31	55.75	55.20	54.60	54.02	53.48	52.91	51.84
3 & 4	53.03	52.72	52.34	51.87	51.35	50.84	50.27	49.76	49.24	48.73	47.74
4, Low	48.79	48.49	48.11	47.66	47.23	46.74	46.24	45.76	45.29	44.80	43.88

Note: Tract work and highly repetitive jobs may reduce the cost 8 to 12%. Add 4% to the square foot cost of floors above the second floor level. Work outside metropolitan areas may cost 2 to 6% less. When the exterior walls are masonry, add 8 to 9% for class 2 and 1 structures and 6 to 7% for class 4 and 3 structures.

The building area includes all full story (7'6" to 8' high) areas within and including the exterior walls of all floor areas of the building, including small inset areas such as entrances outside the exterior wall but under the main roof. For areas with a ceiling height of less than 80", see the section on half-story areas on page 27.

Multi-Family Residences

Quality Classification

	Class 1 Best Quality	Class 2 Good Quality	Class 3 Average Quality	Class 4 Low Quality
Foundation (7%)	Tri-level or built on sloping site	Conventional crawl-space, footing depth over 30"	Conventional crawl-space, footing depth to 30"	Concrete slab
Exterior Wood Frame Walls (17%)	Tri-level or complex wood frame floor and roof plans.	Complex wood frame floor and roof plan	Wood frame with complex roof plan.	Simple wood frame.
Exterior Finish (7.5%)	Good quality wood siding, stucco or masonry with good trim.	Good quality hardboard or plywood siding, some masonry veneer or good stucco	Hardwood or shingle or minimum quality stucco	Composition siding or minimum quality stucco.
Windows, Doors (4% of total Cost)	Large top quality wood or aluminum.	Average quality wood or aluminum.	Average quality wood or aluminum.	Few, smaller low cost aluminum or wood
Roofing (7%)	Heavy shake, slate or tile with a 24" to 36" enclosed soffit.	Wood shingle or built-up with rock cover. Less than 24" enclosed soffit.	Minimum shingle or built-up with rock cover or good quality composition shingle roof.	Minimum quality built-up with gravel surface or composition shingle. Simple open cornice.
Interior Finish (26.5%)	Gypsum board with heavy texture or plaster. Some vinyl-covered walls and paneling. Cathedral ceiling at entry and living room. Several ornamental lighting fixtures. Fireplace with brick or natural stone face.	Taped and textured gypsum wallboard, some wallpaper and plywood paneling. Exposed beams and. over 8' ceiling at entry. At least one ornamental lighting fixture. Concrete or masonry fireplace.	Taped and textured gypsum wallboard, some wallpaper. Good quality lighting fixtures. Insulated prefabricated metal fireplace with masonry veneer.	Minimum quality wallboard. Minimum built-in lighting. Freestanding metal fireplace.
Flooring (5% of total Cost)	Terrazzo or tile entry. Good hardwood or carpet. Good sheet vinyl in kitchen.	Hardwood or average quality carpet. Sheet vinyl in kitchen.	Average quality hard-wood or low cost carpet. Vinyl tile in kitchen.	Composition tile.
Bathrooms (5% of total Cost)	Sheet vinyl floor, tile over tub with glass shower door. 6' marble top vanity.	Sheet vinyl or inlaid tile, tile over tub with glass shower door. 4' plastic vanity.	Tile floor, plastic coated hardboard over tub. Glass or plastic shower door. 3' plastic vanity.	Tile floor, plastic coated hardboard over tub.
Kitchen (8%)	10 linear feet of hard-wood veneer wall and base cabinets, 10 linear feet of ceramic tile or simulated stone counter. Built in oven, range, dish-washer, garbage disposer, range hood oven, range, dishwasher, with fan, all good quality.	8 linear feet of average quality hardwood veneer cabinets and ceramic tile or good quality laminated plastic counter. Built-in oven, range, dishwasher, garbage disposer, range hood range, dishwasher, and fan, all average quality.	6 linear feet of low cost hardwood veneer cabinets and laminated plastic counter. Low cost built-in oven, appliances, garbage disposer and range hood.	5 linear feet of low cost cabinets with laminated face. Minimum built-in appliances.
Plumbing (13% of total Cost)	Copper tubing, good quality fixtures.	Galvanized pipe, good quality fixtures.	Average fixtures.	Plastic pipe, low cost fixtures.

Plumbing costs assume 1 bathroom per unit. See page 27 for the costs of additional bathrooms.

	Class 1	Class 2	Class 3	Class 4
***For Masonry Walls**	Textured block, tile or brick with masonry facing.	Colored or detailed concrete block, brick or tile with facing.	Colored reinforced concrete block, clay tile or common brick.	Reinforced concrete block block or clay tile.

*When masonry walls are used in lieu of wood frame walls, add 5% to the appropriate S.F. cost.

Note: Use the percent of total cost to help identify the correct quality classification.

Multi-Family Residences

2 or 3 Units

Estimating Procedure

1. Establish the structure quality class by applying the information on page 16.
2. Multiply the average unit area by the appropriate square foot cost below. The average unit area is found by dividing the building area on all floors by the number of units in the building. The building area should include office and utility rooms, interior hallways and interior stairways.
3. Multiply the total from step 2 by the correct location factor listed on page 7.
4. Add, when appropriate, the cost of balconies, porches, garages, heating and cooling equipment, basements, fireplaces, carports, appliances and plumbing fixtures beyond that listed in the quality classification. See the cost of these items on pages 24 to 28.
5. Costs assume one bathroom per unit. Add the cost of additional bathrooms from page 27.

Multi-Family, Class 1

Multi-Family, Class 3 & 4

Average Unit Area in Square Feet

Quality Class	400	450	500	550	600	650	700	750	800	900	1,000
1, Best	93.91	90.57	87.98	85.88	84.21	82.84	81.65	80.72	79.88	78.55	77.52
1 & 2	87.85	84.75	82.33	80.39	78.81	77.52	76.45	75.53	74.77	73.48	72.53
2, Good	83.03	80.09	77.79	75.94	74.45	73.24	72.22	71.34	70.60	69.44	68.54
2 & 3	77.70	74.95	72.80	71.09	69.71	68.54	67.63	66.79	66.10	64.99	64.16
3, Average	73.43	70.80	68.77	67.16	65.84	64.76	63.86	63.10	62.43	61.41	60.61
3 & 4	68.23	65.82	63.93	62.42	61.20	60.20	59.35	58.65	58.08	57.07	56.33
4, Low	63.15	60.89	59.15	57.77	56.64	55.69	54.93	54.27	53.74	52.81	52.12

Average Unit Area in Square Feet

Quality Class	1,100	1,200	1,300	1,400	1,500	1,600	1,700	1,800	1,900	2,000	2,200
1, Best	76.45	75.61	74.92	74.34	73.85	73.42	73.07	72.75	72.49	72.24	71.85
1 & 2	71.57	70.79	70.13	69.57	69.14	68.73	68.41	68.13	67.84	67.64	67.29
2, Good	67.62	66.87	66.25	65.74	65.28	64.93	64.63	64.35	64.11	63.90	63.55
2 & 3	63.28	62.58	62.00	61.50	61.11	60.76	60.47	60.20	59.98	59.78	59.44
3, Average	59.79	59.13	58.59	58.14	57.73	57.43	57.12	56.90	56.65	56.51	56.18
3 & 4	55.56	54.97	54.45	54.03	53.66	53.37	53.11	52.89	52.68	52.52	52.23
4, Low	51.44	50.87	50.41	50.01	49.67	49.39	49.18	48.94	48.77	48.60	48.33

Note: Work outside metropolitan areas may cost 2 to 6% less. Add 2% to the costs for second floor areas and 4% for third floor areas. Add 5% when the exterior walls are masonry.

Multi-Family Residences

4 to 9 Units

Estimating Procedure

1. Establish the structure quality class by applying the information on page 16.
2. Multiply the average unit area by the appropriate square foot cost below. The average unit area is found by dividing the building area on all floors by the number of units in the building. The building area should include office and utility rooms, interior hallways and interior stairways.
3. Multiply the total from step 2 by the correct location factor listed on page 7.
4. Add, when appropriate, the cost of balconies, porches, garages, heating and cooling equipment, basements, fireplaces, carports, appliances and plumbing fixtures beyond that listed in the quality classification. See the cost of these items on pages 24 to 28.
5. Costs assume one bathroom per unit. Add the cost of additional bathrooms from page 27.

Multi-Family, Class 3

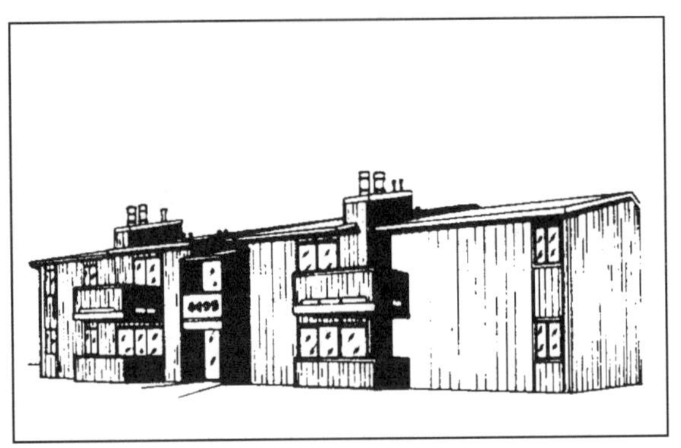

Multi-Family, Class 2 & 3

Average Unit Area in Square Feet

Quality Class	400	450	500	550	600	650	700	750	800	900	1,000
1, Best	88.46	85.26	82.78	80.81	79.22	77.95	76.88	75.99	75.20	73.97	73.06
1 & 2	82.97	79.96	77.65	75.81	74.32	73.10	72.08	71.24	70.53	69.38	68.52
2, Good	78.37	75.57	73.36	71.61	70.25	69.08	68.13	67.31	66.61	65.55	64.76
2 & 3	73.34	70.72	68.66	67.03	65.73	64.67	63.75	63.01	62.33	61.35	60.58
3, Average	69.65	67.10	65.17	63.62	62.39	61.35	60.51	59.78	59.19	58.20	57.51
3 & 4	64.17	61.85	60.05	58.63	57.48	56.54	55.76	55.10	54.53	53.64	52.99
4, Low	59.64	57.48	55.80	54.47	53.44	52.56	51.84	51.20	50.68	49.87	49.28

Average Unit Area in Square Feet

Quality Class	1,100	1,200	1,300	1,400	1,500	1,600	1,700	1,800	1,900	2,000	2,200
1, Best	72.14	71.35	70.71	70.14	69.68	69.28	68.94	68.65	68.39	68.16	67.79
1 & 2	67.51	66.78	66.16	65.66	65.21	64.87	64.50	64.25	63.99	63.81	63.42
2, Good	63.72	63.04	62.42	61.98	61.52	61.17	60.87	60.63	60.39	60.22	59.85
2 & 3	59.71	59.03	58.48	58.03	57.64	57.32	57.02	56.80	56.58	56.40	56.08
3, Average	56.40	55.78	55.27	54.84	54.47	54.18	53.90	53.66	53.49	53.30	53.02
3 & 4	52.43	51.84	51.37	50.98	50.62	50.34	50.08	49.90	49.69	49.52	49.28
4, Low	48.50	48.00	47.53	47.18	46.83	46.57	46.34	46.16	45.97	45.85	45.57

Note: Work outside metropolitan areas may cost 2 to 6% less. Add 2% to the costs for second floor areas and 4% for third floor areas. Add 5% when the exterior walls are masonry.

Multi-Family Residences

10 or More Units

Estimating Procedure

1. Establish the structure quality class by applying the information on page 16.
2. Multiply the average unit area by the appropriate square foot cost below. The average unit area is found by dividing the building area on all floors by the number of units in the building. The building area should include office and utility rooms, interior hallways and interior stairways.
3. Multiply the total from step 2 by the correct location factor listed on page 7.
4. Add, when appropriate, the cost of balconies, porches, garages, heating and cooling equipment, basements, fireplaces, carports, appliances and plumbing fixtures beyond that listed in the quality classification. See the cost of these items on pages 24 to 28.
5. Costs assume one bathroom per unit. Add the cost of additional bathrooms from page 27.

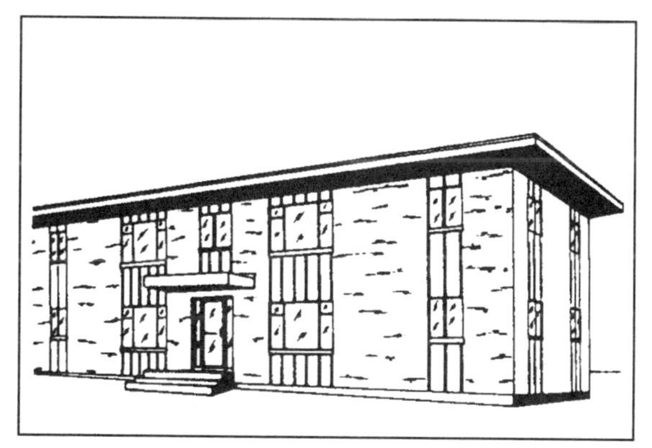

Multi-Family, Class 3 & 4

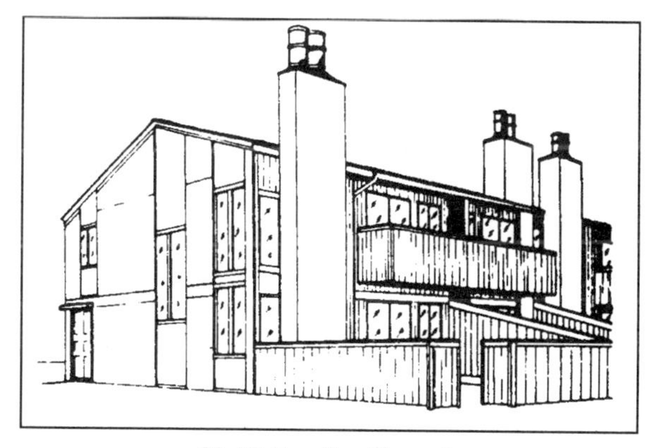

Multi-Family, Class 3

Average Unit in Square Feet

Quality Class	400	450	500	550	600	650	700	750	800	900	1,000
1, Best	83.55	80.58	78.26	76.44	74.95	73.73	72.69	71.84	71.09	69.89	68.97
1 & 2	78.18	75.42	73.25	71.54	70.14	68.99	68.06	67.24	66.53	65.40	64.54
2, Good	73.88	71.24	69.23	67.59	66.27	65.19	64.27	63.53	62.84	61.78	61.01
2 & 3	69.14	66.69	64.78	63.27	62.02	61.01	60.12	59.42	58.82	57.83	57.09
3, Average	65.31	63.04	61.21	59.79	58.62	57.66	56.87	56.18	55.60	54.65	53.94
3 & 4	60.76	58.60	56.92	55.58	54.48	53.60	52.88	52.23	51.69	50.84	50.17
4, Low	56.20	54.22	52.67	51.43	50.44	49.58	48.91	48.32	47.83	47.02	46.42

Average Unit in Square Feet

Quality Class	1,100	1,200	1,300	1,400	1,500	1,600	1,700	1,800	1,900	2,000	2,200
1, Best	68.03	67.30	66.69	66.17	65.75	65.36	65.04	64.74	64.48	64.28	63.90
1 & 2	63.71	63.04	62.45	61.97	61.56	61.15	60.95	60.63	60.39	60.20	59.84
2, Good	60.19	59.53	58.99	58.54	58.15	57.82	57.53	57.27	57.05	56.87	56.53
2 & 3	56.32	55.69	55.19	54.76	54.42	54.10	53.84	53.60	53.38	53.20	52.89
3, Average	53.19	52.64	52.14	51.72	51.41	51.10	50.85	50.61	50.46	50.26	49.99
3 & 4	49.47	48.93	48.47	48.10	47.78	47.52	47.28	47.07	46.87	46.74	46.47
4, Low	45.76	45.27	44.84	44.50	44.21	43.95	43.76	43.54	43.38	43.23	42.96

Note: Work outside metropolitan areas may cost 2 to 6% less. Add 2% to the costs for second floor areas and 4% for third floor areas. Add 5% when the exterior walls are masonry.

Motels

Quality Classification

	Class 1 Best Quality	Class 2 Good Quality	Class 3 Average Quality	Class 4 Low Quality
Foundation (4%) Foundation costs will vary greatly with substrate, type, and location.	Concrete slab	Concrete slab	Concrete slab	Concrete slab
Framing* (20% of total Cost)	Wood frame.	Wood frame.	Wood frame.	Wood frame.
Windows (2% of total Cost)	Large, good quality.	Average number and quality.	Average number and quality.	Small, few, low cost.
Roofing (8% of total Cost)	Heavy, shake, tile or slate.	Medium shake or good built-up with large rock, inexpensive tile.	Wood or good composition shingle, light shake, or good built-up with rock.	Inexpensive shingles or built-up with rock.
Overhang (2% of total Cost)	36" open or 24" closed.	30" open or small closed.	16" open.	12" to 16" open.
Exterior Walls (10% of total Cost)	Good wood or stucco, masonry veneer on front.	Good wood siding or stucco with some veneer.	Hardboard, wood shingle, plywood or stucco.	Low cost stucco, hardboard or plywood.
Flooring (5% of total Cost)	Good carpet, good sheet vinyl.	Good carpet, sheet vinyl or inlaid resilient.	Average carpet, average resilient tile in bath.	Minimum tile or low cost carpet.
Interior Finish (23% of total cost including finish carpentry, wiring, lighting, etc.)	Gypsum board with heavy texture or plaster with putty coat. Some good sheet wall cover or paneling.	Gypsum board, taped, textured and painted or plaster. Some wall-paper.	Gypsum board taped and textured or colored interior stucco.	Minimum gyp-sum board.
Baths (15% of total Cost)	Vinyl or foil wall cover, ceramic tile over tub with glass shower door, ample mirrors.	Ceramic tile over tub with glass shower door.	Plastic coated hard-board with low cost glass shower door.	Plastic coated hardboard with one small mirror.
Plumbing** (9% of total Cost)	Copper tube, good quality fixtures.	Galvanized pipe, good fixtures.	Average cost fixtures.	Plastic pipe, low cost fixtures.
Special Features (2% of total Cost)	8' sliding glass door, 8' to 10' tile pullman in bath.	8' sliding glass door, good tile or plastic top pullman in bath.	Small tile or plastic pullman in bath.	None.
***For Masonry Walls**	8" textured face reinforced masonry.	8" colored or detailed reinforced masonry.	8" colored block or common brick, reinforced.	8" painted concrete block.

Note: When masonry walls are used in lieu of wood frame walls add 8% to the appropriate cost

****Add the Following Amounts per Kitchen Unit**

	Class 1	Class 2	Class 3	Class 4
Kitchens	Good sink, 8' to 10' of good cabinets and drainboard - $3,220	Average sink and 6' to 8' average cabinet and drainboard - $2,990	Low cost sink, and 5' of cabinets and drainboard - $2,160	Minimum sink, cabinets and drainboard - $1,870

Add the cost of built-in kitchen fixtures from the table of costs for built-in appliances on page 26.

Note: Use the percent of total cost to help identify the correct quality classification.

Motels

9 Units or Less

Estimating Procedure

1. Establish the structure quality class by applying the information on page 20.
2. Multiply the average unit area by the appropriate cost below. The average unit area is found by dividing the total building area on all floors (including office and manager's area, utility rooms, interior hallways and stairway area) by the number of units in the building.
3. Multiply the total from step 2 by the correct location factor listed on page 7.
4. Add, when appropriate, the cost of heating and cooling equipment, porches, balconies, exterior stairs, garages, kitchens, built-in kitchen appliances and fireplaces. See pages 20 and 24 to 28.

Motel, Class 3 & 4

Average Unit Area in Square Feet

Quality Class	200	225	250	275	300	330	375	425	500	600	720
1, Best	104.20	100.49	97.53	95.06	93.02	91.01	88.57	86.43	84.03	81.72	79.85
1 & 2	95.72	92.28	89.58	87.34	85.47	83.58	81.31	79.38	77.16	75.09	73.34
2, Good	88.83	85.66	83.14	81.06	79.33	77.58	75.49	73.70	71.61	69.68	68.07
2 & 3	81.62	78.73	76.40	74.48	72.89	71.29	69.35	67.71	65.82	64.06	62.57
3, Average	75.74	73.04	70.90	69.11	67.63	66.13	64.36	62.81	61.07	59.42	58.05
3 & 4	69.52	67.04	65.07	63.43	62.08	60.70	59.06	57.67	56.04	54.54	53.26
4, Low	63.54	61.25	59.45	57.99	56.72	55.50	53.98	52.69	51.20	49.83	48.67

Note: Add 2% for work above the first floor. Work outside metropolitan areas may cost 2 to 6% less. Add 8% when the exterior walls are masonry. Deduct 2% for area built on a concrete slab.

Motels

10 to 24 Units

Estimating Procedure

1. Establish the structure quality class by applying the information on page 20.
2. Multiply the average unit area by the appropriate cost below. The average unit area is found by dividing the total building area on all floors (including office and manager's area, utility rooms, interior hallways and stairway area) by the number of units in the building.
3. Multiply the total from step 2 by the correct location factor listed on page 7.
4. Add, when appropriate, the cost of heating and cooling equipment, porches, balconies, exterior stairs, garages, kitchens, built-in kitchen appliances and fireplaces. See pages 20 and 24 to 28.

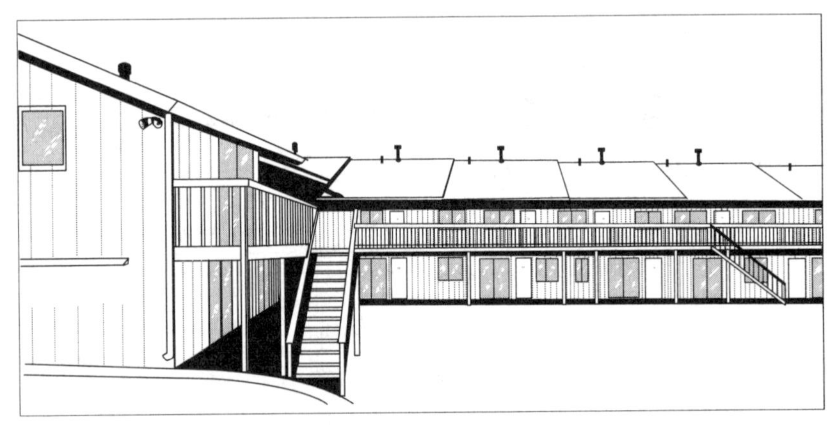

Motel, Class 3

Average Unit Area in Square Feet

Quality Class	200	225	250	275	300	330	375	425	500	600	720
1, Best	100.96	97.38	94.49	92.15	90.17	88.18	85.80	83.74	81.44	79.19	77.39
1 & 2	92.75	89.45	86.79	84.62	82.80	81.00	78.81	76.93	74.79	72.77	71.05
2, Good	86.14	83.08	80.58	78.57	76.91	75.23	73.19	71.43	69.44	67.56	66.02
2 & 3	79.10	76.28	73.99	72.15	70.61	69.02	67.21	65.58	63.75	62.03	60.59
3, Average	73.38	70.77	68.69	66.94	65.50	64.09	62.34	60.85	59.16	57.58	56.22
3 & 4	67.35	64.95	63.04	61.46	60.14	58.81	57.21	55.88	54.28	52.84	51.62
4, Low	61.55	59.34	57.61	56.18	54.96	53.78	52.30	51.05	49.65	48.31	47.19

Note: Add 2% for work above the first floor. Work outside metropolitan areas may cost 2 to 6% less. Add 8% when the exterior walls are masonry. Deduct 2% for area built on a concrete slab.

Motels

Over 24 Units

Estimating Procedure

1. Establish the structure quality class by applying the information on page 20.
2. Multiply the average unit area by the appropriate cost below. The average unit area is found by dividing the total building area on all floors (including office and manager's area, utility rooms, interior hallways and stairway area) by the number of units in the building.
3. Multiply the total from step 2 by the correct location factor listed on page 7.
4. Add, when appropriate, the cost of heating and cooling equipment, porches, balconies, exterior stairs, garages, kitchens, built-in kitchen appliances and fireplaces. See pages 20 and 24 to 28.

Motel, Class 2 & 3

Average Unit Area in Square Feet

Quality Class	200	225	250	275	300	330	375	425	500	600	720
1, Best	97.65	94.16	91.36	89.10	87.16	85.26	82.95	80.99	78.74	76.61	74.86
1 & 2	89.73	86.52	83.94	81.86	80.09	78.33	76.19	74.40	72.34	70.10	68.76
2, Good	83.29	80.36	77.96	76.02	74.39	72.75	70.82	69.09	67.18	65.36	63.86
2 & 3	76.50	73.76	71.58	69.80	68.28	66.79	65.00	63.41	61.67	60.03	58.64
3, Average	70.96	68.43	66.41	64.74	63.33	61.97	60.30	58.87	57.24	55.67	54.41
3 & 4	65.13	62.78	60.91	59.38	58.13	56.87	55.30	53.99	52.51	51.10	49.92
4, Low	59.54	57.40	55.68	54.29	53.13	51.98	50.56	49.36	48.01	46.72	45.61

Note: Add 2% for work above the first floor. Work outside metropolitan areas may cost 2 to 6% less. Add 8% when the exterior walls are masonry. Deduct 2% for area built on a concrete slab.

Additional Costs for Residential Structures

Covered Porches

Estimate covered porches by applying a fraction of the main building square foot cost.

Porch Description	Suggested Fraction
Ground level floor (usually concrete) without banister, with no ceiling and shed-type roof.	1/4 to 1/3
High (house floor level) floor (concrete or wood) with light banister, no ceiling and shed-type roof.	1/3 to 1/2
Same as above with a finished ceiling and roof like the residence (most typical).	1/2
Same as above but partially enclosed with screen or glass.	1/2 to 2/3
Enclosed lean-to (sleeping porch, etc.) with lighter foundation, wall structure, interior finish or roof than that of house to which it is attached.	1/2 to 3/4
Roofed, enclosed, recessed porch, under the same roof as the main building and with the same type and quality foundation (includes shape costs).	3/4
Roofed, enclosed, recessed porch with the same type roof and foundation as the main building (includes shape costs).	4/4
Good arbor or pergola with floor.	1/4 to 1/3

Uncovered Concrete Decks, cost per square foot, 4" thick, no reinforcing

	On Grade	1' High	2' High	3' High	4' High
Less than 100 square feet	$4.48	$6.41	$10.14	$13.60	$16.94
100 to 200 square feet	4.03	5.80	7.69	10.14	13.14
200 to 400 square feet	3.47	4.23	6.58	8.80	10.70
Over 400 square feet	3.36	4.02	6.24	8.14	9.48

Uncovered Wood Decks, cost per square foot

1" board floor on 2" x 6" joists 24" o.c.	$13.09 to $14.97
2" board floor on 4" x 6" girders 48" o.c.	14.87 to 17.30

Porch Roofs, cost per square foot

Type	Cost per Square Foot
Unceiled shed roof	$4.22 to $4.89
Ceiled shed roof	8.05 to 8.61
Unceiled gable roof	4.65 to 5.82
Ceiled gable roof	7.21 to 10.14

(Assumes 3 ply built-up roof & gravel)

Alternate Roof Covers	Cost Difference per S.F.
Corrugated aluminum	Deduct $.22 to $.43
Roll asphalt	Deduct .28 to .38
Fiberglass shingles	Deduct .33 to .43
Wood shakes	Add .63 to .99
Clay or concrete tile	Add 4.77 to 5.33
Slate	Add 5.66 to 8.21

Residential Basements, cost per square foot

Size	Unfinished Basements	Finished Basements
Less than 400 square feet	$39.34 to $56.55	$44.91 to $66.28
400 - 1,000 square feet	28.41 to 34.04	36.22 to 46.38
Over 1,000 square feet	17.59 to 24.86	26.66 to 34.96

All basement costs assume normal soil conditions, 7'6" headroom, no plumbing, partitions, or windows. Unfinished basements are based on reinforced concrete floors and walls, open ceilings and minimum lighting. Finished basement costs are based on reinforced concrete floors with resilient tile cover, reinforced concrete walls with plywood panel finish, acoustical tile ceiling and lighting similar to average quality residences. Basement costs will be lower in areas where normal footing depths approach 8 feet. These figures assume a normal footing depth of 2 feet below grade.

Additional Costs for Residential Structures

Balconies, Standard Wood Frame, cost per square foot

Supported by 4" x 4" posts, 1" wood floor, open on underside, open 2" x 4" railing.	$11.54 to $12.76
Supported by 4" x 4" posts, 1" wood floor, sealed on underside, solid stucco or wood siding on railing.	12.86 to 14.41
Supported by steel columns, lightweight concrete floors, sealed on underside, solid stucco or open grillwork railing	15.64 to 17.74

Heating and Cooling Equipment

Prices include wiring and minimum duct work.
Use the higher figures for smaller residences and in more extreme climates where greater heating and cooling density is required. Cost per square foot of heated or cooled area.

Type	Perimeter Outlets	Overhead Outlets
Central Ducted Air Systems, Single Family		
Forced air heating	$4.09 to $4.53	$3.39 to $4.09
Forced air heating and cooling	5.16 to 5.54	4.63 to 4.85
Gravity heat	3.12 to 4.20	—
Central Ducted Air Systems, Multi-Family		
Forced air heating	3.71 to 4.09	3.56 to 4.09
Forced air heating and cooling	5.16 to 5.66	4.53 to 4.85
Motel Units		
Forced air heating	3.12 to 3.39	4.30 to 4.63
Forced air heating and cooling	5.28 to 5.66	5.11 to 5.28
Circulating hot and cold water system	10.34 to 12.17	10.34 to 12.17

Floor and Wall Furnaces, cost ea.

Single floor unit	$912 to$1,045
Dual floor unit	1,631 to 1,756
Single wall unit	618 to 757
Dual wall unit	704 to 827
Thermostat control, add	154 to 213

Electric Baseboard Units, cost ea.

500 watts, 3'	$277 to $293
1,000 watts, 4'	373 to 415
1,500 watts, 6'	415 to 438
2,000 watts, 8'	473 to 534
2,500 watts, 10'	587 to 618
3,000 watts, 12'	671 to 747

Outside Stairways, cost per square foot of horizontal area

Standard wood frame, wood steps with open risers, open on underside, open 2" x 4" railing, unpainted.	$10.97 to $11.75
Standard wood frame w/ solid wood risers, sealed on underside, solid stucco or wood siding on railing.	13.31 to 14.97
Precast concrete steps with open risers, steel frame, pipe rail with ornamental grillwork.	27.85 to 32.16

Window Type or Thru-the-Wall Refrigerated Room Coolers, cost ea.

1/3 ton	$580 to $681
1/2	651 to 783
3/4	704 to 1,077
1	816 to 937
1-1/2	981 to 1,183
2	1,183 to 1,450

Ton = 12,000 Btu

Electric Wall Heaters, cost ea.

1,000 watts	$341 to $405
2,000	405 to 458
3,000	415 to 534
3,500	426 to 554
4,000	554 to 666
4,500	661 to 821
Add for circulating fan	71 to 93
Add for thermostat	77 to 87

Additional Costs for Residential Structures

Built-In Equipment

Add these costs only when the item is not included in the specification for the applicable quality class. These costs include plumbing, wiring and installation.

Appliances, cost each

Drop-in type, gas or electric	
Range with single oven below	$ 862 to $1,255
Built-in type, gas or electric	
Range with single oven below	1,034 to 1,348
Range with oven above and below	1,328 to 1,631
Surface cooking units, four elements without grill	323 to 539
Oven units, single gas or electric	668 to 888
Oven units, double, gas or electric	862 to 1,506
Electronic ovens, with range and microwave	1,506 to 1,830
Microwave, built-in	$ 339to $ 668
Range hood and fan	281 to 377
Refrigerators, built-in	
Under counter, 5 cubic feet	447 to 668
Single door, 12 cubic feet	764 to 926
Double door, 12 cubic feet	1,151 to 1,307
Double door, 17 cubic feet	1,892 to 2,248
Refrigerator, range, oven and sink	2,007 to 2,300
Dishwashers	458 to 792
Garbage disposers	162 to 350
Trash compactors	350 to 641

Fireplaces, cost each

	1 Story	2 Story
Freestanding wood burning heat circulating prefab metal fireplace with interior flue, base and cap.	$1,161	$1,547
Zero-clearance insulated prefab metal fireplace, brick face	1,652	2,185
5' base, common brick on interior face	2,175	2,384
6' base, common brick, used brick or natural stone on interior face.	3,439	3,555
8' base, used brick or natural stone on interior face, raised hearth.	4,799	5,291

Residential Garages and Carports

Attached and detached garages for single family dwellings are usually classified in the same quality class as the main structure. The attached garage assumes a 20 foot wall in common with the main structure. The costs are per square foot of floor area and include only light excavation for foundations. Multiply the square foot cost below by the correct location factor on page 7 to find the square foot cost for any building site. Costs include no interior finish. Where the interior finish is similar to the main dwelling and the garage walls are in vertical alignment with second floor walls, the square foot cost will be about 80% of the main dwelling square foot cost.

Square Foot Area for Attached Garages for Single Family Dwellings

Quality Class	220	260	280	320	360	400	440	480	540	600	720
1, Best	30.52	29.01	28.46	27.50	26.76	26.21	25.76	25.38	24.90	24.53	23.98
1 & 2	28.91	27.50	26.94	26.04	25.36	24.84	24.38	24.01	23.59	23.26	22.72
2, Good	28.60	26.31	25.80	24.94	24.27	23.74	23.34	22.96	22.56	22.24	21.75
2 & 3	26.05	24.78	24.29	23.49	22.88	22.39	21.99	21.66	21.27	20.95	20.46
3, Average	24.79	23.59	23.11	22.35	21.76	21.28	20.93	20.59	20.24	19.95	19.50
3 & 4	22.53	21.47	21.00	20.32	19.80	19.36	19.01	18.76	18.39	18.14	17.72
4, Low	20.35	19.35	18.95	18.32	17.86	17.45	17.16	16.88	16.62	16.33	16.00

Square Foot Area for Detached Garages for Single Family Dwellings

Quality Class	220	260	280	320	360	400	440	480	540	600	720
1, Best	39.22	36.36	35.25	33.46	32.12	31.02	30.16	29.45	28.59	27.91	26.94
1 & 2	37.09	34.40	33.34	31.63	30.39	29.34	28.51	27.85	27.02	26.41	25.48
2, Good	35.23	32.68	31.69	30.08	28.87	27.89	27.11	26.47	25.67	25.08	24.19
2 & 3	33.39	30.96	30.02	28.51	27.35	26.44	25.67	25.06	24.35	23.76	22.93
3, Average	31.60	29.29	28.39	26.96	25.86	24.97	24.28	23.72	23.05	22.49	21.70
3 & 4	28.58	26.52	25.69	24.40	23.42	22.59	21.99	21.48	20.86	20.35	19.62
4, Low	25.63	23.77	23.04	21.87	20.98	20.31	19.73	19.27	18.70	18.25	17.61

Carports, including asphalt floor, cost between $6.58 and $7.91 per square foot, depending on the type of roof cover.

Additional Costs for Residential Structures

Costs for Multi-Family Residential Bathrooms beyond 1 per unit

	Class 1 Best Quality	Class 2 Good Quality	Class 3 Average Quality	Class 4 Low Quality
2 or 3 units				
2 fixture bath	$4,416	$4,125	$3,564	$2,833
3 fixture bath	6,677	6,106	5,094	4,641
4 fixture bath	8,367	8,142	7,237	5,880
4 to 9 units				
2 fixture bath	4,189	3,844	3,166	2,714
3 fixture bath	6,331	5,708	4,749	3,844
4 fixture bath	8,254	7,581	6,163	5,145
10 or more units				
2 fixture bath	4,070	3,736	3,112	2,434
3 fixture bath	5,880	5,201	4,641	3,511
4 fixture bath	8,142	7,183	5,998	4,976

Half Story Areas

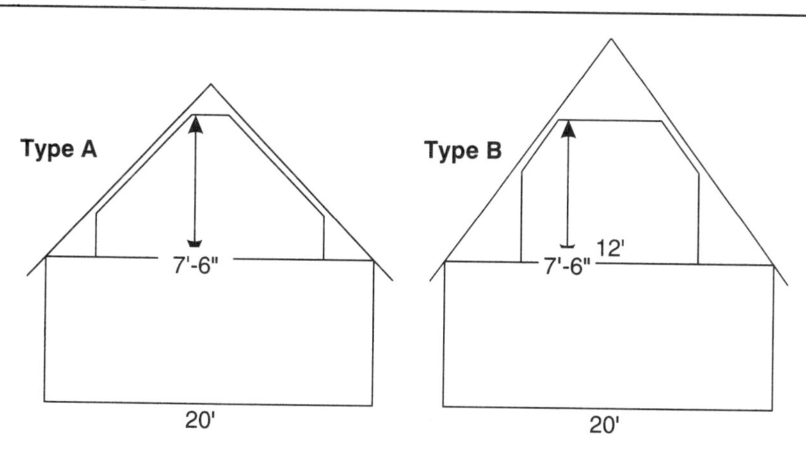

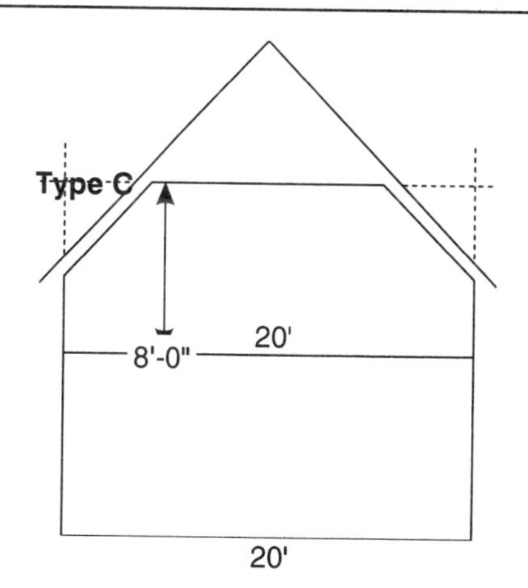

Use a fraction of the basic square foot cost for figuring the reduced headroom floor area Type "C" includes typical dormers.

Type	Same Finish As Main Area	Lesser Quality Finish
A	1/3	1/4
B	1/2	1/3
C	2/3	1/2

Split Level Construction

All classes, all quality add 3% to base costs

Hydraulic Elevators, per shaft cost for car and machinery, 2 stops

Capacity	100 F.P.M.	125 F.P.M.	150 F.P.M.
1,200 lbs	$34,300	—	—
2,000	39,600	$50,700	$64,700
2,500	45,800	57,700	67,500
3,000	49,200	63,750	71,900
3,500	—	67,100	74,300
4,000	—	70,500	79,050

Add for deluxe car, $7,100. Add for each additional stop over 2: baked enamel doors $7,650, stainless steel doors $8,250.

Multi-Family and Motel Garages Cost Per Square Foot

Garages built at ground level under a multi-family or motel unit. The costs below include the following components:

1. A reinforced concrete floor in all areas.
2. Exterior walls, on one long side and two short sides, made up of a wood frame and good quality stucco, wood siding or masonry veneer.
3. A finished ceiling in all areas.
4. The difference between the cost of a standard wood frame floor structure at second floor level and one at ground level.
5. An inexpensive light fixture for each 600 square feet.

Where no exterior walls enclose the two short sides, use $10.00 per square foot.

Garages built as separate structures for multi-family or motel units. The costs below include the following components:

1. Foundations.
2. A reinforced concrete floor in all areas.
3. Exterior walls on one long side and two short sides, made up of a wood frame and good quality stucco, wood siding or masonry veneer.
4. Steel support columns supporting the roof.
5. A wood frame roof structure with composition tar and gravel, wood shingle or light shake cover. No interior ceiling finish.
6. An inexpensive light fixture for each 600 square feet.

Use the location modifiers on page 7 to adjust garage costs to any area.

Basement Garages

Costs are listed below for basement garages built on one level, approximately 5 feet below grade, directly below 2 to 4 story dwelling structures with perimeter walls in vertical alignment. These costs include:

1. Excavation to 5' below ground line.
2. Full wall enclosure.
3. Typical storage facilities.
4. Minimum lighting.
5. Concrete floors.

Access stairways and driveways ramps outside the perimeter walls should be included when calculating the area of the garage.

Use the location modifiers on page 7 to adjust garage costs to any area.

Ground Level Garages

Area	400	800	1,200	2,000	3,000	5,000	10,000	20,000
Cost	23.26	20.71	17.95	17.24	16.08	16.47	14.79	11.16

Separate Structure Garages

Area	400	800	1,200	2,000	3,000	5,000	10,000	20,000
Cost	26.17	24.13	21.32	20.60	20.46	18.67	17.82	17.45

Basement Garages

Type	5,000	7,500	10,000	15,000	20,000	30,000	40,000	60,000
Reinforced concrete exterior walls and columns. Flat concrete roof slab.	33.61	32.40	30.37	28.80	28.34	27.97	27.35	27.24
Concrete block exterior walls, reinforced concrete columns. Flat concrete roof slab .	33.06	31.56	28.44	26.89	26.72	26.29	26.16	25.92
Concrete block exterior walls, steel posts and beams, light concrete/metal roof fireproofed with spray plaster.	29.67	28.10	26.42	25.82	24.72	24.49	24.01	23.29
Concrete block exterior walls, wood posts and beams, light concrete/metal roof fireproofed with spray plaster.	27.99	26.79	24.72	23.29	23.03	22.19	21.73	21.34

Cabins and Recreational Dwellings

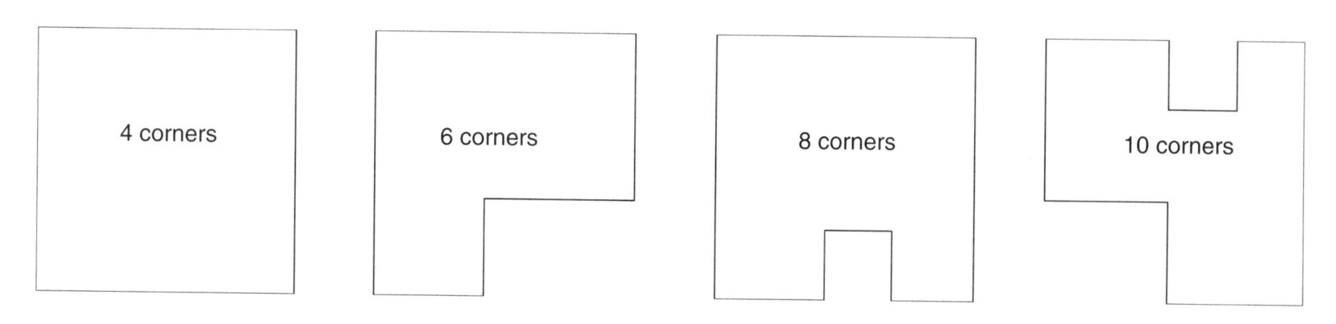

Example of Dwelling Shapes

Cabins and recreational dwellings are designed for single family occupancy, usually on an intermittent basis. These structures are characterized by a more rustic interior and exterior finish and often have construction details which would not meet building requirements in metropolitan areas. Classify these structures into either "conventional type" or "A-frame" construction. Conventional dwellings have an exterior wall which is approximately 8 feet high on all sides. A-frame cabins have a sloping roof which reduces the horizontal area 8 feet above the first floor to between 50% and 75% of the first floor area.

The shape of the outside perimeter is an important consideration in estimating the construction cost. Generally, the more complex the shape, the more expensive the structure per square foot of floor area. The shape classification is based on the outline formed by the outer-most exterior walls, including the garage area, regardless of the varying level. Most cabins and recreational dwellings have 4, 6, 8 or 10 corners as illustrated.

Cabins and recreational dwellings are often built under difficult working conditions and in remote sites. Allowance is made here for transportation of materials to typical recreational locations and for the higher labor cost which is usually associated with greater travel distances and adverse conditions. Individual judgments may be necessary in evaluating the cost impact of the dwelling location. The costs assume construction by skilled professional craftsmen. Where non-professional labor or second quality materials are used, use the next lower quality classification than might otherwise apply. If the structure is assembled from prefabricated components, use costs for the next lower half class.

Conventional Recreational Dwellings

Quality Classification

	Class 1 Best Quality	Class 2 Good Quality	Class 3 Average Quality	Class 4 Low Quality
Foundation (4%)	Concrete slab	Concrete slab	Concrete slab	Concrete slab
Foundation costs will vary greatly with substrate, type, and location.				
Framing * (10%)	Wood frame.	Wood frame.	Wood frame.	Wood frame.
Floor Framing (5%)	4" x 8" girders 48" o.c. with 2" T&G subfloor, or 2" x 8" to 2" x 10" joists 16" o.c. with 1" plywood subfloor.	4" x 8" girders 48" o.c. with 1-1/4" plywood or 2" T&G subfloor, or 2" x 8" to 10" joists 16" o.c. with 1" plywood subfloor.	4" x 6" girders 48" o.c. with 1-1/4" plywood or 2" T&G subfloor, or 2" x 6" to 8" joists 16" o.c. with 1" plywood subfloor.	4" x 6" girders 48" o.c. with 1-1/4" plywood or 2" x 6" joists 16" o.c. with 1" plywood subfloor.
Roof Framing (5%)	Same as floor or heavy glu-lam beams with 2" x 8" or 2" x 10" purlins, 3" T&G flat or low pitch deck and built-up roof.	4" x 8" girders 48" o.c. with 2" or 3" T&G sheathing, or 2" x 6" to 10" rafters 12" to 16" o.c. with 1" sheathing.	4" x 8" girders 48" o.c. with 2" T&G sheathing, or 2" x 6" to 8" rafters 12" to 24" o.c. with 1" sheathing.	4" x 8" girders 48" o.c. with 1-1/4" plywood or 2" T&G sheathing, or 2" x 6" to 8" rafters up to 24" o.c. with 1" sheathing.
Exterior Finish (10%)	Good plywood, lap or board and batt siding. Good trim.	Average to good plywood or board siding or good shingles. Some trim.	Average plywood composition or board siding or shingles. Little or no trim.	Low cost plywood or wood siding or low cost shingle or composition siding.
Windows (2%)	Larger area of good insulated wood or metal windows.	Adequate number of good wood or metal windows.	Average quality wood or metal windows.	Few low cost windows.
Roofing (8%)	Heavy shakes, usually with steep pitch and several roof planes.	Medium shakes, usually with medium to steep pitch.	Wood or composition shingles with medium to steep pitch.	Low cost composition shingles with medium pitch.
Flooring (5%)	Good carpet or hardwood. Sheet vinyl in kitchen and baths.	Average carpet with sheet vinyl or tile in kitchen and baths.	Low cost carpet with resilient tile in kitchen and baths.	Composition tile.
Interior Finish (23% of total cost including finish carpentry, wiring, lighting, fireplace, etc.)	Good quality hardwood veneer paneling, with some decorative exposed structural members.	Good textured gypsum wallboard with some plywood or knotty pine paneling.	Textured gypsum wallboard or average quality plywood or hardboard paneling.	Low cost hardboard paneling or gypsum wallboard.
Bathrooms (5%)	Two 3-fixture and one 2-fixture baths. Good fixtures.	Two 3-fixture baths. Good fixtures.	Two 3-fixture baths.	One 3-fixture bath.
Kitchen (5%)	15' to 18' of good hardwood veneer base and wall cabinet. 15' to 18' of good plastic or ceramic drainboard.	12' to 16' of hardwood veneer base cabinet with matching wall cabinets. 12' to 16' of plastic or ceramic drainboard.	8' to 12' of plywood veneer or painted base cabinet with matching wall cabinets. 8' to 12' of plastic drainboard.	8' to 10' of minimum base cabinets with matching wall cabinets. 8' to 10' of minimum plastic drainboard.
Plumbing (14%)	Nine good quality fixtures with one large or two 30 gallon water heaters, copper supply piping.	Seven good quality fixtures and one water heater.	Seven average quality fixtures and one water heater.	Four low cost fixtures and one water heater. Plastic supply piping.
Special Features (4%)	Built-in oven, range, dishwasher, disposer, range hood with good insulation, good lighting fixtures, insulated sliding glass door and ornate entry door.	Built-in range, oven and range hood, some insulation, 8' sliding glass door, average electric fixtures.	Drop-range and hood, some insulation, low cost electric fixtures.	Minimum electric fixtures.
***For Masonry Walls**	Textured block, tile or brick with masonry facing.	Colored or detailed concrete block, brick or tile with facing.	Colored reinforced concrete block, clay tile or common brick.	Reinforced concrete block block or clay tile.

When masonry walls are used in lieu of wood frame walls, add 5% to the appropriate S.F. cost.

Note: Use the percent of total cost to help identify the correct quality classification.

Conventional Recreational Dwellings

Four Corners

Estimating Procedure

1. Establish the structure quality class by applying the information on page 30.
2. Multiply the structure floor area by the appropriate cost listed below.
3. Multiply the total from step 2 by the correct location factor listed on page 7.
4. Add, when appropriate, the cost of a deck or porch, paving, fireplace, garage or carport, heating, extra plumbing fixtures, supporting walls, half story areas, construction on hillside lots, and construction in remote areas. See page 39.

Conventional Recreational Dwelling, Class 3

Square Foot Area

Quality Class	400	500	600	700	800	900	1,000	1,100	1,200	1,300	1,400
1, Best	124.50	111.81	103.06	96.61	91.69	87.74	84.51	81.86	79.55	77.59	75.86
1 & 2	115.08	103.33	95.26	89.33	84.75	81.09	78.13	75.63	73.52	71.71	70.14
2, Good	106.44	95.58	88.12	82.63	78.38	75.02	72.27	69.95	68.02	66.32	64.86
2 & 3	101.00	90.70	83.61	78.37	74.37	71.18	68.55	66.38	64.49	62.90	61.51
3, Average	96.32	86.51	79.73	74.77	70.94	67.90	65.40	63.30	61.55	60.04	58.69
3 & 4	89.81	80.66	74.33	69.70	66.14	63.30	60.98	59.02	57.38	55.98	54.72
4, Low	82.62	74.21	68.39	64.12	60.83	58.21	56.08	54.28	52.76	51.48	50.33

Square Foot Area

Quality Class	1,500	1,600	1,700	1,800	2,000	2,200	2,400	2,600	2,800	3,000	3,200
1, Best	73.86	72.58	71.44	70.38	68.62	67.13	65.88	64.80	63.87	63.05	62.29
1 & 2	68.17	66.97	65.95	64.97	63.36	61.98	60.80	59.81	58.96	58.17	57.52
2, Good	63.40	62.28	61.34	60.44	58.92	57.63	56.57	55.62	54.83	54.10	53.46
2 & 3	60.19	59.16	58.19	57.36	55.92	54.72	53.67	52.81	52.04	51.37	50.76
3, Average	57.57	56.57	55.67	54.86	53.49	52.34	51.37	50.52	49.77	49.14	48.57
3 & 4	54.22	53.27	52.45	51.68	50.38	49.31	48.37	47.58	46.88	46.28	45.73
4, Low	49.71	48.94	48.26	47.03	45.99	45.14	44.40	43.77	43.18	42.70	42.06

Note: Add 4% to the square foot cost for floors above the second floor level.

Conventional Recreational Dwellings

Six Corners

Estimating Procedure

1. Establish the structure quality class by applying the information on page 30.
2. Multiply the structure floor area by the appropriate cost listed below.
3. Multiply the total from step 2 by the correct location factor listed on page 7.
4. Add, when appropriate, the cost of a deck or porch, paving, fireplace, garage or carport, heating, extra plumbing fixtures, supporting walls, half story areas, construction on hillside lots, and construction in remote areas.
   See page 39.

Conventional Recreational Dwelling, Class 4

Square Foot Area

Quality Class	400	500	600	700	800	900	1,000	1,100	1,200	1,300	1,400
1, Best	125.98	113.20	104.43	98.01	93.05	89.15	85.97	83.28	81.05	79.10	77.41
1 & 2	116.51	104.67	96.55	90.60	86.05	82.42	79.47	77.01	74.93	73.14	71.56
2, Good	107.84	96.90	89.37	83.90	79.68	76.31	73.58	71.31	69.36	67.70	66.25
2 & 3	102.42	91.97	84.87	79.63	75.62	72.43	69.85	67.70	65.85	64.30	62.90
3, Average	97.75	87.81	81.01	76.05	72.20	69.20	66.67	64.62	62.87	61.38	60.06
3 & 4	91.19	81.93	75.57	70.92	67.36	64.50	62.20	60.28	58.66	57.25	56.02
4, Low	83.73	75.20	69.36	65.12	61.83	59.24	57.14	55.36	53.86	52.56	51.45

Square Foot Area

Quality Class	1,500	1,600	1,700	1,800	2,000	2,200	2,400	2,600	2,800	3,000	3,200
1, Best	75.54	74.24	73.09	72.02	70.23	68.72	67.48	66.41	65.47	64.67	63.93
1 & 2	69.87	68.68	67.61	66.62	64.96	63.59	62.42	61.44	60.58	59.81	59.14
2, Good	64.75	63.64	62.65	61.73	60.22	58.93	57.84	56.91	56.12	55.44	54.80
2 & 3	61.60	60.54	59.56	58.72	57.27	56.03	55.03	54.14	53.38	52.71	52.12
3, Average	58.92	57.89	56.98	56.15	54.76	53.60	52.63	51.80	51.05	50.40	49.86
3 & 4	55.58	54.64	53.76	53.00	51.67	50.56	49.66	48.87	48.17	47.57	47.05
4, Low	51.01	50.22	49.48	48.27	47.22	46.38	45.63	44.98	44.41	43.91	43.25

Note: Add 4% to the square foot cost for floors above the second floor level.

Conventional Recreational Dwellings

Eight Corners

Estimating Procedure

1. Establish the structure quality class by applying the information on page 30.
2. Multiply the structure floor area by the appropriate cost listed below.
3. Multiply the total from step 2 by the correct location factor listed on page 7.
4. Add, when appropriate, the cost of a deck or porch, paving, fireplace, garage or carport, heating, extra plumbing fixtures, supporting walls, half story areas, construction on hillside lots, and construction in remote areas.
   See page 39.

Conventional Recreational Dwelling, Class 1 & 2

Square Foot Area

Quality Class	400	500	600	700	800	900	1,000	1,100	1,200	1,300	1,400
1, Best	127.14	114.47	105.78	99.41	94.50	90.60	87.43	84.77	82.51	80.56	78.87
1 & 2	117.66	105.93	97.86	91.97	87.43	83.83	80.90	78.42	76.35	74.53	73.00
2, Good	108.95	98.11	90.65	85.18	80.98	77.65	74.91	72.64	70.71	69.04	67.61
2 & 3	103.52	93.21	86.12	80.93	76.94	73.76	71.18	68.99	67.18	65.60	64.23
3, Average	98.80	88.95	82.22	77.24	73.45	70.40	67.94	65.88	64.12	62.62	61.31
3 & 4	91.88	82.74	76.45	71.85	68.27	65.47	63.17	61.24	59.64	58.22	57.00
4, Low	85.32	76.84	70.99	66.69	63.40	60.80	58.65	56.87	55.37	54.07	52.94

Square Foot Area

Quality Class	1,500	1,600	1,700	1,800	2,000	2,200	2,400	2,600	2,800	3,000	3,200
1, Best	76.84	75.57	74.45	73.43	71.66	70.18	68.92	67.77	66.81	65.96	65.19
1 & 2	71.31	70.17	69.08	68.17	66.53	65.13	63.96	62.95	62.03	61.22	60.54
2, Good	66.13	65.07	64.10	63.22	61.70	60.41	59.31	58.36	57.55	56.78	56.14
2 & 3	62.84	61.79	60.89	60.06	58.63	57.40	56.39	55.46	54.66	53.94	53.33
3, Average	60.18	59.19	58.31	57.54	56.11	54.97	53.97	53.09	52.34	51.66	51.05
3 & 4	56.87	55.94	55.10	54.37	53.05	51.93	51.00	50.20	49.47	48.82	48.27
4, Low	52.69	52.30	51.53	50.84	49.59	48.58	47.69	46.94	46.26	45.66	45.11

Note: Add 4% to the square foot cost for floors above the second floor level.

Conventional Recreational Dwellings

Ten Corners

Estimating Procedure

1. Establish the structure quality class by applying the information on page 30.
2. Multiply the structure floor area by the appropriate cost listed below.
3. Multiply the total from step 2 by the correct location factor listed on page 7.
4. Add, when appropriate, the cost of a deck or porch, paving, fireplace, garage or carport, heating, extra plumbing fixtures, supporting walls, half story areas, construction on hillside lots, and construction in remote areas.
   See page 39.

Conventional Recreational Dwelling, Class 1

Square Foot Area

Quality Class	400	500	600	700	800	900	1,000	1,100	1,200	1,300	1,400
1, Best	128.58	115.97	107.29	100.87	95.98	92.04	88.83	86.15	83.90	81.94	80.24
1 & 2	119.02	107.33	99.31	93.38	88.82	85.19	82.24	79.76	77.66	75.85	74.26
2, Good	110.28	99.48	92.04	86.53	82.29	78.95	76.18	73.91	71.97	70.29	68.82
2 & 3	104.89	94.59	87.51	82.27	78.25	75.05	72.45	70.28	68.44	66.81	65.43
3, Average	100.30	90.45	83.67	78.71	74.85	71.77	69.31	67.23	65.42	63.90	62.60
3 & 4	93.77	84.58	78.23	73.59	69.98	67.15	64.80	62.84	61.17	59.75	58.52
4, Low	86.65	78.16	72.31	68.01	64.68	62.05	59.87	58.08	56.55	55.24	54.08

Square Foot Area

Quality Class	1,500	1,600	1,700	1,800	2,000	2,200	2,400	2,600	2,800	3,000	3,200
1, Best	78.37	77.08	75.93	74.88	73.10	71.60	70.33	69.29	68.32	67.49	66.78
1 & 2	72.63	71.42	70.34	69.39	67.73	66.34	65.18	64.20	63.31	62.57	61.88
2, Good	67.32	66.21	65.21	64.32	62.77	61.49	60.41	59.49	58.67	57.97	57.35
2 & 3	64.22	63.13	62.19	61.35	59.89	58.65	57.62	56.74	55.97	55.28	54.71
3, Average	61.52	60.51	59.58	58.77	57.35	56.20	55.21	54.39	53.62	52.98	52.39
3 & 4	58.16	57.20	56.34	55.57	54.26	53.15	52.20	51.43	50.71	50.10	49.56
4, Low	53.60	52.81	52.10	50.87	49.80	48.94	48.17	47.53	46.96	46.46	45.76

Note: Add 4% to the square foot cost for floors above the second floor level.

"A-Frame" Cabins

Quality Classification

	Class 1 Best Quality	Class 2 Good Quality	Class 3 Average Quality	Class 4 Low Quality
Framing (10% of total cost)	Wood frame.	Wood frame.	Wood frame.	Wood frame.
Floor Framing (5% of total cost)	4" x 8" girders 48" o.c. with 2" T&G subfloor, or 2" x 6" to 2" x 8" joists 16" o.c. with 1" subfloor.	4" x 8" girders 48" o.c. with 1-1/4" plywood or 2" T&G subfloor, or 2" x 6" to 2" x 8" joists 16" o.c. with 1" subfloor.	4" x 6" girders 48" o.c. with 1-1/4" plywood or 2" T&G subfloor, or 2" x 6" joists 16" o.c. with 1" subfloor.	4" x 6" girders 48" o.c. with 1-1/4" plywood or 2" T&G subfloor, or 2" x 6" joists 16" o.c. with 1" subfloor.
Roof Framing (8% of total cost)	4" x 8" at 48" o.c. with 2" or 3" T&G sheathing.	4" x 8" at 48" o.c. with 2" or 3" T&G sheathing.	4" x 8" at 48" o.c. with 2" T&G sheathing.	4" x 8" at 48" o.c. with 1-1/4" plywood or 2" T&G sheathing.
Gable End Finish (5% of total cost)	Good plywood, lap board or board and batt.	Average to good plywood, or boards.	Average plywood, board or wood shingle.	Low cost plywood, shingle or composition siding.
Windows (2% of total cost)	Good quality large insulated wood or metal windows.	Average quality insulated wood or metal windows.	Average quality wood or metal windows.	Small glass area of low cost windows.
Roofing (10% of total cost)	Heavy wood shakes.	Medium wood or aluminum shakes.	Wood or composition shingles.	Low cost composition shingles.
Flooring (5% of total cost)	Good carpet or hardwood with sheet vinyl in kitchen and baths.	Average to good quality carpet with good tile or sheet vinyl in kitchen and baths.	Average quality carpet with resilient tile in kitchen and baths.	Composition tile.
Interior Finish (25% of total cost including finish carpentry, wiring, lighting, fireplace, etc.)	Good quality hardwood veneer paneling.	Good textured gypsum wallboard, good plywood or knotty pine paneling.	Textured gypsum wallboard or plywood paneling.	Low cost paneling or wallboard.
Bathrooms (5% of total cost)	Two 3-fixture baths and one 2-fixture bath, good fixtures.	Two 3-fixture baths, good fixtures.	Two 3-fixture baths, average fixtures.	One 3-fixture bath.
Kitchen (5% of total cost)	15' to 18' good quality hardwood veneer base cabinet with matching wall cabinets. 15' to 18' of good quality plastic or ceramic tile drain board.	12' to 16' of hardwood veneer base cabinet with matching wall cabinets. 12' to 16' of plastic or ceramic tile drainboard.	8' to 12' of average quality veneer or painted base cabinets with matching wall cabinets. 8' to 12' of plastic drainboard.	6' to 8' of minimum base cabinets with matching wall cabinets. 6' to 8' of minimum plastic drainboard.
Plumbing (15% of total cost)	Nine good quality fixtures and one larger or two 30 gallon water heaters. Copper supply piping.	Seven good quality fixtures and one water heater.	Seven average quality fixtures and one water heater.	Four low cost fixtures and one water heater. Plastic supply pipe.
Special Features (5% of total cost)	Built-in oven, range, dishwasher, disposer, range hood with good insulation, good lighting fixtures, insulated sliding glass door and ornate entry door.	Built-in range, oven and range hood, some insulation, 8' sliding glass door, average electric fixtures.	Drop-in range and hood, some insulation, low cost electric fixtures.	Minimum electric fixtures.

Note: Use the percent of total cost to help identify the correct quality classification.

"A-Frame" Cabins

Four Corners

Estimating Procedure

1. Establish the structure quality class by applying the information on page 35.
2. Multiply the structure floor area by the appropriate cost listed below.
3. Multiply the total from step 2 by the correct location factor listed on page 7.
4. Add, when appropriate, the cost of a deck or porch, paving, fireplace, garage or carport, heating, extra plumbing fixtures, supporting walls, half story areas, construction on hillside lots, and construction in remote areas.
   See page 39.

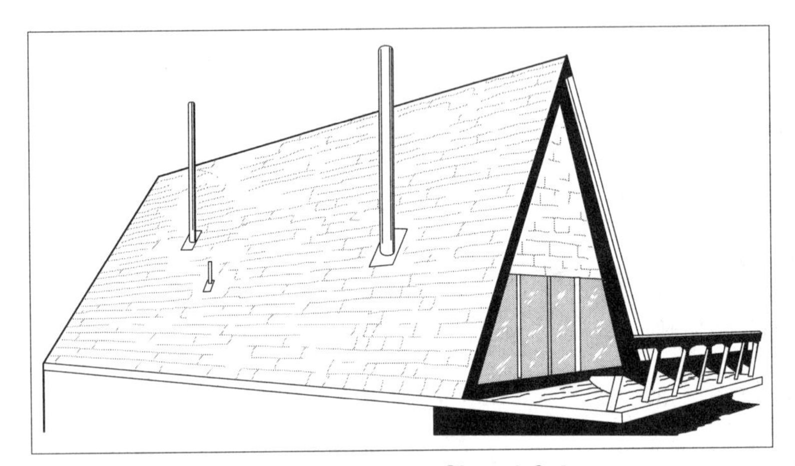

"A-Frame" Cabin, Class 3 & 4

Square Foot Area

Quality Class	400	500	600	700	800	900	1,000	1,100	1,200	1,300	1,400
1, Best	142.43	128.52	118.86	111.74	106.21	101.81	98.18	95.17	92.58	90.33	88.39
1 & 2	130.86	118.09	109.21	102.65	97.59	93.48	90.22	87.43	85.05	83.02	81.21
2, Good	120.08	108.36	100.24	94.21	89.59	85.84	82.80	80.21	78.06	76.16	74.52
2 & 3	113.36	102.28	94.61	88.94	84.54	81.03	78.13	75.75	73.68	71.91	70.34
3, Average	107.38	96.89	89.62	84.24	80.06	76.74	74.01	71.73	69.81	68.12	66.64
3 & 4	97.48	87.95	81.31	76.46	72.69	69.67	67.20	65.12	63.34	61.81	60.50
4, Low	87.43	78.90	72.98	68.62	65.22	62.51	60.29	58.44	56.82	55.46	54.27

Square Foot Area

Quality Class	1,500	1,600	1,700	1,800	2,000	2,200	2,400	2,600	2,800	3,000	3,200
1, Best	85.25	83.87	82.62	81.48	79.50	77.87	76.49	75.28	74.22	73.29	72.46
1 & 2	78.68	77.40	76.20	75.20	73.38	71.86	70.59	69.47	68.51	67.65	66.89
2, Good	72.60	71.39	70.32	69.36	67.70	66.31	65.13	64.10	63.18	62.39	61.70
2 & 3	68.90	67.77	66.77	65.85	64.25	62.96	61.80	60.85	59.98	59.25	58.56
3, Average	65.46	64.36	63.40	62.58	61.07	59.79	58.73	57.80	57.01	56.29	55.65
3 & 4	60.22	59.24	58.34	57.57	56.18	55.02	54.03	53.18	52.45	51.78	51.19
4, Low	53.86	52.80	52.28	51.54	50.88	49.83	48.93	48.14	47.49	46.90	46.37

"A-Frame" Cabins

Six Corners

Estimating Procedure

1. Establish the structure quality class by applying the information on page 35.
2. Multiply the structure floor area by the appropriate cost listed below.
3. Multiply the total from step 2 by the correct location factor listed on page 7.
4. Add, when appropriate, the cost of a deck or porch, paving, fireplace, garage or carport, heating, extra plumbing fixtures, supporting walls, half story areas, construction on hillside lots, and construction in remote areas.
   See page 39.

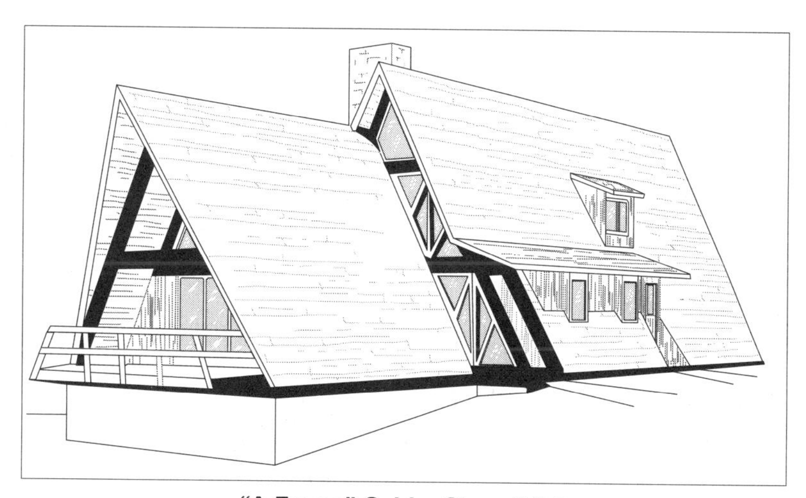

"A-Frame" Cabin, Class 2 & 3

Square Foot Area

Quality Class	400	500	600	700	800	900	1,000	1,100	1,200	1,300	1,400
1, Best	144.72	130.59	120.88	113.72	108.22	103.84	100.26	97.26	94.75	92.50	90.60
1 & 2	132.73	119.79	110.87	104.32	99.28	95.25	91.94	89.20	86.87	84.86	83.02
2, Good	121.93	110.04	101.86	95.80	91.15	87.47	84.46	81.93	79.79	77.95	76.36
2 & 3	115.14	103.87	96.16	90.45	86.10	82.62	79.76	77.37	75.32	73.61	72.07
3, Average	108.51	97.92	90.62	85.28	81.12	77.84	75.17	72.92	71.02	69.35	67.93
3 & 4	99.01	89.38	82.71	77.82	74.04	71.04	68.58	66.56	64.81	63.31	62.00
4, Low	88.63	79.99	74.03	69.67	66.31	63.61	61.42	59.57	58.01	56.64	55.49

Square Foot Area

Quality Class	1,500	1,600	1,700	1,800	2,000	2,200	2,400	2,600	2,800	3,000	3,200
1, Best	87.63	86.18	84.89	83.74	81.77	80.11	78.68	77.45	76.40	75.41	74.58
1 & 2	80.69	79.37	78.20	77.12	75.31	73.77	72.47	71.33	70.33	69.47	68.69
2, Good	74.46	73.23	72.14	71.19	69.49	68.09	66.86	65.83	64.91	64.12	63.39
2 & 3	70.65	69.49	68.47	67.51	65.95	64.59	63.43	62.48	61.61	60.82	60.18
3, Average	67.33	66.22	65.27	64.35	62.84	61.56	60.50	59.53	58.70	57.98	57.31
3 & 4	61.80	60.82	59.90	59.09	57.69	56.52	55.53	54.66	53.88	53.21	52.62
4, Low	55.00	54.19	53.47	52.18	51.11	50.24	49.41	48.75	48.13	47.59	46.87

"A-Frame" Cabins

Eight Corners

Estimating Procedure

1. Establish the structure quality class by applying the information on page 35.
2. Multiply the structure floor area by the appropriate cost listed below.
3. Multiply the total from step 2 by the correct location factor listed on page 7.
4. Add, when appropriate, the cost of a deck or porch, paving, fireplace, garage or carport, heating, extra plumbing fixtures, supporting walls, half story areas, construction on hillside lots, and construction in remote areas.
   See page 39.

"A-Frame" Cabin, Class 2

Square Foot Area

Quality Class	400	500	600	700	800	900	1,000	1,100	1,200	1,300	1,400
1, Best	147.09	133.03	123.32	116.15	110.64	106.18	102.56	99.51	96.95	94.75	92.77
1 & 2	134.88	122.02	113.11	106.53	101.45	97.38	94.07	91.29	88.92	86.87	85.07
2, Good	123.66	111.84	103.71	97.65	93.02	89.29	86.23	83.68	81.52	79.65	78.00
2 & 3	116.71	105.57	97.86	92.18	87.79	84.27	81.38	78.99	76.94	75.17	73.62
3, Average	110.39	99.86	92.54	87.17	83.04	79.70	76.96	74.73	72.76	71.09	69.64
3 & 4	100.32	90.75	84.09	79.19	75.42	72.41	69.95	67.88	66.11	64.62	63.28
4, Low	89.93	81.32	75.38	71.00	67.64	64.89	62.70	60.85	59.27	57.90	56.72

Square Foot Area

Quality Class	1,500	1,600	1,700	1,800	2,000	2,200	2,400	2,600	2,800	3,000	3,200
1, Best	89.85	88.41	87.11	85.97	83.97	82.30	80.85	79.67	78.59	77.66	76.85
1 & 2	81.24	79.93	78.74	77.70	75.89	74.39	73.10	72.01	71.02	70.21	69.44
2, Good	76.34	75.07	73.99	73.02	71.31	69.88	68.69	67.65	66.75	65.95	65.25
2 & 3	72.28	71.09	70.07	69.14	67.50	66.17	65.06	64.07	63.18	62.46	61.77
3, Average	68.83	67.72	66.73	65.83	64.30	63.05	61.97	61.04	60.22	59.48	58.84
3 & 4	63.15	62.13	61.19	60.39	58.98	57.81	56.84	55.97	55.24	54.58	53.98
4, Low	56.29	55.46	54.72	53.46	52.38	51.47	50.68	50.01	49.41	48.92	48.17

Additional Costs

Half-Story Costs

For conventional recreational dwellings, use the suggested fractions found on page 27 in the section "Additional Costs for Residential Structures." For "A-Frame" cabins, use one of the following costs: A simple platform with low cost floor cover, minimum partitions, and minimum lighting costs $24.0 to $32.50 per square foot. Average quality half story area with average quality carpet, average number of partitions finished with gypsum wallboard or plywood veneer and average lighting costs $33.75 to $45.50 per square foot. A good quality half story area with good carpet, decorative rustic partitions, ceiling beams and good lighting costs $53.50 to $67.60 per square foot.

Decks and Porches, per square foot

Small (10 to 50 S.F.) uncovered wood porch with steps and railing	
1' above ground level	$18.36 to $20.22
4' above ground level	20.10 to 23.36
2" wood deck with steps and railing (300 S.F. base)	
1' above ground level	13.30 to 19.60
4' above ground level	16.95 to 21.65

Fireplaces, cost each

Metal hood with concrete slab	$1,234 to $1,673
Simple concrete block	2,227 to 2,907
Simple block with stone facing	2,927 to 3,712
Simple natural stone	3,586 to 3,816
Prefabricated, zero clearance	1,778 to 2,352

Extra Plumbing, cost each

Lavatory	$969 to $1,228
Water closet	1,161 to 1,270
Tub	1,223 to 1,485
Stall shower	883 to 1,161
Sink	969 to 1,213

Heating, cost each

Wall furnaces	
35,000 Btu	$ 1,015
50,000 Btu	1,138
Central heating, perimeter ducts per S.F.	8.02

Garages and Carports, per S.F.

Average carport, no slab (single)	$12.50 to $17.78
Average single garage with slab	27.54 to 31.21
Average double garage with slab	25.95 to 30.05

Flatwork, per square foot

Asphalt paving	$2.91 to $3.12
4" concrete	3.56 to 4.73
6" concrete	4.20 to 5.93

Reinforced concrete walls, per C.F.

Formed one side only	$13.24 to $15.30
Formed both sides	16.90 to 19.17

Supporting Wall Costs

Cabins and recreational dwellings built on sloping lots cost more than if they are built on level lots. The cost of supporting walls of a building that do not enclose any living area should be estimated by using the figures below. These costs include everything above a normal foundation (12" to 18" above ground) up to the bottom of the next floor structure where square foot costs can be applied. In addition to the cost of supporting walls, add the cost of any extra structural members and the higher cost of building on a slope. A good rule of thumb for this is to add $560 for each foot of vertical distance between the highest and the lowest points of intersection of foundation and ground level.

Wood posts, per foot of height

4" x 4"	$1.67 to $2.66
4" x 6"	2.66 to 4.54
6" x 6"	3.43 to 6.46
8" x 8"	7.72 to 12.02
10" x 10"	14.32 to 20.50
12" x 12"	21.53 to 29.95

Brick, per square foot of wall

8" common brick	$30.29 to $37.03
12" common brick	46.53 to 57.78
8" common brick, 1 side face brick	38.41 to 47.39
12" common brick, 1 side face brick	60.08 to 74.89

Reinforced concrete block, per square foot of wall

8" natural	$7.04 to $8.50
8" colored	9.69 to 11.37
8" detailed blocks, natural	8.01 to 10.51
8" detailed blocks, colored	10.91 to 12.34
8" sandblasted	8.50 to 9.96
8" splitface, natural	7.25 to 8.55
8" splitface, colored	11.36 to 12.83
8" slump block, natural	8.01 to 9.69
8" slump block, colored	10.81 to 12.51
12" natural	13.74 to 15.37

Typical Physical Lives in Years for Residential Structures by Quality Class

	Masonry				Wood Frame			
Quality Class	**1**	**2**	**3**	**4**	**1**	**2**	**3**	**4**
1 or 2 family	70	60	60	55	70	60	60	55
3 or more units	60	55	55	50	60	55	55	50
Motels	60	55	55	50	60	55	55	50

Raise half classes to the next higher whole class

Average Life in Years (Percent Good) for Residential and Agricultural Structures

	20 Years		25 Years		30 Years		40 Years	
Age	Rem. Life	% Good	Rem. Life	% Good	Rem. Life	% Good	Rem. Life	% Good
0	20	100	25	100	30	100	40	100
1	19	94	24	95	29	96	39	98
2	18	88	23	90	28	93	38	96
3	17	81	22	86	27	89	37	94
4	16	75	21	81	26	86	36	92
5	15	69	20	77	25	82	35	90
6	14	63	19	72	24	79	34	87
7	13	59	18	68	23	75	33	84
8	12	57	17	63	22	71	32	82
9	11	55	16	60	21	67	31	80
10	11	53	16	58	20	64	30	77
11	10	50	15	56	19	60	29	74
12	9	48	14	54	19	59	28	72
13	8	46	13	53	18	57	27	70
14	7	44	12	51	17	56	27	67
15	7	42	11	49	16	54	26	65
16	6	40	11	48	15	53	25	62
17	5	38	10	46	14	52	24	60
18	5	36	9	44	13	50	23	59
19	4	33	8	43	13	49	22	58
20	4	31	7	41	12	47	21	58
21	3	29	7	39	11	46	21	55
22	3	27	6	37	11	44	20	54
23	3	25	6	35	10	43	19	53
24	3	23	5	34	9	42	18	52
25	2	21	5	32	9	40	17	51
26	2	19	4	30	8	39	17	50
27	2	16	4	29	7	37	16	49
28	2	14	4	27	7	36	15	48
29	2	12	3	25	6	34	14	47
30	1	10	3	24	6	33	14	46
31	–	–	3	22	5	31	13	45
32	–	–	3	20	5	30	12	44
33	–	–	2	18	5	29	12	43
34	–	–	2	17	4	27	11	42
35	–	–	2	15	4	26	11	41
36	–	–	2	13	4	24	10	40
38	–	–	1	10	3	21	9	38
40	–	–	–	–	2	19	7	35
42	–	–	–	–	2	16	6	33
46	–	–	–	–	1	10	5	29
50	–	–	–	–	–	–	4	25
55	–	–	–	–	–	–	3	20
60	–	–	–	–	–	–	2	14
64	–	–	–	–	–	–	1	10

	45 Years		50 Years		55 Years		60 Years		70 Years	
Age	Rem. Life	% Good	Rem. Life	% Good	Rem. Life	% Good	Rem. Life	% Good	Rem. Life	% Good
0	45	100	50	100	55	100	60	100	70	100
2	43	97	48	97	53	98	58	98	68	99
4	41	93	46	94	51	96	56	96	66	98
6	39	89	44	91	49	94	54	94	64	97
8	37	85	42	88	47	91	52	92	62	96
10	35	81	39	85	45	88	50	90	60	95
12	33	77	38	82	43	85	48	88	58	93
14	32	73	36	78	41	82	46	86	56	92
16	30	69	35	74	40	79	45	83	54	90
18	28	65	33	70	38	76	43	80	52	89
20	26	60	31	67	36	73	41	77	50	87
22	24	58	29	63	34	69	39	74	48	86
24	23	56	28	60	32	65	37	71	46	84
26	22	54	26	58	31	62	35	68	44	82
28	20	52	24	56	29	60	34	65	42	80
30	18	50	23	54	27	58	32	63	40	78
32	17	48	21	53	26	56	30	60	38	76
34	15	47	20	51	24	55	29	58	36	73
36	14	45	18	49	23	53	27	57	34	71
38	12	43	17	47	21	51	26	55	32	68
40	11	41	16	45	20	50	24	54	30	65
42	10	39	14	44	19	48	23	52	28	62
44	9	37	13	42	17	46	21	51	26	59
46	8	35	12	40	16	45	20	49	25	56
48	7	33	11	38	15	43	19	47	23	54
50	6	31	10	37	14	41	18	46	21	49
52	5	29	9	35	12	40	16	44	19	45
54	5	28	8	33	11	38	15	43	18	44
56	4	26	7	31	10	36	14	41	16	42
58	4	24	6	30	9	35	13	40	15	38
60	3	22	5	28	8	33	12	38	14	36
62	3	20	4	26	7	31	11	37	12	31
64	3	17	4	24	6	30	10	35	11	30
66	2	16	3	22	5	28	9	33	10	27
68	2	14	3	21	5	27	8	32	9	25
70	2	12	3	19	4	25	7	30	9	24
72	1	10	2	17	4	23	6	29	8	21
74	–	–	2	15	4	21	5	27	7	20
76	–	–	2	14	3	20	5	26	7	19
80	–	–	1	10	2	17	4	23	7	18
82	–	–	–	–	2	15	3	20	6	17
84	–	–	–	–	1	10	2	17	5	16
96	–	–	–	–	–	–	1	10	3	14
98	–	–	–	–	–	–	–	–	2	12
100	–	–	–	–	–	–	–	–	1	10

Commercial Structures Section

Section Contents

Urban Stores

Urban store buildings are designed for retail sales and are usually found in strip or downtown commercial developments. Square foot costs in this section are representative of a building situation where construction activities are restricted to the immediate site. This restriction tends to make the cost slightly higher than suburban type stores where unlimited use of modern machinery and techniques is possible. Do not use these figures for department stores, discount houses or suburban stores. These building types are evaluated in later sections.

Costs are for shell-type buildings without permanent partitions and include all labor, material and equipment costs for the following:

1. Foundations as required for normal soil conditions.
2. Floor, rear wall, side wall and roof structures.
3. A front wall consisting of vertical support columns or pilasters and horizontal beams spanning the area between these members leaving an open space to receive a display front.
4. Interior floor, wall and ceiling finishes.
5. Exterior wall finish on the side and rear walls.
6. Roof cover.
7. Basic lighting and electrical systems.
8. Rough and finish plumbing.
9. Design and engineering fees.
10. Permits and fees.
11. Utility hook-up.
12. Contractor's contingency, overhead and profit.

The in-place costs of the following components should be added to the basic building cost to arrive at the total structure cost. See the section "Additional Costs for Commercial Structures" on page 201.

1. Heating and air conditioning systems.
2. Elevators and escalators.
3. Fire sprinklers and fire escapes.
4. All display front components.
5. Finish materials on the front wall.
6. Canopies, ramps and docks.
7. Interior partitions.
8. Exterior signs.
9. Mezzanines and basements.
10. Communication systems.

For valuation purposes, urban stores are divided into two building types: 1) masonry or concrete frame and, 2) wood or wood and steel frame. Masonry or concrete urban stores vary widely in cost. Consequently, 6 quality classifications are established. Wood or wood and steel frame urban stores are divided into 4 quality classes.

Urban Stores – Masonry or Concrete

Quality Classification

	Class 1 Best Quality	Class 2 Good Quality	Class 3 High Average Quality	Class 4 Low Average Quality	Class 5 Low Quality	Class 6 Minimum Quality
Foundation (15% of total cost)	Reinforced concrete.	Reinforced concrete.	Reinforced concrete.	Reinforced concrete.	Reinforced concrete.	Reinforced concrete.
Floor Structure (15% of total cost)	6" reinforced concrete on 6" rock fill or 2" x 12" joists 16" o.c.	4" to 6" reinforced concrete on 6" rock fill or 2" x 10" joists 16" o.c.	4" reinforced concrete on 6" rock fill or 2" x 10" joists 16" o.c.	4" reinforced concrete on 6" rock fill or 2" x 8" joists 16" o.c.	4" reinforced concrete on 6" rock fill or 2" x 6" joists 16" o.c.	4" reinforced concrete on 4" rock fill or 2" x 6" joists 16" o.c.
Wall Structure (15% of total cost)	Reinforced 8" concrete or 12" common brick or block.	Reinforced 8" concrete or 12" common brick or block.	Reinforced 8" concrete or 12" common brick or block.	Reinforced 8" concrete or 12" common brick or block.	Reinforced 8" concrete block or reinforced 6" concrete.	Reinforced 8" concrete block or reinforced 6" concrete or 8" clay tile or 8" brick.
Roof Covering (10% of total cost)	5 ply composition roof on 1" sheathing with insulation.	5 ply composition roof on 1" sheathing with insulation.	5 ply composition roof on 1" sheathing with insulation.	5 ply composition roof on 1" sheathing with insulation.	4 ply composition roof on 1" sheathing.	4 ply composition roof on 1" sheathing.
Floor Finish (5% of total cost)	Combination solid vinyl tile and terrazzo or very good carpet.	Combination solid vinyl tile and terrazzo or very good carpet.	Solid vinyl tile with some terrazzo or good carpet.	Vinyl tile with small areas of terrazzo, carpet or solid vinyl tile.	Resilient tile.	Composition tile.
Interior Wall Finish (5% of total cost)	Plaster on gypsum or metal lath or 2 layers of 5/8" gypsum wallboard with expensive wallpaper or vinyl wall cover.	Plaster on gypsum or metal lath or 2 layers of 5/8" gypsum wallboard with average wallpaper or vinyl wall cover.	Plaster with putty coat finish on gypsum or metal lath, or 5/8" gypsum wallboard taped, textured and painted, some vinyl wall covering.	Plaster with putty coat finish or gypsum or metal lath, or 5/8" gypsum wallboard taped, textured and painted or with wallpaper.	Lath, 2 coats plaster with putty coat finish or 1/2" gypsum wallboard taped, textured and painted.	Interior plaster on masonry. Colored finish.
Ceiling Finish (5% of total cost)	Acoustical plaster or suspended anodized acoustical metal panels.	Acoustical plaster or suspended anodized acoustical metal panels.	Plaster with putty coat finish and some acoustical plaster or suspended acoustical tile with gypsum wallboard backing.	Plaster with putty coat finish or suspended acoustical tile.	Gypsum wallboard taped and textured or lath, 2 coats of plaster and putty coat finish.	Ceiling tile or gypsum wallboard and paint.
Exterior Wall Finish (5% of total cost)	Waterproofed and painted finish with face brick on exposed walls.	Waterproofed and painted finish with face brick on exposed walls.	Waterproofed and painted finish, face brick on exposed walls.	Painted finish, face brick on exposed wall.	Painted finish.	Unfinished.
Lighting (10% of total cost)	Encased modular units and custom designed chandeliers. Many spotlights.	Encased modular units and stock design chandeliers. Many spotlights.	Encased modular units and stock chandeliers. Many spotlights.	Quad open strip fixtures or triple encased louvered strip fixtures. Average number of spotlights.	Triple open strip fixtures or double encased louvered strip fixtures. Some spotlights.	Double open strip fluorescent fixtures.
Plumbing *(Per each 5,000 S.F.)* (12% of total cost)	6 good fixtures, metal or marble toilet partitions.	6 good fixtures, metal or marble toilet partitions.	6 standard fixtures, metal or marble toilet partitions.	6 standard fixtures, metal toilet partitions.	4 standard commercial fixtures, metal toilet partitions.	4 standard commercial fixtures, wood toilet partitions.
Bath Wall Finish (3% of total cost)	Ceramic tile or marble or custom mosaic tile.	Ceramic tile or marble or custom mosaic tile.	Ceramic tile or marble or plain mosaic tile.	Gypsum wallboard and paint, some ceramic tile or plastic finish wallboard.	Gypsum wallboard and paint.	Gypsum wallboard and paint.

Note: Use the percent of total cost to help identify the correct quality classification

Urban Stores - Masonry or Concrete

First Floor, Length Less than Twice Width

Estimating Procedure

1. Establish the structure quality class by applying the information on page 43.
2. Compute the building ground floor area. This should include everything within the exterior walls and all insets outside the walls but under the main roof.
3. Add to or subtract from the cost below the appropriate amount from the Wall Height Adjustment Table on page 48 if the first floor wall height is more or less than 16 feet for large stores or 12 feet for small stores.
4. Multiply the adjusted square foot cost by the building area.
5. Deduct, if appropriate, for common walls, using the figures on page 48.
6. Multiply the total cost by the location factor on page 7.
7. Add the cost of heating and cooling equipment, elevators, escalators, fire escapes, fire sprinklers, display fronts, canopies, ramps, docks, interior partitions, mezzanines, basements, and communication systems from pages 201 to 213.
8. Add the cost of second and higher floors from page 47.

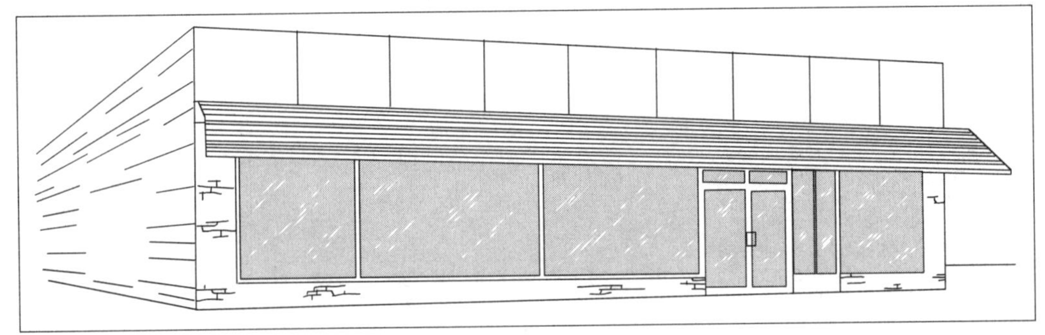

Urban Store, Class 4 & 5

Smaller Stores - Square Foot Area

Quality Class	500	600	700	800	900	1,000	1,200	1,500	1,700	2,000	2,500
4, Low Avg.	99.53	94.57	90.69	87.63	85.04	82.86	79.35	75.45	73.45	71.05	68.02
4 & 5	93.55	88.88	85.25	82.33	79.89	77.87	74.56	70.90	69.02	66.76	63.91
5, Low	88.42	84.00	80.55	77.81	75.51	73.57	70.45	66.99	65.25	63.09	60.42
5 & 6	84.21	80.02	76.77	74.14	71.92	70.09	67.14	63.83	62.14	60.12	57.55
6, Minimum	80.37	76.37	73.25	70.76	68.66	66.90	64.08	60.93	59.33	57.38	54.95

Larger Stores - Square Foot Area

Quality Class	3,000	3,500	4,000	4,500	5,000	6,000	7,500	10,000	15,000	20,000
1, Best	99.00	93.46	91.30	89.57	88.10	85.78	83.28	80.45	77.20	75.29
1 & 2	90.22	87.70	85.70	84.05	82.68	80.50	78.14	75.50	72.44	70.63
2, Good	84.96	82.62	80.72	79.18	77.88	75.81	73.60	71.10	68.23	66.56
2 & 3	81.09	78.84	77.01	74.46	74.31	72.35	70.23	67.84	65.12	63.48
3, Hi. Avg.	76.33	73.90	71.79	70.74	69.46	67.57	66.09	64.11	61.23	59.66
3 & 4	72.04	69.98	68.31	66.95	65.83	63.96	61.94	59.71	57.04	55.46
4, Low Avg.	66.02	64.38	62.85	61.60	60.55	58.87	57.02	54.93	52.51	51.06
4 & 5	62.78	60.99	59.53	58.33	57.35	55.75	54.01	52.03	49.69	48.31
5, Low	59.78	58.04	56.67	55.56	54.61	53.06	51.40	49.55	47.31	46.04
5 & 6	56.66	55.02	53.72	52.64	51.75	51.41	48.56	46.96	44.87	43.64
Minimum	53.65	52.11	50.86	49.86	49.02	47.64	46.14	44.46	42.51	41.31

Urban Stores – Masonry or Concrete

First Floor, Length Between 2 and 4 Times Width

Estimating Procedure

1. Establish the structure quality class by applying the information on page 43.
2. Compute the building ground floor area. This should include everything within the exterior walls and all insets outside the walls but under the main roof.
3. Add to or subtract from the cost below the appropriate amount from the Wall Height Adjustment Table on page 48 if the first floor wall height is more or less than 16 feet for large stores or 12 feet for small stores.
4. Multiply the adjusted square foot cost by the building area.
5. Deduct, if appropriate, for common walls, using the figures on page 48.
6. Multiply the total cost by the location factor on page 7.
7. Add the cost of heating and cooling equipment, elevators, escalators, fire escapes, fire sprinklers, display fronts, canopies, ramps, docks, interior partitions, mezzanines, basements, and communication systems from pages 201 to 213.
8. Add the cost of second and higher floors from page 47.

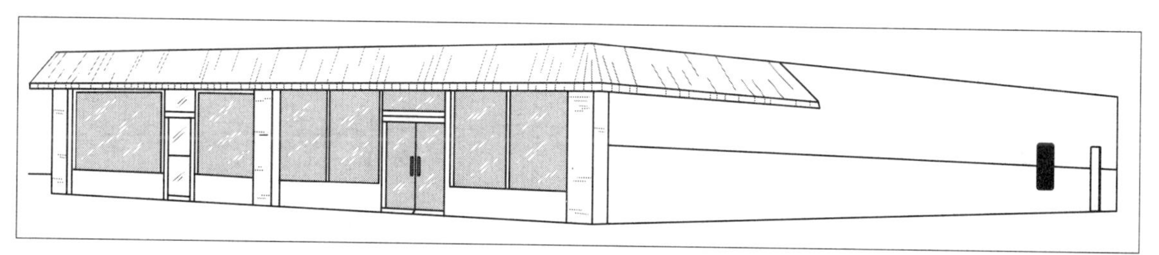

Urban Store, Class 4

Smaller Stores - Square Foot Area

Quality Class	500	600	700	800	900	1,000	1,250	1,500	1,700	2,000	2,500
4, Low Avg.	109.15	104.68	100.09	96.42	93.39	90.85	85.85	82.22	79.91	77.13	73.67
4 & 5	104.19	98.58	94.28	90.78	87.95	85.52	80.85	77.41	75.27	72.59	69.36
5, Low	98.52	93.22	89.12	85.84	83.15	80.86	76.44	73.17	71.14	68.65	65.57
5 & 6	94.18	89.08	85.18	82.05	79.47	77.30	73.06	69.96	68.00	65.64	62.70
6, Minimum	90.25	85.42	81.67	78.68	76.17	74.09	70.04	67.07	65.18	62.92	60.10

Larger Stores - Square Foot Area

Quality Class	3,000	3,500	4,000	4,500	5,000	6,000	7,500	10,000	15,000	20,000
1, Best	102.99	99.97	97.55	95.55	93.87	91.17	88.17	84.84	80.87	78.52
1 & 2	96.93	94.12	91.83	89.95	88.39	85.82	83.04	79.87	76.13	73.89
2, Good	91.15	88.49	86.35	84.57	83.09	80.70	78.06	75.07	71.61	69.50
2 & 3	86.99	84.46	82.41	80.74	79.31	77.02	74.53	71.69	68.34	66.35
3, Hi. Avg.	82.35	81.16	79.21	77.57	76.21	74.02	71.61	68.86	65.64	63.76
3 & 4	77.83	75.36	73.38	71.75	70.37	68.19	65.81	63.08	59.90	58.03
4, Low Avg.	71.76	69.25	67.47	65.93	64.71	62.68	60.44	57.96	55.04	53.34
4 & 5	68.12	65.95	64.22	62.79	61.59	59.68	57.57	55.21	52.42	50.79
5, Low	64.75	62.67	61.01	59.69	58.55	56.71	54.71	52.47	49.82	48.27
5 & 6	61.31	59.34	57.80	56.52	55.41	53.70	51.81	49.67	47.19	45.70
Minimum	58.16	56.31	54.84	53.62	52.60	50.94	49.16	47.15	44.77	43.01

Urban Stores – Masonry or Concrete

First Floor, Length More Than 4 Times Width

Estimating Procedure

1. Establish the structure quality class by applying the information on page 43.
2. Compute the building ground floor area. This should include everything within the exterior walls and all insets outside the walls but under the main roof.
3. Add to or subtract from the cost below the appropriate amount from the Wall Height Adjustment Table on page 48 if the first floor wall height is more or less than 16 feet for large stores or 12 feet for small stores.
4. Multiply the adjusted square foot cost by the building area.
5. Deduct, if appropriate, for common walls, using the figures on page 48.
6. Multiply the total cost by the location factor on page 7.
7. Add the cost of heating and cooling equipment, elevators, escalators, fire escapes, fire sprinklers, display fronts, canopies, ramps, docks, interior partitions, mezzanines, basements, and communication systems from pages 201 to 213.
8. Add the cost of second and higher floors from page 47.

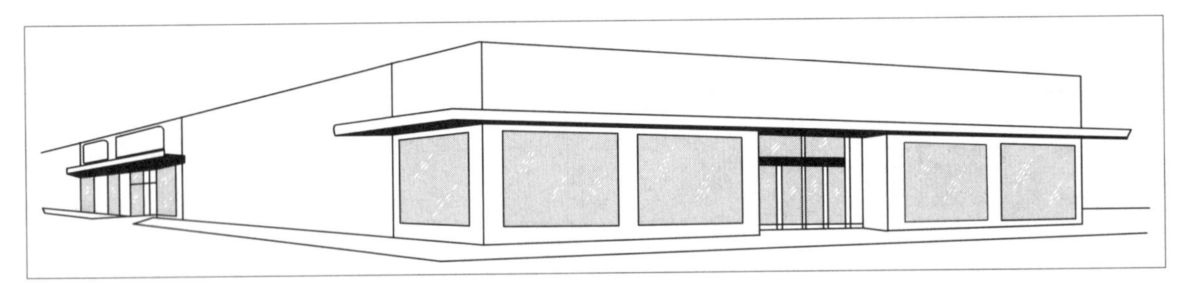

Urban Store, Class 3 & 4

Smaller Stores - Square Foot Area

Quality Class	500	600	700	800	900	1,000	1,200	1,500	1,700	2,000	2,500
4, Low Avg.	120.77	115.97	110.41	105.99	102.35	99.26	94.36	89.01	86.26	82.97	78.90
4 & 5	116.03	109.22	104.00	99.78	96.38	93.49	88.88	83.82	81.26	78.15	70.32
5, Low	109.70	103.27	98.30	94.36	91.12	88.42	84.04	79.25	76.83	73.90	70.27
5 & 6	105.23	99.04	94.31	90.51	87.39	84.79	80.61	76.03	73.68	70.85	67.41
6, Minimum	101.17	95.24	90.68	87.03	84.05	81.51	77.49	73.11	70.88	68.15	64.82

Larger Stores - Square Foot Area

Quality Class	3,000	3,500	4,000	4,500	5,000	6,000	7,500	10,000	15,000	20,000
1, Best	110.82	107.11	104.16	101.76	99.68	96.40	92.81	88.77	84.01	81.18
1 & 2	104.33	100.89	98.10	95.82	93.91	90.77	87.38	83.61	79.11	76.92
2, Good	98.13	94.88	92.24	90.10	88.28	85.40	82.18	78.62	74.37	71.90
2 & 3	90.23	87.23	84.80	82.83	81.16	78.49	75.56	72.25	68.39	66.10
3, Hi. Avg.	89.43	86.97	84.58	82.59	80.93	78.28	75.36	72.08	68.19	65.92
3 & 4	83.72	80.81	78.46	76.52	74.90	72.27	69.38	66.13	62.33	60.07
4, Low Avg.	77.33	74.62	72.47	70.67	69.17	66.77	64.09	61.11	57.57	55.48
4 & 5	73.48	70.92	68.83	67.14	65.71	63.40	60.86	58.05	54.68	52.71
5, Low	69.75	67.33	65.37	63.77	62.39	60.21	57.83	55.12	51.93	50.05
5 & 6	66.17	63.85	61.99	60.46	59.18	57.12	54.85	52.28	49.26	47.49
6, Minimum	62.78	60.59	58.70	57.38	55.53	53.32	51.31	48.98	46.75	45.06

Urban Stores – Masonry or Concrete

Second and Higher Floors

Estimating Procedure

1. Establish the structure quality class. The class for second and higher floors will usually be the same as the first floor.
2. Compute the square foot area of the second floor and each higher floor.
3. Add to or subtract from the square foot cost below the appropriate amount from the Wall Height Adjustment Table on page 48 if the wall height is more or less than 12 feet.
4. Multiply the adjusted square foot cost by the area of each floor.
5. Deduct, if appropriate, for common walls, using the figures on page 48.
6. Add 2% to the cost of each floor above the second floor. For example, the third floor cost would be 102% of the second floor cost and the fourth floor cost would be 104% of the second floor cost.
7. Multiply the total cost for each floor by the location factor on page 7.
8. Add the cost of heating and cooling equipment, escalators, fire escapes, fire sprinklers, and interior partitions from pages 201 to 213.

Length less than twice width - Square Foot Area

Quality Class	2,500	3,000	3,500	4,000	4,500	5,000	6,000	7,500	10,000	15,000	20,000
1, Best	88.89	86.37	84.40	82.83	81.52	80.43	78.68	76.73	74.51	71.88	70.33
1 & 2	82.94	80.59	78.76	77.30	76.07	75.06	73.42	71.58	69.53	67.06	65.64
2, Good	77.26	75.08	73.38	72.01	70.85	69.92	68.38	66.70	64.76	62.48	61.13
2 & 3	73.00	70.95	69.33	68.03	66.96	66.06	64.61	63.01	61.18	59.02	57.76
3, High Avg.	69.55	67.58	66.05	64.81	63.81	62.92	61.56	60.03	58.29	56.25	55.02
3 & 4	64.19	62.20	60.63	59.40	58.36	57.48	56.06	54.51	52.77	50.68	49.44
4, Low Avg.	58.41	56.63	55.23	54.08	53.15	52.33	51.08	49.64	48.06	46.15	45.02
4 & 5	54.74	53.05	51.73	50.66	49.75	49.02	47.81	46.51	45.02	43.22	42.17
5, Low	51.10	49.52	48.27	47.27	46.44	45.77	44.66	43.41	42.00	40.33	39.34
5 & 6	48.24	46.73	45.58	44.61	43.85	43.19	42.13	40.98	39.65	38.08	37.14
6, Minimum	45.46	44.06	42.96	42.08	41.36	40.71	39.73	38.62	37.37	35.90	35.02

Length between 2 and 4 times width - Square Foot Area

Quality Class	2,500	3,000	3,500	4,000	4,500	5,000	6,000	7,500	10,000	15,000	20,000
1, Best	92.33	89.57	87.41	85.66	84.23	83.01	81.07	78.88	76.43	73.50	71.77
1 & 2	86.17	83.59	81.56	79.94	78.58	77.49	75.63	73.58	71.30	68.58	66.94
2, Good	80.21	77.81	75.93	74.42	73.17	72.11	70.42	68.53	66.40	63.86	62.34
2 & 3	75.90	73.63	71.85	70.44	69.27	68.24	66.61	64.85	62.80	60.41	58.98
3, Hi. Avg.	72.27	70.11	68.42	67.08	65.96	65.01	63.44	61.77	59.82	57.54	56.18
3 & 4	66.82	64.60	62.86	61.49	60.33	59.37	57.80	56.07	54.13	51.84	50.46
4, Low Avg.	60.82	58.78	57.26	55.94	54.92	54.06	52.63	51.08	49.27	47.17	45.95
4 & 5	57.05	55.15	53.69	52.50	51.52	50.69	49.34	47.88	46.22	44.23	43.08
5, Low	53.39	51.58	50.19	49.08	48.20	47.43	46.18	44.79	43.24	41.39	40.31
5 & 6	50.39	48.70	47.40	46.35	45.50	44.77	43.58	43.33	41.46	39.08	38.04
6, Minimum	47.48	45.93	44.68	43.72	42.89	42.20	41.07	39.85	38.48	36.85	35.86

Length over 4 times width - Square Foot Area

Quality Class	2,500	3,000	3,500	4,000	4,500	5,000	6,000	7,500	10,000	15,000	20,000
1, Best	96.04	93.54	91.45	89.72	88.22	86.92	84.76	82.25	79.28	75.59	73.28
1 & 2	89.57	87.24	85.32	83.70	82.29	81.08	79.07	76.74	73.95	70.51	68.35
2, Good	83.37	81.22	79.43	77.91	76.61	75.47	73.60	71.41	68.84	65.62	63.62
2 & 3	70.90	70.85	75.15	73.72	72.50	71.42	69.65	67.58	65.14	62.09	60.22
3, Hi. Avg.	75.10	73.16	71.57	70.18	69.00	67.99	66.32	64.36	62.01	59.13	57.33
3 & 4	69.86	67.78	66.08	64.64	63.45	62.39	60.64	58.66	56.31	53.42	51.63
4, Low Avg.	63.19	61.73	60.19	58.89	57.79	56.86	55.26	53.42	51.31	48.67	47.05
4 & 5	59.66	57.86	56.41	55.21	54.17	53.27	51.79	50.10	48.11	45.62	44.09
5, Low	55.89	54.22	52.85	51.72	50.73	49.93	48.53	46.91	45.06	42.73	41.29
5 & 6	52.76	51.19	49.91	48.83	47.90	47.12	45.83	44.31	42.54	40.36	38.98
6, Minimum	49.90	48.44	47.20	46.19	45.33	44.58	43.34	41.93	40.23	38.19	36.88

Urban Stores – Masonry or Concrete Wall Adjustments

Wall Height Adjustment

The square foot costs for urban stores are based on the wall heights listed on each page. The main or first floor height is the distance from the bottom of the floor slab or joists to the top of the roof slab or ceiling joists. Second and higher floors are measured from the top of the floor slab or floor joists to the top of the roof slab or ceiling joists. Add or subtract the amount listed to the square foot cost for each foot more or less than the standard wall height in the tables. For second and higher floors use only 75% of the wall height adjustment cost.

Area	**500**	**600**	**700**	**800**	**900**	**1,000**	**1,250**	**1,500**	**1,750**	**2,000**	**2,500**
Adjustment	2.85	2.60	2.40	2.28	2.15	2.03	1.85	1.64	1.54	1.46	1.30

Area	**3,000**	**3,500**	**4,000**	**4,500**	**5,000**	**6,000**	**7,500**	**10,000**	**15,000**	**20,000**
Adjustment	1.17	1.10	1.05	.96	.92	.82	.74	.66	.50	.45

Perimeter Wall Adjustment

A common wall exists when two buildings share one wall. Adjust for common walls by deducting the linear foot costs below from the total structure cost. In some structures one or more walls are not owned at all. In this case, deduct the "No Ownership" cost per linear foot of wall not owned.

Small Urban Stores (to 2,500 S.F.)
Common wall, deduct $90.00 per linear foot
No ownership, deduct $181.00 per linear foot

Large Urban Stores (over 2,500 S.F.)
Common wall, deduct $173 per linear foot
No ownership, deduct $347 per linear foot

Urban Stores – Wood or Wood and Steel Frame

Quality Classification

	Class 1 Best Quality	Class 2 Good Quality	Class 3 Average Quality	Class 4 Low Quality
Foundation (15% of total cost)	Reinforced concrete.	Reinforced concrete.	Reinforced concrete.	Reinforced concrete.
Floor Structure (15% of total cost)	4" reinforced concrete on 6" rock fill or 2" x 10" joists 16" o.c.	4" reinforced concrete on 6" rock fill or 2" x 8" joists 16" o.c.	4" reinforced concrete on 6" rock fill or 2" x 6" joists 16" o.c.	4" reinforced concrete on 4" rock fill or 2" x 6" joists 16" o.c.
Wall Structure (15% of total cost)	2" x 6" studs 16" o.c.	2" x 4" or 2" x 6" studs 16" o.c.	2" x 4" studs 16" o.c. up to 14' high, 2" x 6" studs 16" o.c. over 14' high.	2" x 4" studs 16" o.c. up to 14' high, 2" x 6" studs 16" o.c. over 14' high.
Roof Covering (10% of total cost)	5 ply composition roof on 1" x 6" sheathing with insulation.	5 ply composition roof on 1" x 6" sheathing with insulation.	4 ply composition roof on 1" x 6" sheathing.	4 ply composition roof on 1" x 6" sheathing.
Floor Finish (5% of total cost)	Sheet vinyl with some terrazzo or good carpet.	Resilient tile with small areas of terrazzo, carpet or solid vinyl tile.	Composition tile.	Minimum grade tile.
Interior Wall Finish (5% of total cost)	Plaster with putty coat finish on gypsum or metal lath, or 5/8" gypsum wallboard taped, textured and painted or some vinyl wall cover.	Plaster with putty coat finish on gypsum or metal lath, or 5/8" gypsum wallboard taped, textured and painted or with wallpaper.	Lath, 2 coats plaster with putty coat finish or 1/2" gypsum wallboard taped, textured and painted.	1/2" gypsum wallboard taped, textured and painted.
Ceiling Finish (5% of total cost)	Plaster with putty coat finish and some acoustical plaster or suspended acoustical tile with gypsum wallboard backing.	Plaster with putty coat finish or suspended acoustical tile with exposed grid.	Gypsum wallboard taped and textured or lath, 2 coats of plaster and putty coat finish.	Ceiling tile or gypsum wallboard and paint.
Exterior Wall Finish (5% of total cost)	Good wood siding.	Average wood siding.	Stucco or average wood siding.	Stucco or inexpensive wood siding.
Lighting (10% of total cost)	Encased modular units and stock chandeliers. Many spotlights.	Quad open strip fixtures or triple encased louvered strip fixtures. Average number of spotlights.	Triple open strip fixtures or double encased louvered strip fixtures. Some spotlights.	Double open strip fixtures.
Plumbing (12% of total cost) *(Per 5,000 S.F.)*	6 standard fixtures, metal or marble toilet partitions.	6 standard fixtures, metal toilet partitions.	4 standard commercial fixtures, metal toilet partitions.	4 standard commercial fixtures, wood toilet partitions.
Bath Wall Finish (3% of total cost)	Ceramic tile, marble or plain mosaic tile.	Gypsum wallboard and paint, some ceramic tile or plastic finish wallboard.	Gypsum wallboard and paint.	Gypsum wallboard and paint.

Note: Use the percent of total cost to help identify the correct quality classification

Urban Stores – Wood or Wood and Steel Frame

First Floor, Length Less Than Twice Width

Estimating Procedure

1. Establish the structure quality class by applying the information on page 49.
2. Compute the building ground floor area. This should include everything within the exterior walls and all insets outside the walls but under the main roof.
3. Add to or subtract from the cost below the appropriate amount from the Wall Height Adjustment Table on page 54 if the first floor wall height is more or less than 16 feet for large stores or 12 feet for small stores.
4. Multiply the adjusted square foot cost by the building area.
5. Deduct, if appropriate, for common walls, using the figures on page 54.
6. Multiply the total cost by the location factor on page 7.
7. Add the cost of heating and cooling equipment, elevators, escalators, fire escapes, fire sprinklers, display fronts, canopies, ramps, docks, interior partitions, mezzanines, basements, and communication systems from pages 201 to 213.
8. Add the cost of second and higher floors from page 53.

Smaller Stores - Square Foot Area

Quality Class	500	600	700	800	900	1,000	1,250	1,500	1,750	2,000	2,500
2, Good	74.16	71.49	69.43	67.73	66.33	65.13	62.79	61.05	59.70	58.60	56.90
2 & 3	69.47	66.98	65.05	63.45	62.15	61.01	58.82	57.17	55.95	54.90	53.30
3, Average	65.24	62.87	61.06	59.55	58.35	57.29	55.22	53.71	52.49	51.53	50.05
3 & 4	59.86	57.71	56.04	54.68	53.54	52.58	50.68	49.29	48.16	47.31	45.94
4, Low	54.67	52.73	51.18	49.96	48.91	48.02	46.31	45.03	44.01	43.21	41.95

Larger Stores - Square Foot Area

Quality Class	3,000	3,500	4,000	4,500	5,000	6,000	7,500	10,000	15,000	20,000
1, Best	69.19	67.77	66.64	65.72	64.89	63.65	62.24	60.65	58.74	57.63
1 & 2	62.63	61.36	60.33	59.50	58.78	57.64	56.36	54.90	53.20	52.17
2, Good	56.70	55.53	54.60	53.85	53.17	52.16	51.00	49.69	48.15	47.23
2 & 3	53.10	52.00	51.14	50.44	49.82	48.87	47.79	46.54	45.09	44.24
3, Average	49.70	48.69	47.87	47.20	46.64	45.72	44.73	43.57	42.21	41.40
3 & 4	45.00	44.08	43.37	42.74	42.23	41.42	40.48	39.44	38.22	37.51
4, Low	41.21	40.39	39.70	39.17	38.67	37.91	37.10	36.12	35.01	34.33

Urban Stores - Wood or Wood and Steel Frame

First Floor, Length Between 2 and 4 Times Width

Estimating Procedure

1. Establish the structure quality class by applying the information on page 49.
2. Compute the building ground floor area. This should include everything within the exterior walls and all insets outside the walls but under the main roof.
3. Add to or subtract from the cost below the appropriate amount from the Wall Height Adjustment Table on page 54 if the first floor wall height is more or less than 16 feet for large stores or 12 feet for small stores.
4. Multiply the adjusted square foot cost by the building area.
5. Deduct, if appropriate, for common walls, using the figures on page 54.
6. Multiply the total cost by the location factor on page 7.
7. Add the cost of heating and cooling equipment, elevators, escalators, fire escapes, fire sprinklers, display fronts, canopies, ramps, docks, interior partitions, mezzanines, basements, and communication systems from pages 201 to 213.
8. Add the cost of second and higher floors from page 53.

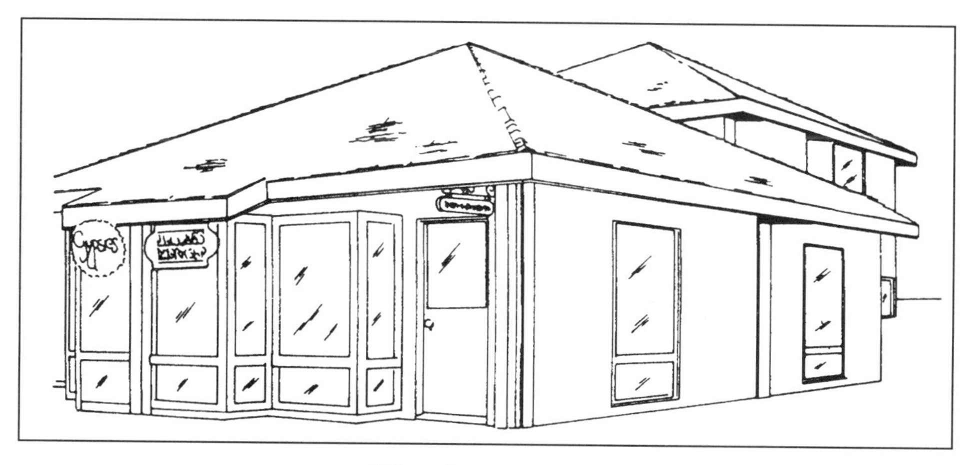

Urban Store, Class 3

Smaller Stores - Square Foot Area

Quality Class	500	600	700	800	900	1,000	1,250	1,500	1,750	2,000	2,500
2, Good	80.20	76.99	74.50	72.51	70.86	69.46	66.73	64.73	63.18	61.92	59.99
2 & 3	74.70	71.71	69.42	67.56	66.01	64.72	62.17	60.29	58.83	57.67	55.91
3, Average	70.45	67.70	65.49	63.73	62.30	61.08	58.66	56.90	55.53	54.42	52.73
3 & 4	64.72	62.15	60.13	58.52	57.18	56.06	53.86	52.24	50.99	49.96	48.43
4, Low	58.97	56.60	54.79	53.31	52.12	51.07	49.09	47.60	46.46	45.52	44.13

Larger Stores - Square Foot Area

Quality Class	3,000	3,500	4,000	4,500	5,000	6,000	7,500	10,000	15,000	20,000
1, Best	71.98	70.84	69.50	67.96	67.23	65.87	64.35	62.46	60.25	58.92
1 & 2	66.18	64.68	63.46	62.46	61.62	60.27	58.77	57.05	55.02	53.78
2, Good	59.20	57.85	56.78	55.90	55.13	53.90	52.56	51.04	49.20	48.12
2 & 3	55.25	54.01	53.01	52.17	51.45	50.31	49.03	47.62	45.93	44.91
3, Average	51.97	50.79	49.85	49.09	48.40	47.33	46.13	44.79	43.21	42.24
3 & 4	47.69	46.60	45.75	45.04	44.42	43.43	42.34	41.11	39.66	38.79
4, Low	43.66	42.67	41.87	41.22	40.65	39.74	38.75	37.62	36.27	35.49

Urban Stores – Wood or Wood and Steel Frame

First Floor, Length More Than 4 Times Width

Estimating Procedure

1. Establish the structure quality class by applying the information on page 49.
2. Compute the building ground floor area. This should include everything within the exterior walls and all insets outside the walls but under the main roof.
3. Add to or subtract from the cost below the appropriate amount from the Wall Height Adjustment Table on page 54 if the first floor wall height is more or less than 16 feet for large stores or 12 feet for small stores.
4. Multiply the adjusted square foot cost by the building area.
5. Deduct, if appropriate, for common walls, using the figures on page 54.
6. Multiply the total cost by the location factor on page 7.
7. Add the cost of heating and cooling equipment, elevators, escalators, fire escapes, fire sprinklers, display fronts, canopies, ramps, docks, interior partitions, mezzanines, basements, and communication systems from pages 201 to 213.
8. Add the cost of second and higher floors from page 53.

Urban Store, Class 3

Smaller Stores - Square Foot Area

Quality Class	500	600	700	800	900	1,000	1,250	1,500	1,750	2,000	2,500
2, Good	88.39	84.04	80.76	78.19	76.10	74.36	71.03	68.66	66.85	65.41	63.25
2 & 3	82.57	78.51	75.46	73.06	71.09	69.44	66.39	64.15	62.43	61.12	59.08
3, Average	77.42	73.62	70.75	68.49	66.66	65.15	62.23	60.13	58.55	57.30	55.44
3 & 4	70.83	67.36	64.73	62.67	61.00	59.58	56.96	55.03	53.57	52.39	50.68
4, Low	64.75	61.56	59.19	57.29	55.75	54.47	52.07	50.32	48.96	47.94	46.36

Larger Stores - Square Foot Area

Quality Class	3,000	3,500	4,000	4,500	5,000	6,000	7,500	10,000	15,000	20,000
1, Best	78.23	76.18	74.54	73.19	72.04	70.23	68.21	65.96	63.30	61.74
1 & 2	71.46	69.60	68.12	66.86	65.81	64.17	62.32	60.26	57.83	56.42
2, Good	65.41	63.70	62.32	61.19	60.23	58.71	57.04	55.15	52.95	51.63
2 & 3	58.03	56.52	55.29	54.30	53.44	52.10	50.62	48.93	46.97	45.82
3, Average	55.27	53.83	52.66	51.69	50.90	49.62	48.20	46.60	44.74	43.64
3 & 4	49.80	48.48	47.45	46.57	45.89	44.70	43.43	41.96	40.29	39.32
4, Low	45.07	43.89	42.95	42.17	41.50	40.46	39.32	38.02	36.47	35.57

Urban Stores - Wood or Wood and Steel Frame

Second and Higher Floors

Estimating Procedure

1. Establish the structure quality class. The class for second and higher floors will usually be the same as the first floor.
2. Compute the square foot area of the second floor and each higher floor.
3. Add to or subtract from the square foot cost below the appropriate amount from the Wall Height Adjustment Table on page 54 if the wall height is more or less than 12 feet.
4. Multiply the adjusted square foot cost by the area of each floor.
5. Deduct, if appropriate, for common walls, using the figures on page 54.
6. Add 2% to the cost for each floor above the second floor. For example, the third floor cost would be 102% of the second floor cost and the fourth floor cost would be 104% of the second floor cost.
7. Multiply the total cost by the location factor on page 7.
8. Add the cost of heating and cooling equipment, escalators, fire escapes, fire sprinklers, and interior partitions from pages 201 to 213.

Length less than twice width - Square Foot Area

Quality Class	2,500	3,000	3,500	4,000	4,500	5,000	6,000	7,500	10,000	15,000	20,000
1, Best	60.15	58.92	57.90	57.11	56.47	55.92	55.00	54.04	52.91	51.57	50.78
1 & 2	54.63	53.47	52.58	51.87	51.28	50.78	49.94	49.04	48.03	46.82	46.10
2, Good	49.56	48.53	47.74	47.06	46.49	46.06	45.31	44.50	43.58	42.47	41.81
2 & 3	45.95	45.01	44.25	43.65	43.14	42.72	42.03	41.26	40.41	39.40	38.80
3, Average	42.84	41.95	41.27	40.70	40.24	39.83	39.20	38.48	37.69	36.73	36.16
3 & 4	40.36	39.50	38.85	38.32	37.89	37.52	36.93	36.24	35.51	34.60	34.04
4, Low	37.79	37.04	36.39	35.90	35.52	35.13	34.59	33.97	33.23	32.42	31.91

Length between 2 and 4 times width - Square Foot Area

Quality Class	2,500	3,000	3,500	4,000	4,500	5,000	6,000	7,500	10,000	15,000	20,000
1, Best	61.91	60.50	59.43	58.54	57.81	57.19	56.19	55.10	53.84	52.38	51.46
1 & 2	56.18	54.90	53.92	53.12	52.47	51.93	51.01	50.00	48.87	47.54	46.73
2, Good	50.98	49.80	48.90	48.20	47.60	47.09	46.27	45.36	44.33	43.13	42.37
2 & 3	47.33	46.26	45.44	44.77	44.22	43.74	42.98	42.15	41.17	40.04	39.36
3, Average	44.10	43.13	42.35	41.74	41.21	40.78	40.07	39.29	38.38	37.31	36.71
3 & 4	41.50	40.58	39.85	39.27	38.76	38.34	37.69	36.96	36.12	35.11	34.52
4, Low	38.90	38.04	37.34	36.79	36.35	35.96	35.32	34.63	33.86	32.91	32.27

Length over 4 times width - Square Foot Area

Quality Class	2,500	3,000	3,500	4,000	4,500	5,000	6,000	7,500	10,000	15,000	20,000
1, Best	64.55	62.89	61.62	60.60	59.76	59.01	57.84	56.56	55.11	53.38	52.37
1 & 2	58.63	57.11	55.98	55.03	54.25	53.58	52.53	51.37	50.04	48.48	47.55
2, Good	53.17	51.83	50.77	49.93	49.24	48.64	47.67	46.60	45.41	43.96	43.16
2 & 3	49.44	48.16	47.19	46.42	45.73	45.18	44.30	43.32	42.21	40.88	40.09
3, Average	46.03	44.87	43.96	43.23	42.63	42.10	41.28	40.34	39.31	38.07	37.32
3 & 4	43.38	42.26	41.40	40.71	40.14	39.67	38.86	38.03	37.04	35.87	35.16
4, Low	40.53	39.49	38.71	38.06	37.53	37.08	36.35	35.54	34.61	33.53	32.88

Urban Stores - Wood or Wood and Steel Frame

Wall Height Adjustment

The square foot costs for urban stores are based on the wall heights listed on each page. The main or first floor height is the distance from the bottom of the floor slab or joists to the top of the roof slab or ceiling joists. Second and higher floors are measured from the top of the floor slab or floor joists to the top of the roof slab or ceiling joists. Add or subtract the amount listed to the square foot cost for each foot more or less than the standard wall height in the tables. For second and higher floors use only 75% of the wall height adjustment cost.

Area	500	600	700	800	900	1,000	1,250	1,500	1,750	2,000	2,500
Adjustment	1.19	1.09	1.02	.94	.89	.84	.77	.70	.65	.61	.54

Area	3,000	3,500	4,000	4,500	5,000	6,000	7,500	10,000	15,000	20,000
Adjustment	.48	.45	.44	.42	.40	.38	.36	.34	.32	.31

Perimeter Wall Adjustment

A common wall exists when two buildings share one wall. Adjust for common walls by deducting the linear foot costs below from the total structure cost. In some structures one or more walls are not owned at all. In this case, deduct the "No Ownership" cost per linear foot of wall not owned.

Small Urban Stores (2,500 S.F. or less)
Common wall, deduct $60.20 per linear foot
No ownership, deduct $119.00 per linear foot

Large Urban Stores (over 2,500 S.F.)
Common wall, deduct $87.00 per linear foot
No ownership, deduct $177.00 per linear foot

Suburban Stores

Suburban stores are usually built as part of shopping centers. They differ from urban stores in that they are built in open areas where modern construction techniques, equipment and more economical designs can be used. They are also subject to greater variations in size and shape than are urban stores. Do not use the figures in this section for department stores, discount houses or urban stores. These building types are evaluated in other sections.

Costs identified "building shell only" do not include permanent partitions, display fronts or finish materials on the front of the building. Costs for "multi-unit buildings" include partitions, display fronts and finish materials on the front of the building. All figures include the following costs:

1. Foundations as required for normal soil conditions.
2. Floor, rear wall, side wall and roof structures.
3. A front wall consisting of vertical support columns or pilasters and horizontal beams spanning the area between these members, leaving an open space to receive a display front.
4. Interior floor, wall and ceiling finishes.
5. Exterior wall finish on the side and rear walls.
6. Roof cover.
7. Basic lighting and electrical systems.
8. Rough and finish plumbing.
9. A usual or normal parapet wall.
10. Design and engineering fees.
11. Permits and fees.
12. Utility hook-ups.
13. Contractor's contingency, overhead and mark-up.

The in-place costs of these extra components should be added to the basic building cost to arrive at total structure cost. See the section "Additional Costs for Commercial and Industrial Structures" on page 201.

1. Heating and air conditioning systems.
2. Fire sprinklers.
3. All display front components (shell-type buildings only).
4. Finish materials on the front wall of the building (shell-type building only).
5. Canopies.
6. Interior partitions (shell-type buildings only).
7. Exterior signs.
8. Mezzanines and basements.
9. Loading docks and ramps.
10. Miscellaneous yard improvements.
11. Communications systems.

For valuation purposes suburban stores are divided into two building types: masonry or concrete frame, or wood or wood and steel frame. Each building type is divided into four shape classes:

1. Buildings in which the depth is greater than the front.
2. Buildings in which the front is between one and two times the depth.
3. Buildings in which the front is between two and four times the depth.
4. Buildings in which the front is greater than four times the depth. Angular buildings should be classed by comparing the sum of the length of all wings to the width of the wings. All areas should be included, but no area should be included as part of two different wings. Note the example at the right.

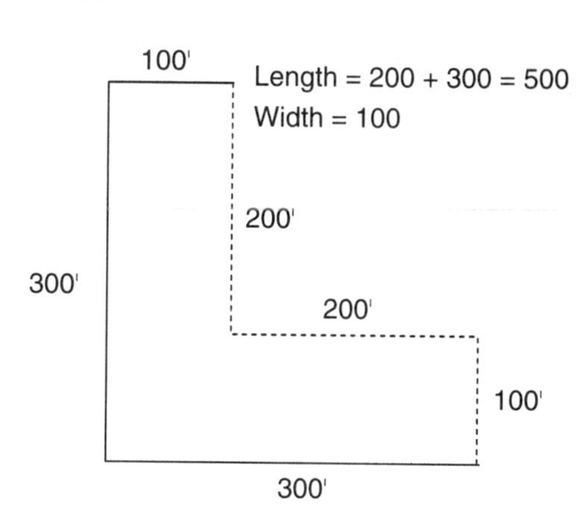

Suburban Stores – Masonry or Concrete

Quality Classification

	Class 1 Best Quality	Class 2 Good Quality	Class 3 Average Quality	Class 4 Low Quality
Foundation (15% of total cost)	Reinforced concrete.	Reinforced concrete.	Reinforced concrete.	Reinforced concrete.
Floor Structure (15% of total cost)	6" reinforced concrete on 6" rock base.	6" reinforced concrete on 6" rock base.	6" reinforced concrete on 6" rock base.	4" reinforced concrete on 6" rock base.
Wall Structure (15% of total cost)	8" reinforced decorative concrete block, 6" concrete tilt-up or 8" reinforced brick.	8" reinforced decorative concrete block, 6" concrete tilt-up or 8" reinforced brick.	8" reinforced concrete block, 6" concrete tilt-up or 8" reinforced common brick.	8" reinforced concrete block or 6" concrete tilt-up.
Roof (15% of total cost)	Glu-lam or steel beams on steel intermediate columns. Panelized roof system, 1/2" plywood sheathing, 5 ply built-up roof with insulation.	Glu-lam or steel beams on steel intermediate columns. Panelized roof system, 1/2" plywood sheathing, 5 ply built-up roof with insulation.	Glu-lam beams on steel intermediate columns. Panelized roof system, 1/2" plywood sheathing, 4 ply built-up roof.	Glu-lam beams on steel intermediate columns. Panelized roof system, 1/2" plywood sheathing, 4 ply built-up roof.
Floor Finish (5% of total cost)	Terrazzo, sheet vinyl or very good carpet.	Resilient tile with 50% solid vinyl tile, terrazzo, or good carpet.	Composition tile.	Minimum grade tile.
Interior Wall Finish (5% of total cost)	Inside of exterior walls furred out with gypsum wallboard or lath and plaster cover. Exterior walls and partitions finished with vinyl wall covers and hardwood veneers.	Interior stucco on inside of exterior walls, gypsum wallboard and texture or paper on partitions, some vinyl wall cover and plywood paneling.	Interior stucco on inside of exterior walls, gypsum wallboard and texture and paint on partitions.	Paint on inside of exterior walls, gypsum wallboard with texture and paint on partitions.
Ceiling Finish (5% of total cost)	Suspended good grade acoustical tile with gypsum wallboard backing.	Suspended acoustical tile with concealed grid system.	Suspended acoustical tile with exposed grid system.	Exposed beams with ceiling tile or paint.
Lighting (5% of total cost)	Recessed fluorescent lighting in modular plastic panels.	Continuous recessed 3 tube fluorescent strips with egg crate diffusers, 8' o.c.	Continuous 3 tube fluo-rescent strips with egg crate diffusers, 8' o.c.	Continuous exposed 2 tube fluorescent strips, 8' o.c.
Exterior (8% of total cost)	Face brick or stone veneer.	Exposed aggregate, some stone veneer.	Paint on exposed areas, some exposed aggregate.	Paint on exposed areas.
Plumbing (12% of total cost)	6 good fixtures per 5,000 S.F. of floor area, metal toilet partitions.	6 standard fixtures per 5,000 S.F. of floor area, metal toilet partitions.	4 standard fixtures per 5,000 S.F. of floor area, metal toilet partitions.	4 standard fixtures per 5,000 S.F. of floor area wood toilet partitions.

Note: Use the percent of total cost to help identify the correct quality classification.

The costs on page 59 and 60, and 66 to 69 include display fronts. The quality of the display front will help establish the quality class of the building as a whole. Display fronts are classified as follows:

	Class 1 Best Quality	Class 2 Good Quality	Class 3 Average Quality	Class 4 Low Quality
Bulkhead *(0 to 4' high)*	Vitrolite, domestic marble or stainless steel.	Black flagstone, terrazzo or good ceramic tile.	Average ceramic tile, Roman brick or imitation flagstone.	Stucco, wood or common brick.
Window Frame	Bronze or stainless steel.	Heavy aluminum.	Aluminum.	Light aluminum with with stops.
Glass	1/4" float glass with mitered joints.	1/4" float glass, some mitered joints.	1/4" float glass.	Crystal or 1/4" float glass.
Sign Area *(4' high)*	Vitrolite, domestic marble or stainless steel.	Black flagstone, terrazzo, or good ceramic tile.	Average ceramic tile, Roman brick or imitation flagstone.	Stucco.
Pilasters	Vitrolite, domestic marble.	Black flagstone, terrazzo or good ceramic tile.	Average ceramic tile, Roman brick or imitation flagstone.	Stucco.

Suburban Stores - Masonry or Concrete

Building Shell Only

Estimating Procedure

1. Use these figures to estimate the cost of shell-type buildings without permanent partitions, display fronts or finish material on the front wall of the building.
2. Establish the structure quality class by applying the information on page 56.
3. Compute the building floor area. This should include everything within the exterior walls and all inset areas outside the main walls but under the main roof.
4. Add to or subtract from the cost below the appropriate amount from the Wall Height Adjustment Table (at the bottom of this page) if the wall height is more or less than 16 feet.
5. Multiply the adjusted square foot cost by the building area.
6. Deduct, if appropriate, for common walls, using the figures at the bottom of page 58.
7. Multiply the total cost by the location factor on page 7.
8. Add the cost of the appropriate additional components from page 201: heating and cooling equipment, fire sprinklers, display fronts, finish materials on the front wall, canopies, interior partitions, exterior signs, mezzanines and basements, loading docks and ramps, yard improvements, and communication systems.

Depth greater than length of front - Square Foot Area

Quality Class	500	1,000	2,000	2,500	3,000	4,000	5,000	7,500	10,000	12,500	15,000
1, Best	114.57	95.56	82.15	78.75	76.23	72.70	70.27	66.53	64.28	62.77	61.63
1 & 2	106.03	88.44	76.06	72.88	70.55	67.27	65.03	61.58	59.50	58.10	57.05
2, Good	99.28	82.80	71.19	68.23	66.06	63.00	60.91	57.65	55.70	54.39	53.40
2 & 3	94.01	78.43	67.42	64.63	62.58	59.66	57.69	54.60	52.77	51.50	50.58
3, Average	90.33	75.36	64.77	62.07	60.12	57.32	55.41	52.45	50.71	49.53	48.62
3 & 4	85.15	71.03	61.09	58.51	56.64	54.00	52.21	49.43	47.79	46.65	45.81
4, Low	81.09	67.63	58.14	55.71	53.94	51.42	49.71	47.09	45.48	44.42	43.61

Length of front between 1 and 2 times depth - Square Foot Area

Quality Class	500	1,000	2,000	3,000	5,000	7,500	10,000	15,000	20,000	25,000	35,000
1, Best	109.46	91.95	79.63	74.19	68.75	65.31	63.24	60.82	59.40	58.38	57.11
1 & 2	101.14	84.99	73.57	68.55	63.50	60.36	58.45	56.21	54.87	53.96	52.73
2, Good	94.71	79.55	68.91	64.19	59.45	56.49	54.71	52.63	51.36	50.49	49.39
2 & 3	89.62	75.30	65.20	60.75	56.27	53.42	51.77	49.76	48.59	47.81	46.74
3, Average	86.25	72.48	62.77	58.47	54.16	51.48	49.86	47.93	46.78	46.03	44.98
3 & 4	81.74	68.70	59.47	55.38	51.32	48.78	47.24	45.41	44.33	43.60	42.64
4, Low	77.11	64.79	56.08	52.23	48.40	46.00	44.54	42.83	41.82	41.10	40.21

Wall Height Adjustment: Costs above are based on a 16' wall height, measured from the bottom of the floor slab or floor joists to the top of the roof cover. Add or subtract the amount listed to the square foot cost for each foot more or less than 16 feet.

Area	500	1,000	2,000	2,500	3,000	4,000	5,000	7,500	10,000	12,500
Cost	1.98	1.47	1.07	.93	.84	.74	.68	.49	.43	.39

Area	15,000	20,000	25,000	30,000	35,000	50,000	70,000	75,000	100,000	150,000
Cost	.38	.37	.35	.33	.31	.27	.23	.22	.16	.14

Suburban Stores – Masonry or Concrete

Building Shell Only

Estimating Procedure

1. Use these figures to estimate the cost of shell-type buildings without permanent partitions, display fronts or finish material on the front wall of the building.
2. Establish the structure quality class by applying the information on page 56.
3. Compute the building floor area. This should include everything within the exterior walls and all inset areas outside the main walls but under the main building roof.
4. Add to or subtract from the cost below the appropriate amount from the Wall Height Adjustment Table on page 57 if the wall height is more or less than 16 feet.
5. Multiply the adjusted square foot cost by the building area.
6. Deduct, if appropriate, for common walls, using the figures at the bottom of this page.
7. Multiply the total cost by the location factor on page 7.
8. Add the cost of the appropriate additional components from page 201: heating and cooling equipment, fire sprinklers, display fronts, finish materials on the front wall, canopies, interior partitions, exterior signs, mezzanines and basements, loading docks and ramps, yard improvements, and communication systems.

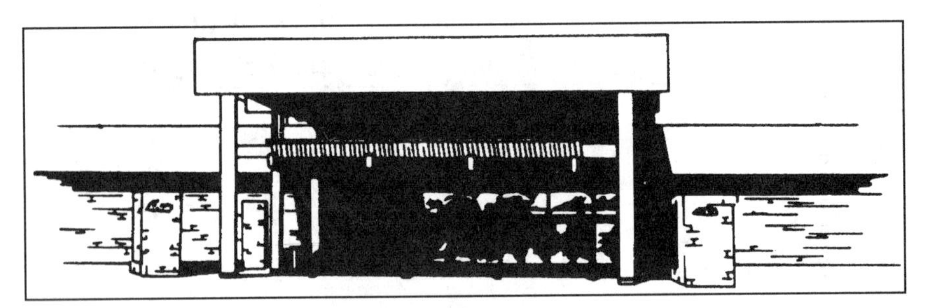

Suburban Store, Class 1

Length between 2 and 4 times depth - Square Foot Area

Quality Class	1,000	2,000	3,000	5,000	7,500	10,000	15,000	20,000	30,000	50,000	70,000
1, Best	95.36	82.26	76.47	70.66	66.97	64.82	62.24	60.53	58.85	57.01	56.02
1 & 2	87.82	75.73	70.40	65.04	61.67	59.68	57.28	55.88	54.17	52.51	51.59
2, Good	81.76	70.52	65.56	60.56	57.43	55.58	53.34	52.03	50.45	48.88	48.05
2 & 3	77.45	66.80	62.08	57.38	54.41	52.63	50.52	49.28	47.80	46.30	45.48
3, Average	74.38	64.19	59.70	55.11	52.29	50.55	48.54	47.37	45.90	44.46	43.72
3 & 4	70.66	60.96	56.65	52.35	49.64	48.04	46.09	44.95	43.60	42.24	41.49
4, Low	66.71	57.55	53.48	49.42	46.88	45.33	43.55	42.47	41.16	39.88	39.22

Length greater than 4 times depth - Square Foot Area

Quality Class	2,000	3,000	5,000	10,000	15,000	20,000	30,000	50,000	75,000	100,000	150,000
1, Best	86.78	80.24	73.67	67.06	64.15	62.41	60.38	58.30	57.00	56.23	55.33
1 & 2	79.04	73.07	67.07	61.05	58.40	56.85	54.96	53.08	51.90	51.22	50.37
2, Good	73.54	67.98	62.39	56.83	54.36	52.88	51.14	49.39	48.29	47.64	46.84
2 & 3	69.61	64.33	59.07	53.79	51.46	50.05	48.42	46.76	45.72	45.07	44.36
3, Average	66.86	61.85	56.76	51.69	49.42	48.09	46.51	44.93	43.94	43.34	42.62
3 & 4	63.22	58.46	53.68	48.85	46.72	45.47	43.96	42.49	41.51	40.96	40.31
4, Low	59.84	55.34	50.78	46.25	44.23	43.05	41.62	40.21	39.30	38.79	38.12

Perimeter Wall Adjustment: A common wall exists when two buildings share one wall. Adjust for common walls by deducting $166.00 per linear foot of common wall from the total structure cost. In some structures one or more walls are not owned at all. In this case, deduct $326.00 per linear foot of wall not owned.

Suburban Stores – Masonry or Concrete

Multi-Unit Buildings

Estimating Procedure

1. Use these square foot costs to estimate the cost of stores designed for multiple occupancy. These costs include all components of shell buildings plus the cost of display fronts, finish materials on the front of the building and normal interior partitions.
2. Establish the structure quality class by applying the information on page 56. Evaluate the quality of the display front to help establish the correct quality class of the building as a whole. See also pages 207 to 209.
3. Compute the building floor area. This should include everything within the building exterior walls and all inset areas outside the main walls but under the main building roof.
4. Add to or subtract from the cost below the appropriate amount from the Wall Height Adjustment Table (at the bottom of this page) if the wall height is more or less than 16 feet.
5. Multiply the adjusted square foot cost by the building area.
6. Deduct, if appropriate, for common walls, using the figures at the bottom of page 60.
7. Multiply the total cost by the location factor on page 7.
8. Add the cost of the appropriate additional components from page 201: heating and cooling equipment, fire sprinklers, canopies, exterior signs, mezzanines and basements, loading docks and ramps, yard improvements, and communications systems.

Depth greater than length of front - Square Foot Area

Quality Class	500	1,000	2,000	2,500	3,000	4,000	5,000	7,500	10,000	12,500	15,000
1, Best	163.12	133.17	112.01	106.62	102.63	97.08	93.28	87.34	83.80	81.42	79.62
1 & 2	146.12	119.31	100.36	95.53	91.96	86.96	83.53	78.25	75.10	72.93	71.34
2, Good	131.32	107.19	90.18	85.83	82.64	78.15	75.07	70.30	67.48	65.54	64.09
2 & 3	124.24	98.47	82.81	78.84	75.89	71.77	68.96	64.55	61.97	60.18	58.89
3, Average	111.58	91.07	76.63	72.93	70.24	66.39	63.78	59.75	57.33	55.70	54.48
3 & 4	103.46	84.46	71.04	67.64	65.10	61.58	59.14	55.38	53.15	51.63	50.49
4, Low	95.31	77.81	65.46	62.30	59.99	56.70	54.51	51.05	48.96	47.55	46.54

Length of front between 1 and 2 times depth - Square Foot Area

Quality Class	500	1,000	2,000	3,000	5,000	7,500	10,000	15,000	20,000	25,000	35,000
1, Best	179.47	145.31	120.99	110.14	99.24	92.34	88.22	83.35	80.44	78.43	75.83
1 & 2	159.49	129.13	107.49	97.88	88.18	82.06	78.41	74.05	71.47	69.70	67.40
2, Good	141.83	114.83	95.61	87.04	78.40	72.97	69.72	65.88	63.56	61.99	59.92
2 & 3	129.43	104.80	87.25	79.44	71.57	66.59	63.64	60.13	58.00	56.59	54.68
3, Average	123.18	99.74	83.05	75.59	68.13	63.40	60.57	57.21	55.24	53.84	52.05
3 & 4	116.56	94.38	78.57	71.55	64.47	59.97	57.33	54.14	52.24	50.95	49.25
4, Low	106.34	86.12	71.69	65.27	58.81	54.74	52.30	49.39	47.67	46.48	44.95

Wall Height Adjustment: Costs above are based on a 16' wall height, measured from the bottom of the floor slab or floor joists to the top of the roof cover. Add or subtract the amount listed to the square foot cost for each foot more or less than 16 feet.

Square Foot Area

Class	2,000	3,000	5,000	10,000	15,000	25,000	35,000	50,000	75,000	100,000	150,000
1, Best	4.64	3.77	2.92	2.04	1.71	1.30	1.12	.98	.84	.79	.73
2, Good	3.45	2.82	2.17	1.51	1.29	.98	.84	.75	.64	.59	.53
3, Average	2.65	2.16	1.66	1.18	.99	.78	.67	.55	.48	.46	.42
4, Low	2.09	1.70	1.30	.90	.79	.60	.49	.44	.39	.36	.35

Suburban Stores – Masonry or Concrete

Multi-Unit Buildings

Estimating Procedure

1. Use these square foot costs to estimate the cost of stores designed for multiple occupancy. These costs include all components of shell buildings plus the cost of display fronts, finish materials on the front of the building and normal interior partitions.
2. Establish the structure quality class by applying the information on page 56. Evaluate the quality of the display front to help establish the correct quality class of the building as a whole. See also pages 207 to 209.
3. Compute the building floor area. This should include everything within the building exterior walls and all inset areas outside the main walls but under the main building roof.
4. Add to or subtract from the cost below the appropriate amount from the Wall Height Adjustment Table (at the bottom of this page) if the wall height is more or less than 16 feet.
5. Multiply the adjusted square foot cost by the building area.
6. Deduct, if appropriate, for common walls, using the figures at the bottom of this page.
7. Multiply the total cost by the location factor on page 7.
8. Add the cost of the appropriate additional components from page 201: heating and cooling equipment, fire sprinklers, canopies, exterior signs, mezzanines and basements, loading docks and ramps, yard improvements, and communications systems .

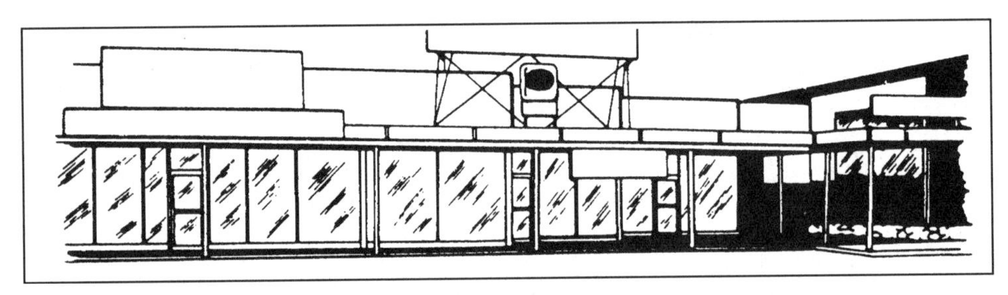

Suburban Store, Class 3 & 4

Length of front between 2 and 4 times depth - Square Foot Area

Quality Class	1,000	2,000	3,000	5,000	7,500	10,000	15,000	20,000	30,000	50,000	70,000
1, Best	163.85	134.42	121.30	108.06	99.68	94.69	88.78	85.22	81.02	76.77	74.54
1 & 2	146.07	119.85	108.15	96.36	88.88	84.43	79.13	75.99	72.21	68.46	66.44
2, Good	129.41	106.14	95.78	85.33	78.72	74.77	70.09	67.29	63.97	60.62	58.84
2 & 3	117.43	96.34	86.94	77.46	71.45	67.88	63.63	61.10	58.05	55.04	53.40
3, Average	110.60	88.02	79.44	70.78	65.25	62.01	58.12	55.79	53.06	50.27	48.79
3 & 4	98.98	81.22	73.26	65.30	60.22	57.23	53.63	51.49	48.93	46.39	45.03
4, Low	90.59	74.32	67.06	59.76	55.13	52.37	49.05	47.12	44.78	42.47	41.21

Length of front greater than 4 times depth - Square Foot Area

Quality Class	2,000	3,000	5,000	10,000	15,000	20,000	30,000	50,000	75,000	100,000	150,000
1, Best	150.25	134.07	117.94	101.85	94.74	90.52	85.54	80.55	77.42	75.53	73.34
1 & 2	132.89	118.57	104.29	90.07	83.80	80.09	75.67	71.27	68.47	66.81	64.86
2, Good	117.52	104.85	92.26	79.65	74.14	70.83	66.91	63.02	60.55	59.10	57.36
2 & 3	106.47	94.99	83.60	72.18	67.15	64.17	60.61	57.07	54.86	53.54	51.96
3, Average	97.45	86.95	76.49	66.05	61.46	58.73	55.48	52.24	50.19	48.97	47.55
3 & 4	89.96	80.26	70.62	60.99	56.75	54.21	51.22	48.24	46.36	45.21	43.91
4, Low	82.10	73.25	64.44	55.65	51.77	49.46	46.72	44.01	42.29	41.27	40.08

Perimeter Wall Adjustment: A common wall exists when two buildings share one wall. Adjust for common walls by deducting $171.00 per linear foot of common wall from the total structure cost. In some structures one or more walls are not owned at all. In this case, deduct $342.00 per linear foot of wall not owned.

Suburban Stores – Wood or Wood and Steel Frame

Quality Classification

	Class 1 Best Quality	Class 2 Good Quality	Class 3 Average Quality	Class 4 Low Quality
Foundation (15% of total cost)	Reinforced concrete.	Reinforced concrete.	Reinforced concrete.	Reinforced concrete.
Floor Structure (15% of total cost)	6" reinforced concrete on 6" rock base.	6" reinforced concrete on 6" rock base.	6" reinforced concrete on 6" rock base.	4" reinforced concrete on 6" rock base.
Wall Structure (15% of total cost)	2" x 6" - 16" o.c.	2" x 6" - 16" o.c.	2" x 6" - 16" o.c.	2" x 4" - 16" o.c.
Roof (15% of total cost)	Glu-lams or steel beams on steel intermediate columns. Panelized roof system, 1/2" plywood sheathing, 5 ply built-up roof with insulation.	Glu-lams or steel beams on steel intermediate columns. Panelized roof system, 1/2" plywood sheathing, 5 ply built-up roof with insulation.	Glu-lams on steel intermediate columns. Panelized roof system, 1/2" plywood sheathing, 4 ply built-up roof.	Glu-lams on steel intermediate columns. Panelized roof system, 1/2" plywood sheathing, 4 ply built-up roof.
Floor Finish (5% of total cost)	Terrazzo, sheet vinyl, or very good carpet.	Resilient tile with 50% solid vinyl tile, terrazzo, or good carpet.	Composition tile.	Minimum grade tile.
Interior Wall Finish (5% of total cost)	Gypsum wallboard or lath and plaster on exterior walls and partitions, finished with vinyl wall covers and hardwood veneers.	Gypsum wallboard, texture and paper on exterior walls and partitions, some vinyl wall cover and plywood paneling.	Gypsum wallboard, texture and paint on interior walls and partitions.	Gypsum wallboard, texture and paint on interior walls and partitions.
Ceiling Finish (5% of total cost)	Suspended good grade acoustical tile with gypsum board backing.	Suspended acoustical tile with concealed grid system.	Suspended acoustical tile with exposed grid system.	Exposed beams with ceiling tile or painted.
Lighting (5% of total cost)	Recessed fluorescent lighting in modular plastic panels.	Continuous recessed 3 tube fluorescent strips with egg crate diffusers, 8' o.c.	Continuous 3 tube fluorescent strips with egg crate diffusers, 8' o.c.	Continuous exposed 2 tube fluorescent strips, 8' o.c.
Exterior (8% of total cost)	Face brick or stone veneer.	Wood siding, some stone veneer.	Stucco on exposed areas, some brick trim.	Stucco on exposed areas.
Plumbing (12% of total cost)	6 good fixtures per 5,000 S.F. of floor area, metal toilet partitions.	6 standard fixtures per 5,000 S.F. of floor area, metal toilet partitions.	4 standard fixtures per 5,000 S.F. of floor area, metal toilet partitions.	4 standard fixtures per 5,000 S.F. of floor area wood toilet partitions.

Note: Use the percent of total cost to help identify the correct quality classification

Strip and Island Suburban Stores

For estimating purposes, wood frame suburban stores should be divided into strip type units or island type units. Strip type buildings have a front wall made up of display fronts. The side and rear walls, except for delivery or walk-through doors, are made up of solid, continuous wood frame walls. If there are any display areas in the sides or rear of these buildings, the cost of the display front must be added to the building cost and the cost of the wall that it replaces must be deducted from the building costs.

Island type suburban store buildings have display fronts on the major portion of all four sides. Stores may be arranged so that one store fronts on two sides or they may be partitioned in such a way that there are two separate stores fronting on each side of the building.

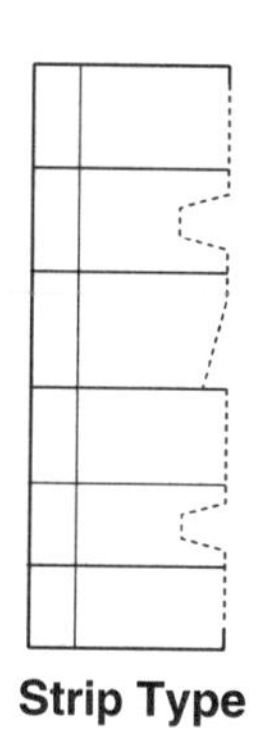

Strip Type

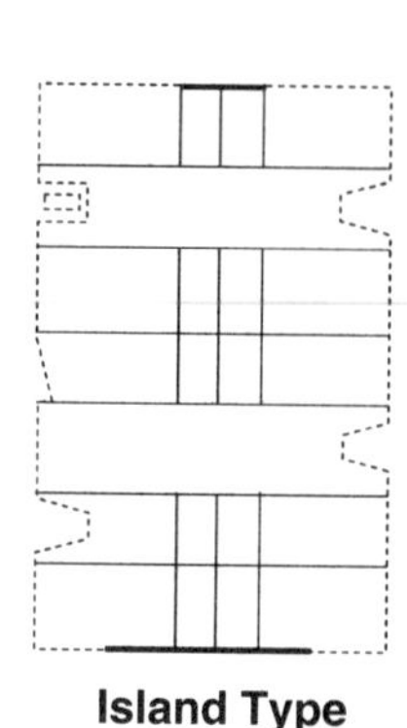

Island Type

Suburban Stores – Wood or Wood and Steel Frame

Building Shell Only, Island Type

Estimating Procedure

1. Use these figures to estimate the cost of shell-type buildings without permanent partitions, display fronts or finish material on the front wall of the building.
2. Establish the structure quality class by applying the information on page 61.
3. Compute the building floor area. This should include everything within the exterior walls and all inset areas outside the main walls but under the main roof.
4. Add to or subtract from the square foot cost below the appropriate amount from the Wall Height Adjustment Table (at the bottom of this page) if the wall height is more or less than 16 feet.
5. Multiply the adjusted square foot cost by the building area.
6. Deduct, if appropriate, for common walls, using the figures at the bottom of this page.
7. Multiply the total cost by the location factor on page 7.
8. Add the cost of the appropriate additional components from page 201: heating and cooling equipment, fire sprinklers, display fronts, finish materials on the front wall, canopies, interior partitions, exterior signs, mezzanines and basements, loading docks and ramps, yard improvements, and communication systems.

Length less than 1-1/2 times depth - Square Foot Area

Quality Class	3,500	5,000	7,500	10,000	12,500	15,000	20,000	30,000	40,000	50,000	75,000
1, Best	64.71	63.63	62.58	61.98	61.56	61.23	60.81	60.27	59.95	59.76	59.43
1 & 2	58.62	57.63	56.69	56.15	55.74	55.48	55.09	54.59	54.32	54.12	53.84
2, Good	53.44	52.56	51.70	51.20	50.85	50.59	50.24	49.80	49.53	49.38	49.08
2 & 3	51.60	49.86	48.22	47.75	47.42	47.21	46.87	46.44	46.19	46.04	45.80
3, Average	47.14	46.38	45.61	45.16	44.86	44.60	44.32	43.95	43.72	43.55	43.30
3 & 4	43.27	42.56	41.85	41.44	41.15	40.97	40.66	40.32	40.09	39.96	39.73
4, Low	39.34	38.66	38.05	37.67	37.40	37.23	36.94	36.65	36.44	36.33	36.11

Length between 1-1/2 and 2 times depth - Square Foot Area

Quality Class	4,500	5,000	7,500	10,000	15,000	20,000	30,000	40,000	50,000	75,000	100,000
1, Best	64.41	64.04	62.81	62.14	61.34	60.90	60.34	60.07	59.86	59.54	59.38
1 & 2	58.31	57.97	56.87	56.22	55.52	55.12	54.65	54.38	54.19	53.91	53.77
2, Good	53.19	52.88	51.89	51.33	50.64	50.29	49.84	49.59	49.44	49.18	49.03
2 & 3	49.54	49.26	48.33	47.80	47.21	46.86	46.44	46.20	46.06	45.83	45.69
3, Average	46.86	46.57	45.68	45.21	44.58	44.29	43.92	43.71	43.54	43.31	43.21
3 & 4	43.00	42.73	41.94	41.48	40.96	40.65	40.33	40.09	39.97	39.77	39.63
4, Low	38.99	38.77	38.05	37.62	37.16	36.88	36.55	36.39	36.23	36.08	35.96

Wall Height Adjustment: Add or subtract the amount listed to the square foot costs above for each foot of wall height more or less than 16 feet.

Area	3,500	5,000	7,500	10,000	12,500	15,000	Over 20,000
Cost	.15	.13	.11	.09	.07	.06	.05

Perimeter Wall Adjustment: For common wall deduct $35.90 per linear foot of common wall. For no wall ownership, deduct $71.90 per linear foot of wall.

Suburban Stores – Wood or Wood and Steel Frame

Building Shell Only, Island Type

Estimating Procedure

1. Use these figures to estimate the cost of shell-type buildings without permanent partitions, display fronts or finish material on the front wall of the building.
2. Establish the structure quality class by applying the information on page 61.
3. Compute the building floor area. This should include everything within the exterior walls and all inset areas outside the main walls but under the main roof.
4. Add to or subtract from the square foot cost below the appropriate amount from the Wall Height Adjustment Table (at the bottom of this page) if the wall height is more or less than 16 feet.
5. Multiply the adjusted square foot cost by the building area.
6. Deduct, if appropriate, for common walls, using the figures at the bottom of this page.
7. Multiply the total cost by the location factor on page 7.
8. Add the cost of the appropriate additional components from page 201: heating and cooling equipment, fire sprinklers, display fronts, finish materials on the front wall, canopies, interior partitions, exterior signs, mezzanines and basements, loading docks and ramps, yard improvements, and communication systems.

Length between 2 and 3 times depth - Square Foot Area

Quality Class	10,000	15,000	20,000	30,000	40,000	50,000	75,000	100,000	150,000	200,000	250,000
1, Best	63.07	62.15	61.59	60.95	60.57	60.30	59.89	59.69	59.42	59.25	59.11
1 & 2	57.06	56.22	55.73	55.15	54.83	54.58	54.24	54.02	53.78	53.60	53.52
2, Good	51.81	51.04	50.57	50.06	49.74	49.52	49.22	49.00	48.79	48.66	48.56
2 & 3	48.28	47.57	47.14	46.65	46.38	46.14	45.88	45.66	45.47	45.33	45.27
3, Average	45.40	44.71	44.32	43.83	43.57	43.37	43.10	42.93	42.76	42.63	42.53
3 & 4	41.63	40.99	40.63	40.21	39.98	39.77	39.54	39.35	39.21	39.06	38.98
4, Low	37.74	37.21	36.88	36.48	36.24	36.10	35.87	35.72	35.55	35.46	35.39

Length greater than 3 times depth - Square Foot Area

Quality Class	7,500	10,000	15,000	20,000	30,000	40,000	50,000	75,000	100,000	150,000	200,000
1, Best	62.89	62.50	61.76	61.33	60.74	60.39	60.12	59.73	59.45	59.10	58.94
1 & 2	58.83	58.45	57.77	55.52	55.02	54.67	54.44	54.05	53.84	53.53	53.37
2, Good	51.63	51.33	50.68	50.34	49.86	49.56	49.35	49.00	48.81	48.55	48.37
2 & 3	48.14	47.83	47.24	46.91	46.48	46.19	46.02	45.67	45.49	45.25	45.08
3, Average	45.29	45.02	44.46	44.13	43.75	43.47	43.29	42.98	42.80	42.56	42.43
3 & 4	41.50	41.26	40.78	40.48	40.09	39.84	39.71	39.40	39.26	39.02	38.90
4, Low	37.71	37.49	37.04	36.76	36.43	36.20	36.07	35.81	35.65	35.45	35.34

Wall Height Adjustment: Add or subtract the amount listed to the square foot costs above for each foot of wall height more or less than 16 feet.

Area	3,500	5,000	7,500	10,000	12,500	15,000	Over 20,000
Cost	.15	.13	.11	.09	.07	.06	.05

Perimeter Wall Adjustment: For common wall, deduct $35.90 per linear foot of common wall. For no wall ownership, deduct $71.90 per linear foot of wall.

Suburban Stores – Wood or Wood and Steel Frame

Building Shell Only, Strip Type

Estimating Procedure

1. Use these figures to estimate the cost of shell-type buildings without permanent partitions, display fronts or finish material on the front wall of the building.
2. Establish the structure quality class by applying the information on page 61.
3. Compute the building floor area. This should include everything within the exterior walls and all inset areas outside the main walls but under the main roof.
4. Add to or subtract from the square foot cost below the appropriate amount from the Wall Height Adjustment Table (at the bottom of this page) if the wall height is more or less than 16 feet.
5. Multiply the adjusted square foot cost by the building area.
6. Deduct, if appropriate, for common walls, using the figures at the bottom of this page.
7. Multiply the total cost by the location factor on page 7.
8. Add the cost of the appropriate additional components from page 201: heating and cooling equipment, fire sprinklers, display fronts, finish materials on the front wall, canopies, interior partitions, exterior signs, mezzanines and basements, loading docks and ramps, yard improvements, and communication systems.

Length less than 1-1/2 times depth - Square Foot Area

Quality Class	500	1,000	2,000	2,500	3,000	4,000	5,000	7,500	10,000	12,500	15,000
1, Best	101.12	87.35	77.63	75.14	73.31	70.75	68.99	66.27	64.65	63.54	62.71
1 & 2	92.39	79.83	70.93	68.66	66.98	64.66	63.05	60.55	59.07	58.07	57.31
2, Good	84.79	73.27	65.12	63.04	61.47	59.33	57.86	55.58	54.21	53.28	52.58
2 & 3	80.44	69.50	61.76	59.80	58.32	56.29	54.91	52.71	51.43	50.54	49.89
3, Average	77.10	66.61	59.20	57.29	55.92	53.94	52.63	50.52	49.29	48.45	47.80
3 & 4	72.24	62.36	55.44	53.66	52.37	50.52	49.28	47.32	46.18	45.39	44.78
4, Low	67.34	58.16	51.68	50.04	48.85	47.11	45.96	44.13	43.04	42.30	41.76

Length between 1-1/2 and 2 times depth - Square Foot Area

Quality Class	500	1,000	2,000	3,000	5,000	6,000	7,500	10,000	15,000	20,000	35,000
1, Best	97.48	84.58	75.39	71.28	67.17	65.93	64.58	63.02	61.17	60.07	58.32
1 & 2	89.10	77.29	68.88	65.14	61.39	60.26	59.02	57.59	55.90	54.88	53.33
2, Good	81.82	70.96	63.25	59.82	56.37	55.32	54.19	52.89	51.35	50.41	48.93
2 & 3	77.49	67.24	59.90	56.64	53.39	52.41	51.33	50.08	48.62	47.76	46.38
3, Average	74.18	64.35	57.33	54.25	51.10	50.14	49.12	47.93	46.53	45.68	44.39
3 & 4	69.72	60.48	53.89	50.98	48.05	47.13	46.18	45.05	43.76	42.95	41.71
4, Low	65.22	56.59	50.45	47.72	44.92	44.09	43.20	42.15	40.92	40.18	39.02

Wall Height Adjustment: Add or subtract the amount listed to the square foot cost above for each foot of wall height more or less than 16 feet.

Area	500	1,000	2,000	2,500	3,000	4,000	5,000	7,500	10,000	12,500
Cost	1.15	.81	.57	.52	.47	.43	.39	.35	.32	.30

Area	15,000	20,000	25,000	30,000	35,000	50,000	70,000	75,000	100,000	150,000
Cost	.23	.15	.14	.13	.12	.11	.10	.09	.08	.07

Perimeter Wall Adjustment: For a common wall, deduct $122.00 per linear foot of common wall. For no wall ownership, deduct $246.00 per linear foot of wall.

Suburban Stores – Wood or Wood and Steel Frame

Building Shell Only, Strip Type

Estimating Procedure

1. Use these figures to estimate the cost of shell-type buildings without permanent partitions, display fronts or finish material on the front wall of the building.
2. Establish the structure quality class by applying the information on page 61.
3. Compute the building floor area. This should include everything within the exterior walls and all inset areas outside the main walls but under the main roof.
4. Add to or subtract from the square foot cost below the appropriate amount from the Wall Height Adjustment Table (at the bottom of this page) if the wall height is more or less than 16 feet.
5. Multiply the adjusted square foot cost by the building area.
6. Deduct, if appropriate, for common walls, using the figures at the bottom of this page.
7. Multiply the total cost by the location factor on page 7.
8. Add the cost of the appropriate additional components from page 201: heating and cooling equipment, fire sprinklers, display fronts, finish materials on the front wall, canopies, interior partitions, exterior signs, mezzanines and basements, loading docks and ramps, yard improvements, and communication systems.

Length between 2 and 3 times depth - Square Foot Area

Quality Class	1,000	2,000	3,000	5,000	7,500	10,000	15,000	20,000	30,000	50,000	70,000
1, Best	85.85	76.55	72.40	68.25	65.66	64.11	62.23	61.14	59.84	58.54	57.83
1 & 2	78.46	69.93	66.15	62.39	60.02	58.59	56.89	55.88	54.68	53.48	52.85
2, Good	72.01	64.18	60.72	57.25	55.06	53.75	52.20	51.28	50.19	49.10	48.49
2 & 3	68.35	60.96	57.65	54.35	52.28	51.01	49.53	48.66	47.63	46.63	46.05
3, Average	67.91	60.24	55.32	52.18	50.14	48.98	47.56	46.74	45.71	44.74	44.19
3 & 4	61.67	54.97	51.99	49.01	47.15	46.02	44.71	43.92	42.99	42.06	41.53
4, Low	57.65	51.37	48.59	45.80	44.05	43.00	41.79	41.04	40.16	39.30	38.81

Length greater than 3 times depth - Square Foot Area

Quality Class	2,000	3,000	5,000	10,000	15,000	20,000	30,000	50,000	75,000	100,000	150,000
1, Best	79.07	74.48	69.90	65.32	63.29	62.09	60.65	59.24	58.31	57.79	57.14
1 & 2	71.80	67.65	63.48	59.30	57.47	56.39	55.09	53.79	52.95	52.49	51.89
2, Good	65.91	62.08	58.27	54.44	52.76	51.74	50.56	49.36	48.64	48.17	47.63
2 & 3	62.51	58.87	55.26	51.62	50.03	49.04	47.93	46.79	46.09	45.66	45.18
3, Average	59.93	56.45	53.01	49.50	47.99	47.05	45.97	44.88	44.19	43.80	43.31
3 & 4	56.53	53.25	49.98	46.71	45.25	44.40	43.38	42.33	41.70	41.33	40.86
4, Low	53.02	49.94	46.86	43.80	42.43	41.63	40.67	39.73	39.11	38.75	38.31

Wall Height Adjustment: Add or subtract the amount listed to the square foot cost above for each foot of wall height more or less than 16 feet.

Area	500	1,000	2,000	2,500	3,000	4,000	5,000	7,500	10,000	12,500
Cost	1.15	.81	.57	.52	.47	.43	.39	.35	.32	.30

Area	15,000	20,000	25,000	30,000	35,000	50,000	70,000	75,000	100,000	150,000
Cost	.23	.15	.14	.13	.12	.11	.10	.09	.08	.07

Perimeter Wall Adjustment: For a common wall, deduct $122.00 per linear foot of common wall. For no wall ownership, deduct $246.00 per linear foot of wall.

Suburban Stores - Wood or Wood and Steel Frame

Multi-Unit, Island Type

Estimating Procedure

1. Use these figures to estimate the cost of stores designed for multiple occupancy. These costs include all components of shell buildings plus the cost of display fronts, finish materials on the front of the building and normal interior partitions.
2. Establish the structure quality class by applying the information on page 61. Evaluate the quality of the display front to help establish the correct quality class of the building as a whole. The building classes have display fronts as classified on page 56. See also pages 207 to 209.
3. Compute the building floor area. This should include everything within the exterior walls and all inset areas outside the main walls but under the main roof.
4. Add to or subtract from the square foot cost below the appropriate amount from the Wall Height Adjustment Table (at the bottom of this page) if the wall height is more or less than 16 feet.
5. Multiply the adjusted square foot cost by the building area.
6. Deduct, if appropriate, for common walls, using the figures at the bottom of this page.
7. Multiply the total cost by the location factor on page 7.
8. Add the cost of the appropriate additional components from page 201: heating and cooling equipment, fire sprinklers, canopies, exterior signs, mezzanines and basements, loading docks and ramps, yard improvements, and communication systems.

Length less than 1-1/2 times depth - Square Foot Area

Quality Class	3,500	5,000	7,500	10,000	12,500	15,000	20,000	30,000	40,000	50,000	75,000
1, Best	143.97	131.05	119.27	111.09	106.49	102.85	96.97	91.04	87.55	82.60	80.44
1 & 2	125.83	114.73	104.69	97.62	93.57	90.58	85.54	84.84	80.43	75.42	72.01
2, Good	108.65	99.65	91.15	86.07	82.55	79.98	76.37	72.03	69.44	67.68	64.91
2 & 3	95.03	87.18	79.75	75.31	72.23	69.97	66.81	63.03	60.76	59.24	56.80
3, Average	84.32	77.33	70.76	66.80	64.09	62.07	59.25	55.92	53.89	52.55	50.38
3 & 4	75.52	69.27	63.34	59.81	57.38	55.56	53.06	50.04	48.27	47.05	45.12
4, Low	66.79	61.25	56.02	52.91	50.77	49.15	46.94	44.29	42.69	41.60	39.91

Length between 1-1/2 and 2 times depth - Square Foot Area

Quality Class	4,500	5,000	7,500	10,000	15,000	20,000	30,000	40,000	50,000	75,000	100,000
1, Best	137.10	133.22	120.37	113.00	104.47	99.54	93.80	90.45	88.20	84.75	82.71
1 & 2	120.84	117.41	106.08	99.58	92.09	87.74	82.67	79.71	77.74	74.72	72.92
2, Good	105.80	102.81	92.89	87.18	80.62	76.79	72.39	69.83	68.09	65.42	63.90
2 & 3	93.56	90.90	82.14	77.08	71.29	67.91	64.04	61.74	60.20	57.82	56.46
3, Average	82.27	79.93	72.25	67.81	62.71	59.75	56.32	54.28	52.95	50.86	49.65
3 & 4	74.30	72.21	65.24	61.21	56.61	53.92	50.86	49.01	47.80	45.93	44.83
4, Low	66.23	64.36	58.15	54.59	50.48	48.10	45.34	43.71	42.63	40.96	39.95

Wall Height Adjustment: Add or subtract the amount listed to the square foot cost above for each foot of wall height more or less than 16 feet.

Square Foot Area

Quality Class	3,500	5,000	7,500	10,000	12,500	15,000	20,000	30,000	40,000	50,000	75,000	100,000	150,000
1	6.85	5.75	4.73	4.12	3.70	3.38	2.95	2.41	2.12	1.91	1.57	1.37	1.15
2	4.73	3.99	3.26	2.83	2.56	2.34	2.06	1.68	1.47	1.31	1.10	.94	.78
3	3.26	2.75	2.24	1.97	1.75	1.61	1.41	1.15	1.02	.90	.75	.67	.53
4	2.40	2.02	1.65	1.43	1.29	1.17	1.03	.86	.73	.66	.55	.46	.40

Perimeter Wall Adjustment: For a common wall, deduct per linear foot: Class 1, $677, Class 2, $447, Class 3, $317, Class 4, $208. For no wall ownership, deduct per linear foot: Class 1, $1,343, Class 2, $884, Class 3, $611, Class 4, $426

Suburban Stores – Wood or Wood and Steel Frame

Multi-Unit, Island Type

Estimating Procedure

1. Use these figures to estimate the cost of stores designed for multiple occupancy. These costs include all components of shell buildings plus the cost of display fronts, finish materials on the front of the building and normal interior partitions.
2. Establish the structure quality class by applying the information on page 61. Evaluate the quality of the display front to help establish the correct quality class of the building as a whole. The building classes have display fronts as classified on page 56. See also pages 207 to 209.
3. Compute the building floor area. This should include everything within the exterior walls and all inset areas outside the main walls but under the main roof.
4. Add to or subtract from the square foot cost below the appropriate amount from the Wall Height Adjustment Table (at the bottom of this page) if the wall height is more or less than 16 feet.
5. Multiply the adjusted square foot cost by the building area.
6. Deduct, if appropriate, for common walls, using the figures at the bottom of this page.
7. Multiply the total cost by the location factor on page 7.
8. Add the cost of the appropriate additional components from page 201: heating and cooling equipment, fire sprinklers, canopies, exterior signs, mezzanines and basements, loading docks and ramps, yard improvements, and communication systems.

Length between 2 and 3 times depth - Square Foot Area

Quality Class	7,500	10,000	15,000	20,000	30,000	40,000	50,000	75,000	100,000	150,000	200,000
1, Best	121.26	114.23	105.86	100.85	94.86	91.29	88.81	84.99	82.72	80.00	78.39
1 & 2	107.17	100.94	93.52	89.11	83.81	80.66	78.48	75.10	73.10	70.72	69.27
2, Good	94.17	88.72	82.19	78.31	73.65	70.89	68.97	66.02	64.24	62.13	60.88
2 & 3	83.61	78.74	72.94	69.50	65.38	62.90	61.22	58.61	57.01	55.14	54.03
3, Average	74.00	69.70	64.59	61.51	57.87	55.68	54.21	51.88	50.49	48.84	47.83
3 & 4	66.51	62.67	58.07	55.30	52.03	50.05	48.73	46.63	45.39	43.87	43.00
4, Low	58.95	55.55	51.47	49.00	46.10	44.36	43.19	41.29	40.19	38.89	38.12

Length greater than 3 times depth - Square Foot Area

Quality Class	10,000	15,000	20,000	30,000	40,000	50,000	75,000	100,000	150,000	200,000	250,000
1, Best	120.30	110.59	104.87	98.07	94.04	91.30	87.02	84.49	81.48	79.69	78.46
1 & 2	106.36	97.79	92.70	86.69	83.13	80.71	76.94	74.70	72.03	70.43	69.35
2, Good	93.42	85.91	81.46	76.16	73.03	70.91	67.59	65.61	63.26	61.88	60.93
2 & 3	83.08	76.42	72.41	67.75	64.93	63.05	60.09	58.35	56.25	55.04	54.20
3, Average	73.56	67.63	64.11	59.96	57.52	55.81	53.19	51.63	49.81	48.71	47.99
3 & 4	66.15	60.87	57.68	53.94	51.73	50.21	47.85	46.47	44.82	43.82	43.16
4, Low	58.64	53.94	51.13	47.81	45.87	44.50	42.43	41.20	39.74	38.86	38.28

Wall Height Adjustment: Add or subtract the amount listed to the square foot cost above for each foot of wall height more or less than 16 feet.

Square Foot Area

Quality Class	7,500	10,000	12,500	15,000	20,000	30,000	40,000	50,000	75,000	100,000	150,000	200,000	250,000
1	4.95	4.29	3.87	3.51	3.06	2.49	2.19	1.98	1.62	1.41	1.17	1.04	.91
2	3.46	3.00	2.71	2.44	2.16	1.78	1.54	1.37	1.15	.99	.82	.74	.65
3	2.40	2.10	1.88	1.71	1.50	1.21	1.09	.96	.79	.71	.56	.49	.45
4	1.75	1.55	1.36	1.27	1.11	.90	.78	.71	.58	.48	.43	.38	.36

Perimeter Wall Adjustment: For a common wall, deduct per linear foot: Class 1, $677, Class 2, $447, Class 3, $317, Class 4, $208. For no wall ownership, deduct per linear foot: Class 1, $1,343, Class 2, $884, Class 3, $611, Class 4, $426

Suburban Stores – Wood or Wood and Steel Frame

Multi-Unit, Strip Type

Estimating Procedure

1. Use these figures to estimate the cost of stores designed for multiple occupancy. These costs include all components of shell buildings plus the cost of display fronts, finish materials on the front of the building and normal interior partitions.
2. Establish the structure quality class by applying the information on page 61. Evaluate the quality of the display front to help establish the correct quality class of the building as a whole. The building classes have display fronts as classified on page 56. See also pages 207 to 209.
3. Compute the building floor area. This should include everything within the exterior walls and all inset areas outside the main walls but under the main roof.
4. Add to or subtract from the square foot cost below the appropriate amount from the Wall Height Adjustment Table (at the bottom of this page) if the wall height is more or less than 16 feet.
5. Multiply the adjusted square foot cost by the building area.
6. Deduct, if appropriate, for common walls, using the figures at the bottom of this page.
7. Multiply the total cost by the location factor on page 7.
8. Add the cost of the appropriate additional components from page 201: heating and cooling equipment, fire sprinklers, canopies, exterior signs, mezzanines and basements, loading docks and ramps, yard improvements, and communication systems.

Length less than 1-1/2 times depth - Square Foot Area

Quality Class	500	1,000	2,000	2,500	3,000	4,000	5,000	7,500	10,000	12,500	15,000
1, Best	148.13	123.50	106.16	101.73	98.49	93.90	90.79	85.96	83.07	81.08	79.63
1 & 2	131.22	109.44	94.04	90.13	87.25	83.17	80.41	76.12	73.56	71.82	70.52
2, Good	115.70	96.48	82.89	79.43	76.93	73.33	70.92	67.13	64.87	63.31	62.18
2 & 3	105.33	87.80	75.47	72.32	70.02	66.75	64.54	61.10	59.04	57.65	56.60
3, Average	96.48	80.44	69.15	66.30	64.14	61.18	59.14	55.99	54.09	52.81	51.87
3 & 4	88.87	74.09	63.69	63.30	59.08	56.34	54.48	51.56	49.81	48.64	47.79
4, Low	80.77	67.34	57.86	55.46	53.67	51.18	49.49	46.86	45.29	44.23	43.40

Length between 1-1/2 and 2 times depth - Square Foot Area

Quality Class	500	1,000	2,000	3,000	5,000	7,500	10,000	15,000	20,000	25,000	35,000
1, Best	165.47	136.16	115.41	106.22	96.99	91.18	87.71	83.60	81.13	79.47	77.27
1 & 2	145.96	120.13	101.82	93.72	85.58	80.43	77.38	73.75	71.58	70.09	68.17
2, Good	128.21	105.50	89.40	82.31	75.17	70.62	67.98	64.78	62.86	61.59	59.87
2 & 3	116.37	95.76	81.17	74.70	68.22	64.12	61.67	58.78	57.04	55.90	54.34
3, Average	105.97	87.21	73.93	68.03	62.11	58.40	56.18	53.85	51.96	50.91	49.49
3 & 4	96.95	79.79	67.62	62.24	56.82	53.43	51.41	48.96	47.56	46.57	45.29
4, Low	88.08	72.46	61.41	56.53	51.61	48.50	46.68	44.48	43.19	42.29	41.11

Wall Height Adjustment: Add or subtract the amount listed to the square foot of floor cost for each foot of wall height more or less than 16 feet.

Square Foot Area

Quality Class	500	1,000	2,000	3,000	5,000	7,500	10,000	15,000	20,000	25,000	35,000
1, Best	10.07	7.02	4.94	4.03	3.14	2.54	2.23	1.84	1.60	1.44	1.30
2, Good	6.60	4.60	3.24	2.64	2.03	1.67	1.47	1.19	1.07	.93	.87
3, Average	4.83	3.38	2.38	1.95	1.53	1.23	1.09	.88	.77	.71	.65
4, Low	3.51	2.43	1.71	1.41	1.10	.89	.77	.65	.52	.48	.46

Perimeter Wall Adjustment: For a common wall, deduct $122.00 per linear foot. For no wall ownership, deduct $245.00 per linear foot.

Suburban Stores – Wood or Wood and Steel Frame

Multi-Unit, Strip Type

Estimating Procedure

1. Use these figures to estimate the cost of stores designed for multiple occupancy. These costs include all components of shell buildings plus the cost of display fronts, finish materials on the front of the building and normal interior partitions.
2. Establish the structure quality class by applying the information on page 61. Evaluate the quality of the display front to help establish the correct quality class of the building as a whole. The building classes have display fronts as classified on page 56. See also pages 207 to 209.
3. Compute the building floor area. This should include everything within the exterior walls and all inset areas outside the main walls but under the main roof.
4. Add to or subtract from the square foot cost below the appropriate amount from the Wall Height Adjustment Table (at the bottom of this page) if the wall height is more or less than 16 feet.
5. Multiply the adjusted square foot cost by the building area.
6. Deduct, if appropriate, for common walls, using the figures at the bottom of this page.
7. Multiply the total cost by the location factor on page 7.
8. Add the cost of the appropriate additional components from page 201: heating and cooling equipment, fire sprinklers, canopies, exterior signs, mezzanines and basements, loading docks and ramps, yard improvements, and communication systems.

Length between 2 and 3 times depth - Square Foot Area

Quality Class	1,000	2,000	3,000	5,000	7,500	10,000	15,000	20,000	30,000	50,000	70,000
1, Best	155.57	129.20	117.53	105.85	98.51	94.12	88.90	85.81	82.14	78.48	76.50
1 & 2	137.11	113.86	103.58	93.31	86.81	82.96	78.36	75.63	72.41	69.17	67.43
2, Good	120.04	99.67	90.69	81.67	76.00	72.61	68.61	66.22	63.38	60.54	59.03
2 & 3	108.69	90.25	82.11	73.96	68.82	65.76	62.14	59.95	57.40	54.83	53.47
3, Average	98.51	81.80	74.42	67.02	62.34	59.60	56.30	54.33	52.00	49.68	48.45
3 & 4	90.14	74.83	68.11	61.29	57.04	54.52	51.48	49.71	47.58	45.46	44.32
4, Low	81.59	67.74	61.65	55.52	51.63	49.37	46.64	45.01	43.12	41.15	40.13

Length greater than 3 times depth - Square Foot Area

Quality Class	2,000	3,000	5,000	10,000	15,000	20,000	30,000	50,000	75,000	100,000	150,000
1, Best	141.99	127.77	113.70	99.77	93.68	90.03	85.74	81.49	78.80	77.22	75.32
1 & 2	125.02	112.51	100.10	87.82	82.47	79.27	75.51	71.76	69.41	67.98	66.33
2, Good	109.32	98.39	87.56	76.82	72.12	69.34	66.05	62.76	60.68	59.47	58.01
2 & 3	95.45	85.88	76.45	67.07	63.00	60.54	57.65	54.78	53.01	51.93	50.63
3, Average	89.46	80.49	71.63	62.84	58.99	56.74	54.02	51.34	49.65	48.64	47.47
3 & 4	82.33	74.08	65.94	57.84	54.29	52.19	49.74	47.24	45.69	44.77	43.71
4, Low	74.97	67.47	60.05	52.69	49.47	47.56	45.30	43.03	41.63	40.80	39.79

Wall Height Adjustment: Add or subtract the amount listed to the square foot cost above for each foot of wall height more or less than 16 feet.

Square Foot Area

Quality Class	2,000	3,000	5,000	10,000	15,000	20,000	30,000	50,000	75,000	100,000	150,000
1, Best	4.79	3.90	3.05	2.18	1.82	1.58	1.29	1.08	.88	.79	.69
2, Good	3.08	2.51	1.95	1.41	1.15	1.02	.85	.67	.56	.50	.43
3, Average	2.26	1.86	1.45	1.06	.86	.75	.63	.49	.43	.39	.33
4, Low	1.69	1.36	1.08	.75	.64	.57	.45	.38	.32	.30	.24

Perimeter Wall Adjustment: For a common wall, deduct $122.00 per linear foot. For no wall ownership, deduct $241.00 per linear foot.

Supermarkets – Masonry or Concrete

Quality Classification

	Class 1 Best Quality	Class 2 Good Quality	Class 3 Average Quality	Class 4 Low Quality
Foundation (15% of total cost)	Reinforced concrete.	Reinforced concrete.	Reinforced concrete.	Reinforced concrete.
Floor Structure (12% of total cost)	4" reinforced concrete on 6" rock fill.	4" reinforced concrete on 6" rock fill.	4" reinforced concrete on 6" rock fill.	4" reinforced concrete on 6" rock fill.
Wall Structure (15% of total cost)	6" concrete tilt-up or ornamental block or brick.	6" concrete tilt-up, colored concrete block or brick.	6" concrete tilt-up or 8" concrete block.	6" concrete tilt-up or 8" concrete block.
Roof Structure (10% of total cost)	Glu-lams or steel "I" beams on steel intermediate columns, 2" x 12" purlins 16" o.c., 1/2" plywood sheathing.	Glu-lams or steel "I" beams on steel intermediate columns, 2" x 12" purlins 16" o.c., 1/2" plywood sheathing.	Glu-lams or steel "I" beams on steel intermediate columns, 2" x 12" purlins 16" o.c., 1/2" plywood sheathing.	Glu-lams or steel "I" beams on steel intermediate columns, 3" x 12" purlins 3' o.c., 1/2" plywood sheathing.
Floor Finish (5% of total cost)	Terrazzo in sales area. Sheet vinyl or carpet in cashiers' area.	Resilient tile in sales area. Terrazzo, solid vinyl tile or carpet in cashiers' area.	Composition tile in sales area.	Minimum grade tile in sales area.
Interior Wall Finish (5% of total cost)	Inside of exterior walls furred out with gypsum wallboard and paint or interior stucco, interior stucco or gypsum wallboard and vinyl wall cover on partitions.	Paint on inside of exterior walls, gypsum wallboard and paint or vinyl wall cover on partitions.	Paint on inside of exterior walls, wallboard and paint on partitions.	Paint on inside of exterior walls, wallboard and paint on partitions.
Ceiling Finish (5% of total cost)	Suspended acoustical tile, dropped ceiling over meat and produce departments.	Suspended acoustical tile or gypsum board and acoustical texture, dropped ceiling over meat and produce departments.	Ceiling tile on roof purlins, dropped ceiling over meat department.	Open.
Front (7% of total cost)	A large amount of float glass in good aluminum frames (18'-22' high for 3/4 of width), brick or stone veneer on remainder, 1 pair of good automatic doors per 7,000 S.F. of floor area, anodized aluminum sunshade over glass area, 8' canopy across front, 10'-12' raised walk across front.	A large amount of float glass in good aluminum frames (16'-18' high for 2/3 of width), brick or stone veneer on remainder, 1 pair of good automatic doors per 10,000 S.F. of floor area, 8' canopy across front, 10' raised walk across front.	A moderate amount of float glass in average quality aluminum frames (12'-16' high for 2/3 of width), exposed aggregate on remainder, 1 pair average automatic doors per 10,000 S.F. of floor area, 6' canopy across front, 8' raised walk across front.	Stucco or exposed aggregate with a small amount of float glass in an inexpensive aluminum frame (6'-10' high for 1/2 of width), 6' canopy across front, 6' ground level walk across front.
Exterior Finish (8% of total cost)	Large ornamental rock or brick veneer.	Large ornamental rock or brick veneer.	Paint, some exposed aggregate.	Paint.
Roof Cover (5% of total cost)	5 ply built-up roofing with large rock.	5 ply built-up roofing with tar and rock.	4 ply built-up roofing.	4 ply built-up roofing.
Plumbing (8% of total cost)	2 rest rooms with 3 fixtures, floor piping and drains to refrigerated cases, 2 double sinks with drain board.	2 rest rooms with 3 fixtures each, floor piping and drains to refrigerated cases, 2 double sinks with drain board.	2 rest rooms with 2 fixtures each, floor piping and drains to refrigerated cases, 2 double sinks.	1 rest room with 2 fixtures, floor piping and drains to refrigerated cases.
Electrical (5% of total cost)	Conduit wiring, recessed 4 tube fluorescent fixures 8' o.c., 30-40 spotlights.	Conduit wiring, 4 tube fluorescent fixtures with diffusers 8' o.c., 30-40 spotlights.	Conduit wiring, 3 tube fluorescent fixtures, 8' o.c., 5 or 10 spotlights.	Conduit wiring, double tube fluorescent fixtures, 8' o.c.

Note: Use the percent of total cost to help identify the correct quality classification.

Square foot costs include the following components: Foundations as required for normal soil conditions. Floor, wall, and roof structures. Interior floor, wall and ceiling finishes. Exterior wall finish and roof cover. Display fronts. Interior partitions. Entry and delivery doors. A canopy and walk across the front of the building as described in the applicable building specifications. Basic lighting and electrical systems. Rough and finish plumbing. All plumbing, piping and wiring necessary to operate the usual refrigerated cases and vegetable cases. Design and engineering fees. Permits and hook-up fees. Contractor's mark-up.

Supermarkets - Masonry or Concrete

Estimating Procedure

1. Establish the structure quality class by using the information on page 70.
2. Compute the building floor area. This should include everything within the building exterior walls and all insets outside the main walls but under the main building roof.
3. Add to or subtract from the square foot cost below the appropriate amount from the Wall Height Adjustment Row (at the bottom of this page) if the wall height is more or less than 20 feet.
4. Multiply the adjusted square foot cost by the building area.
5. Deduct, if appropriate, for common walls, using the figures at the bottom of this page.
6. Multiply the total cost by the location factor listed on page 7.
7. Add the cost of heating and cooling equipment, fire sprinklers, exterior signs, yard improvements, loading docks, ramps and walk-in boxes if they are an integral part of the building. See pages 201 to 213.

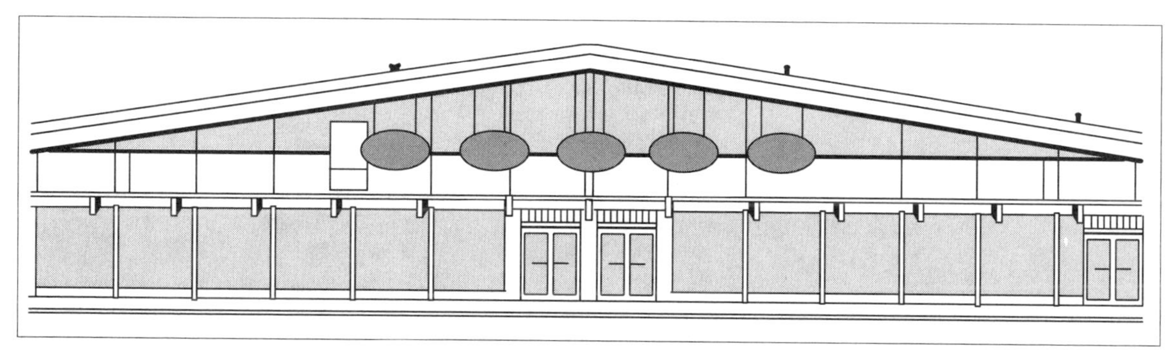

Supermarket, Class 2

Square Foot Area

Quality Class	5,000	7,500	10,000	12,500	15,000	20,000	25,000	30,000	35,000	40,000	50,000
Exceptional	104.22	96.91	92.55	89.53	87.33	84.20	82.04	80.48	79.25	78.25	76.74
1, Best	99.31	92.35	88.18	85.32	83.21	80.25	78.20	76.70	75.52	74.59	73.12
1 & 2	92.08	85.60	81.75	79.09	77.13	74.37	72.49	71.09	69.99	69.13	67.80
2, Good	85.26	79.28	75.71	73.26	71.44	68.91	67.15	65.85	64.85	64.04	62.79
2 & 3	79.54	73.95	70.63	68.35	66.63	64.27	62.64	61.40	60.46	59.73	58.56
3, Average	74.76	69.54	66.39	64.24	62.64	60.43	58.84	57.73	56.85	56.15	55.03
3 & 4	67.81	63.04	60.20	58.26	56.81	54.79	53.40	52.36	51.55	50.92	49.94
4, Low	60.56	56.29	53.75	52.03	50.73	48.92	47.67	46.75	46.04	45.46	44.56
Wall Height Adjustment*	1.12	.89	.79	.72	.67	.55	.48	.43	.41	.39	.36

***Wall Height Adjustment:** Add or subtract the amount listed in this row to the square foot of floor cost for each foot of wall height more or less than 20 feet.

Perimeter Wall Adjustment: A common wall exists when two buildings share one wall. Adjust for common walls by deducting the linear foot costs below from the total structure cost. In some structures one or more walls are not owned at all. In this case, deduct the "No Ownership" cost per linear foot of wall not owned. For common wall, deduct $195 per linear foot. For no wall ownership, deduct $392 per linear foot.

Supermarkets - Wood and Steel Frame

Quality Classification

	Class 1 Best Quality	Class 2 Good Quality	Class 3 Average Quality	Class 4 Low Quality
Foundation (15% of total cost)	Reinforced concrete.	Reinforced concrete.	Reinforced concrete.	Reinforced concrete.
Floor Structure (12% of total cost)	4" reinforced concrete on 6" rock fill.	4" reinforced concrete on 6" rock fill.	4" reinforced concrete on 6" rock fill.	4" reinforced concrete on 6" rock fill.
Wall Structure (10% of total cost)	2" x 6" - 16" o.c.	2" x 6" - 16" o.c.	2" x 4" - 16" o.c.	2" x 4" - 16" o.c.
Roof Structure (10% of total cost)	Glu-lams or steel "I" beams on steel intermediate columns, 2" x 12" purlins 16" o.c., 1/2" plywood sheathing.	Glu-lams or steel "I" beams on steel intermediate columns, 2" x 12" purlins 16" o.c., 1/2" plywood sheathing.	Glu-lams or steel "I" beams on steel intermediate columns, 2" x 12" purlins 16" o.c., 1/2" plywood sheathing.	Glu-lams or steel "I" beams on steel intermediate columns, 3" x 12" purlins 3' o.c., 1/2" plywood sheathing.
Floor Finish (5% of total cost)	Terrazzo in sales area. Sheet vinyl or carpet in cashiers' area.	Resilient tile in sales area. Terrazzo, solid vinyl tile or carpet in cashiers' area.	Composition tile in sales area.	Minimum tile or inexpensive composition tile in sales area.
Interior Wall Finish (7% of total cost)	Gypsum wallboard and vinyl wall cover or interior stucco on inside of exterior walls and on partitions.	Gypsum wallboard, texture and paint or vinyl wall cover on inside of exterior walls, and on partitions.	Gypsum wallboard, texture and paint on inside of exterior walls, and on partitions.	Gypsum wallboard and paint on inside of exterior walls, and on partitions.
Ceiling Finish (5% of total cost)	Suspended acoustical tile, dropped ceiling over meat and produce departments.	Suspended acoustical tile or gypsum board and acoustical texture, dropped ceiling over meat and produce departments.	Ceiling tile on roof purlins, dropped ceiling over meat department.	Open.
Front (10% of total cost)	A large amount of float glass in good aluminum frames (18'-22' high for 3/4 of width), brick or stone veneer on remainder, 1 pair good automatic doors per 7,000 S.F. of floor area, anodized aluminum sunshade over glass area, 8' canopy across front, 10'-12' raised walk across front.	A large amount of float glass in good aluminum frames (16'-18' high for 2/3 of width), brick or stone veneer on remainder, 1 pair good automatic doors per 10,000 S.F. of floor area, 9' canopy across front, 10' raised walk across front.	Moderate amount of float glass in average quality aluminum frames (12'-16' high for 2/3 of width), wood siding on remainder, 1 pair average automatic doors per 10,000 S.F. of floor area, 6' canopy across front, 8' raised walk across front.	Stucco with small amount of float glass in an inexpensive aluminum frame (6'-10' high for 1/2 of width), 6' canopy across front, 6' ground level walk across front.
Exterior Finish (8% of total cost)	Large ornamental rock or brick veneer.	Good wood siding, some masonry veneer.	Stucco or wood siding.	Stucco.
Roof Cover (5% of total cost)	5 ply built-up roofing with large rock.	5 ply built-up roofing with tar and rock.	4 ply built-up roofing.	4 ply built-up roofing.
Plumbing (8% of total cost)	2 rest rooms with 3 fixtures each, floor piping and drains to refrigerated cases, 2 double sinks with drain board.	2 rest rooms with 3 fixtures each, floor piping and drains to refrigerated cases, 2 double sinks with drain board.	2 rest rooms with 2 fixtures each, floor piping and drains to refrigerated cases, 2 double sinks.	1 rest room with 2 fixtures, floor piping and drains to refrigerated cases.
Electrical (5% of total cost)	Conduit wiring, recessed 4 tube fluorescent fixtures, 8' o.c., 30 to 40 spotlights.	Conduit wiring, 4 tube fluorescent fixtures with diffusers, 8' o.c., 30 to 40 spotlights.	Conduit wiring, 3 tube fluorescent fixtures, 8' o.c., 5 or 10 spotlights.	Conduit wiring, double tube fluorescent fixtures, 8' o.c.

Note: Use the percent of total cost to help identify the correct quality classification.

Square foot costs include the following components: Foundations as required for normal soil conditions. Floor, wall, and roof structures. Interior floor, wall and ceiling finishes. Exterior wall finish and roof cover. Display fronts. Interior partitions. Entry and delivery doors. A canopy and walk across the front of the building as described in the applicable building specifications. Basic lighting and electrical systems. Rough and finish plumbing. All plumbing, piping and wiring necessary to operate the usual refrigerated cases and vegetable cases. Design and engineering fees. Permits and hook-up fees. Contractor's mark-up.

Supermarkets - Wood and Steel Frame

Estimating Procedure

1. Establish the structure quality class by using the information on page 72.
2. Compute the building floor area. This should include everything within the building exterior walls and all insets outside the main walls but under the main building roof.
3. Add to or subtract from the square foot cost below the appropriate amount from the Wall Height Adjustment Row (at the bottom of this page) if the wall height is more or less than 20 feet.
4. Multiply the adjusted square foot cost by the building area.
5. Deduct, if appropriate, for common walls, using the figures at the bottom of this page.
6. Multiply the total cost by the location factor listed on page 7.
7. Add the cost of heating and cooling equipment, fire sprinklers, exterior signs, yard improvements, loading docks, ramps and walk-in boxes if they are an integral part of the building. See pages 201 to 213.

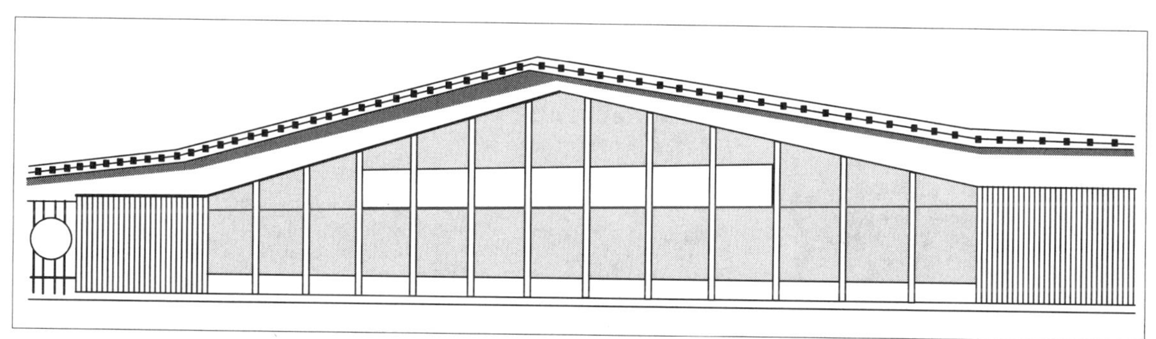

Supermarket, Class 3

Square Foot Area

Quality Class	5,000	7,500	10,000	12,500	15,000	20,000	25,000	30,000	35,000	40,000	50,000
Exceptional	97.55	91.93	88.53	86.22	84.50	82.10	80.46	79.23	78.30	77.54	76.34
1, Best	93.07	87.69	84.46	82.28	80.64	78.33	76.76	75.60	74.68	73.95	72.83
1 & 2	85.50	80.58	77.60	75.59	74.07	71.96	70.52	69.46	68.61	67.93	66.90
2, Good	78.45	73.88	71.19	69.31	67.97	66.02	64.69	63.69	62.94	62.32	61.39
2 & 3	73.31	69.09	66.53	64.80	63.53	61.71	60.48	59.55	58.82	58.26	57.40
3, Average	69.02	65.05	62.63	60.98	59.77	58.09	56.89	56.04	55.39	54.81	53.99
3 & 4	63.22	59.57	57.39	55.87	54.75	53.21	52.14	51.35	50.74	50.25	49.48
4, Low	56.95	53.66	51.69	50.31	49.34	47.93	46.96	46.24	45.68	45.24	44.55
Wall Height Adjustment*	.62	.50	.46	.42	.39	.36	.34	.32	.31	.30	.27

***Wall Height Adjustment:** Add or subtract the amount listed in this row to the square foot of floor cost for each foot of wall height more or less than 20 feet.

Perimeter Wall Adjustment: A common wall exists when two buildings share one wall. Adjust for common walls by deducting the linear foot costs below from the total structure cost. In some structures one or more walls are not owned at all. In this case, deduct the "No Ownership" cost per linear foot of wall not owned. For common wall, deduct $140 per linear foot. For no wall ownership, deduct $280 per linear foot.

Supermarkets - Masonry Construction

Quality Classification

	Class 1 Best Quality	Class 2 Good Quality	Class 3 Average Quality	Class 4 Low Quality
Foundation (15% of total cost)	Reinforced concrete.	Reinforced concrete.	Reinforced concrete.	Reinforced concrete.
Floor Structure (12% of total cost)	4" reinforced concrete on 6" rock fill.	4" reinforced concrete on 6" rock fill.	4" reinforced concrete on 6" rock fill.	4" reinforced concrete on 6" rock fill.
Wall Structure (10% of total cost)	6" concrete tilt-up or ornamental block or brick.	6" concrete tilt-up, colored concrete block or brick.	6" concrete tilt-up or 8" concrete block.	6" concrete tilt-up or 8" concrete block.
Roof Structure (10% of total cost)	Glu-lams or steel "I" beams on steel intermediate columns, 2" x 12" purlins 16" o.c., 1/2" plywood sheathing.	Glu-lams or steel "I" beams, 2" x 12" purlins 16" o.c., 1/2" plywood sheathing.	Glu-lams, 3" x 12" purlins 3' o.c., 1/2" plywood sheathing.	Glu-lams, 3" x 12" purlins 3' o.c., 1/2" plywood sheathing.
Floor Finish (5% of total cost)	Resilient tile in sales area.	Composition tile in sales area.	Minimum grade tile in sales area.	Concrete.
Interior Wall Finish (7% of total cost)	Paint on inside of exterior walls, gypsum wallboard, texture and paint or vinyl wall cover on partitions.	Paint on inside of exterior walls, gypsum wallboard, texture and paint on partitions.	Paint on inside of exterior walls, gypsum wallboard and paint on partitions.	Paint on inside of exterior walls, gypsum wallboard and paint on partitions.
Ceiling Finish (5% of total cost)	Suspended acoustical tile or gypsum board and acoustical texture.	Ceiling tile on roof purlins.	Open.	Open.
Front (10% of total cost)	A large amount of float glass in good aluminum frames (10'-12' high for 2/3 of width), brick or stone veneer on remainder, 1 pair good aluminum and glass doors per 3,000 S.F. of floor area, 8' canopy across front, 10' raised walk across front.	A moderate amount of float glass in average aluminum frames (8' to 10' high for 2/3 of width), exposed aggregate on remainder, 1 pair average aluminum and glass doors per 3,000 S.F. of floor area, 6' canopy across front, 8' raised walk across front.	Painted concrete block with a small amount of float glass in an inexpensive aluminum frame (6' to 8' high for 1/2 of width), 6' canopy across front, 6' ground level walk across front.	Painted concrete block with small amount of crystal glass in wood frames, wood and glass doors, small canopy over entrance, 6' ground level walk at entrances.
Exterior Finish (8% of total cost)	Colored block.	Paint.	Paint.	Paint.
Roof Cover (5% of total cost)	5 ply built-up roofing with tar and rock.	4 ply built-up roofing.	4 ply built-up roofing.	4 ply built-up roofing.
Plumbing (8% of total cost)	2 rest rooms with 3 fixtures each, floor piping and drains to refrigerated cases.	1 rest room with 3 fixtures, floor piping and drains to refrigerated cases.	1 rest room with 2 fixtures, floor piping and drains to refrigerated cases.	1 rest room with 2 fixtures, floor piping and drains to refrigerated cases.
Electrical (5% of total cost)	Conduit wiring, 4 tube fluorescent fixtures with diffusers, 8' o.c., 5 spotlights.	Conduit wiring, 3 tube fluorescent fixtures, 8' o.c.	Conduit wiring, double tube fluorescent fixtures, 8' o.c.	Conduit wiring, incandescent fixtures,10' o.c. or single tube fluorescent fixtures, 8' o.c.

Note: Use the percent of total cost to help identify the correct quality classification.

Square foot costs include the following components: Foundations as required for normal soil conditions. Floor, wall, and roof structures. Interior floor, wall and ceiling finishes. Exterior wall finish and roof cover. Display fronts. Interior partitions. Entry and delivery doors. A canopy and walk across the front of the building as described in the applicable building specifications. Basic lighting and electrical systems. Rough and finish plumbing. All plumbing, piping and wiring necessary to operate the usual refrigerated cases and vegetable cases. Design and engineering fees. Permits and hook-up fees. Contractor's mark-up.

Supermarkets - Masonry Construction

Estimating Procedure

1. Establish the structure quality class by using the information on page 74.
2. Compute the building floor area. This should include everything within the building exterior walls and all insets outside the main walls but under the main building roof.
3. Add to or subtract from the square foot cost below the appropriate amount from the Wall Height Adjustment Row (at the bottom of this page) if the wall height is more or less than 12 feet.
4. Multiply the adjusted square foot cost by the building area.
5. Deduct, if appropriate, for common walls, using the figures at the bottom of this page.
6. Multiply the total cost by the location factor listed on page 7.
7. Add the cost of heating and cooling equipment, fire sprinklers, exterior signs, yard improvements, loading docks, ramps and walk-in boxes if they are an integral part of the building. See pages 201 to 213.

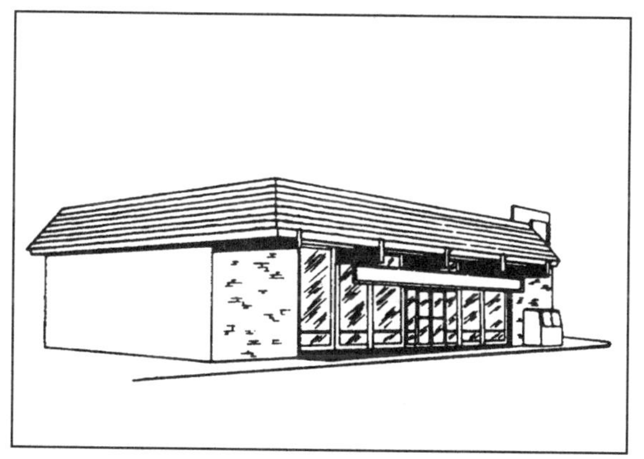

Small Food Store, Class 1 & 2

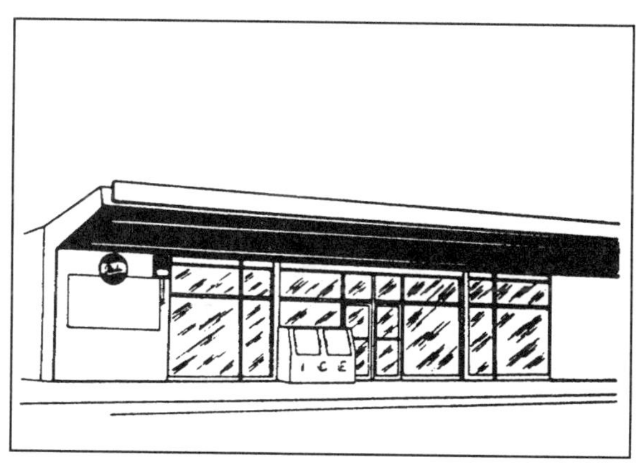

Small Food Store, Class 2

Square Foot Area

Quality Class	500	1,000	1,500	2,000	2,500	3,000	3,500	4,000	4,500	5,000	6,000
1, Best	153.59	116.37	101.75	93.56	88.21	84.38	81.52	79.21	77.33	75.78	73.30
1 & 2	141.48	107.44	94.04	86.56	81.65	78.14	75.49	73.42	71.68	70.25	67.98
2, Good	130.44	99.33	87.11	80.22	75.73	72.50	70.06	68.11	66.52	65.23	63.13
2 & 3	123.22	94.49	82.81	76.17	71.76	68.57	66.12	64.19	62.61	61.26	59.12
3, Average	116.39	89.06	78.05	71.81	67.70	64.75	62.48	60.73	59.23	58.00	56.03
3 & 4	96.40	81.22	72.89	67.52	63.65	60.71	58.31	56.41	54.77	53.37	51.06
4, Low	89.80	75.14	67.33	62.28	58.67	55.92	53.72	51.93	50.40	49.13	47.00
Wall Height Adjustment*	3.53	2.49	2.03	1.76	1.56	1.45	1.33	1.24	1.17	1.12	1.05

***Wall Height Adjustment:** Add or subtract the amount listed in this row to the square foot of floor cost for each foot of wall height more or less than 12 feet.

Perimeter Wall Adjustment: For common wall, deduct $103 per linear foot. For no wall ownership, deduct $205 per linear foot.

Small Food Stores – Wood Frame Construction

Quality Classification

	Class 1 Best Quality	Class 2 Good Quality	Class 3 Average Quality	Class 4 Low Quality
Foundation (15% of total cost)	Reinforced concrete.	Reinforced concrete.	Reinforced concrete.	Reinforced concrete.
Floor Structure (12% of total cost)	4" reinforced concrete on 6" rock fill.	4" reinforced concrete on 6" rock fill.	4" reinforced concrete on 6" rock fill.	4" reinforced concrete on 6" rock fill.
Wall Structure (10% of total cost)	2" x 6" - 16" o.c.	2" x 6" - 16" o.c.	2" x 4" - 16" o.c.	2" x 4" - 16" o.c.
Roof Structure (10% of total cost)	Glu-lams or steel "I" beams on steel intermediate columns, 2" x 12" purlins 16" o.c., 1/2" plywood sheathing.	Glu-lams or steel "I" beams on steel intermediate columns, 2" x 12" purlins 16" o.c., 1/2" plywood sheathing.	Glu-lams, or steel "I" beams, 3" x 12" purlins 3' o.c., 1/2" plywood sheathing.	Glu-lams, 3" x 12" purlins 3' o.c. 1/2" plywood sheathing.
Floor Finish (5% of total cost)	Resilient tile in sales area.	Resilient tile in sales area.	Inexpensive composition tile in sales area.	Concrete.
Interior Wall Finish (7% of total cost)	Gypsum wallboard, texture and paint or vinyl wall cover on inside of exterior walls, and on partitions.	Gypsum wallboard, texture and paint on inside of exterior walls, and on partitions.	Gypsum wallboard and paint on inside of exterior walls, and on partitions.	Gypsum wallboard and paint on inside of exterior walls and on partitions.
Ceiling Finish (5% of total cost)	Suspended acoustical tile or gypsum board and acoustical texture.	Ceiling tile on roof purlins.	Open.	Open.
Front (10% of total cost)	A large amount of float glass in good aluminum frames (10'-12' high for 2/3 of width), brick or stone veneer on remainder, 1 pair good aluminum and glass doors per 3,000 S.F. of floor area, 8' canopy across front, 10' raised walk across front.	A moderate amount of float glass in average quality aluminum frames (8' to 10' high for 2/3 of width), ornamental concrete block on remainder, 1 pair average aluminum and glass doors per 3,000 S.F. of floor area, 6' canopy across front, 8' raised walk across front.	Painted stucco with a small amount of float glass in an inexpensive aluminum frame (6' to 8' high for 1/2 of width) 6' canopy across front. 6' ground level walk across front.	Stucco with small amount of crystal glass in wood frames, wood and glass doors, small canopy over entrance, 6' ground level walk at entrances.
Exterior Finish (8% of total cost)	Stucco and paint or wood siding.	Stucco and paint.	Stucco.	Stucco.
Roof Cover (5% of total cost)	5 ply built-up roofing with tar and rock.	4 ply built-up roofing.	4 ply built-up roofing.	4 ply built-up roofing.
Plumbing (8% of total cost)	2 rest rooms with 3 fixtures each, floor piping and drains to refrigerated cases.	1 rest room with 3 fixtures, floor piping and drains to refrigerated cases.	1 rest room with 2 fixtures, floor piping and drains to refrigerated cases.	1 rest room with 2 fixtures, floor piping and drains to refrigerated cases.
Electrical (5% of total cost)	Conduit wiring, 4 tube fluorescent fixtures with diffusers, 8' o.c., 5 spotlights.	Conduit wiring, 3 tube fluorescent fixtures, 8' o.c.	Conduit wiring, double tube fluorescent fixtures, 8' o.c.	Conduit wiring, incandescent fixtures, 10' o.c. or single tube fluorescent fixtures, 8' o.c.

Note: Use the percent of total cost to help identify the correct quality classification.

Square foot costs include the following components: Foundations as required for normal soil conditions. Floor, wall, and roof structures. Interior floor, wall and ceiling finishes. Exterior wall finish and roof cover. Display fronts. Interior partitions. Entry and delivery doors. A canopy and walk across the front of the building as described in the applicable building specifications. Basic lighting and electrical systems. Rough and finish plumbing. All plumbing, piping and wiring necessary to operate the usual refrigerated cases and vegetable cases. Design and engineering fees. Permits and hook-up fees. Contractor's mark-up.

Small Food Stores – Wood Frame Construction

Estimating Procedure

1. Establish the structure quality class by using the information on page 76.
2. Compute the building floor area. This should include everything within the building exterior walls and all insets outside the main walls but under the main building roof.
3. Add to or subtract from the square foot cost below the appropriate amount from the Wall Height Adjustment Row (at the bottom of this page) if the wall height is more or less than 12 feet.
4. Multiply the adjusted square foot cost by the building area.
5. Deduct, if appropriate, for common walls, using the figures at the bottom of this page.
6. Multiply the total cost by the location factor listed on page 7.
7. Add the cost of heating and cooling equipment, fire sprinklers, exterior signs, yard improvements, loading docks, ramps and walk-in boxes if they are an integral part of the building. See pages 201 to 213.

Small Food Store, Class 1

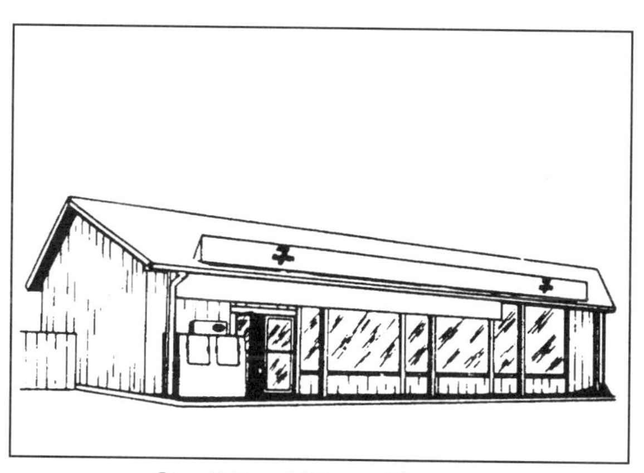

Small Food Store, Class 3

Square Foot Area

Quality Class	500	1,000	1,500	2,000	2,500	3,000	3,500	4,000	4,500	5,000	6,000
1, Best	140.03	106.70	94.12	87.22	82.75	79.64	77.30	75.46	73.94	72.73	70.76
1 & 2	129.44	98.86	87.27	80.93	76.88	73.99	71.83	70.16	68.76	67.63	65.84
2, Good	120.44	92.23	81.53	75.66	71.89	69.24	67.23	65.67	64.38	63.32	61.66
2 & 3	109.72	84.22	74.58	69.31	65.93	63.53	61.71	60.28	59.13	58.17	56.71
3, Average	100.30	77.27	68.53	63.79	60.70	58.54	56.90	55.64	54.63	53.74	52.39
3 & 4	90.90	70.18	62.29	58.02	55.25	53.31	51.83	50.68	49.76	48.98	47.80
4, Low	81.73	62.60	55.58	51.84	49.47	47.81	46.57	45.65	44.88	44.24	43.25
Wall Height Adjustment*	2.17	1.58	1.29	1.13	1.02	.90	.85	.80	.75	.72	.66

***Wall Height Adjustment:** Add or subtract the amount listed in this row to the square foot of floor cost for each foot of wall height more or less than 12 feet.

Perimeter Wall Adjustment: For common wall, deduct $70 per linear foot. For no wall ownership, deduct $140 per linear foot.

Discount Houses - Masonry or Concrete

Quality Classification

	Class 1 Best Quality	Class 2 Good Quality	Class 3 Average Quality	Class 4 Low Quality
Foundation (15% of total cost)	Reinforced concrete.	Reinforced concrete.	Reinforced concrete.	Reinforced concrete.
Floor Structure (12% of total cost)	4" reinforced concrete on 6" rock fill.	4" reinforced concrete on 6" rock fill.	4" reinforced concrete on 6" rock fill.	4" reinforced concrete on 6" rock fill.
Wall Structure (10% of total cost)	6" concrete tilt-up or ornamental block or brick.	6" concrete tilt-up, colored concrete block or brick.	6" concrete tilt-up or 8" concrete block.	6" concrete tilt-up or 8" concrete block.
Roof Structure (10% of total cost)	Glu-lams or steel "I" beams on steel intermediate columns, plywood sheathing.	Glu-lams or steel "I" beams on steel intermediate columns, plywood sheathing.	Glu-lams, or steel "I" beams on steel intermediate columns, plywood sheathing.	Glu-lams, or steel "I" beams on steel intermediate columns, plywood sheathing.
Floor Finish (5% of total cost)	Terrazzo, sheet vinyl or good carpet in sales area.	Resilient tile in sales area with some terrazzo, solid vinyl tile, or carpet.	Composition tile in sales area.	Inexpensive composition tile in sales area.
Interior Wall Finish (7% of total cost)	Inside of exterior walls furred out with gypsum wallboard and paint or interior stucco or gypsum wallboard and vinyl wall cover on partitions.	Paint on inside of exterior walls, gypsum wallboard and paint or vinyl wall cover on partitions.	Paint on inside of exterior walls, gypsum wallboard and paint on partitions.	Paint on inside of exterior walls, gypsum wallboard and paint on partitions.
Ceiling Finish (5% of total cost)	Suspended acoustical tile, dropped ceilings in some areas.	Suspended acoustical tile or gypsum board and acoustical texture, dropped ceilings in some areas.	Ceiling tile on roof purlins, dropped ceiling over some areas.	Open.
Front (10% of total cost)	A large amount of float glass in good aluminum frames (18'-22' high for 3/4 of width), brick or stone veneer on remainder, 1 pair good automatic doors per 7,000 S.F. of floor area, anodized aluminum sunshade over glass area, 8' canopy across front, 10'-12' raised walk across front.	A large amount of float glass in good aluminum frames (16'-18' high for 2/3 of width), brick or stone veneer on remainder, 1 pair good automatic doors per 10,000 S.F. of floor area, 8' canopy across front, 10' raised walk across front.	A moderate amount of float glass in average quality aluminum frames (12'-16' high extending 20' on each side of entrances), exposed aggregate on remainder, 1 pair average automatic doors per 20,000 S.F. of floor area, 6' canopy over glass area.	Stucco or exposed aggregate with a small amount of float glass in inexpensive aluminum frames (6'-10' high for 1/2 of width), 6' canopy at entrances, 6' ground level walk across front.
Exterior Finish (8% of total cost)	Large ornamental rock or brick veneer on exposed walls.	Exposed aggregate on exposed walls.	Paint, some exposed aggregate.	Paint.
Roof Cover (5% of total cost)	5 ply built-up roofing with large rock.	5 ply built-up roofing with tar and rock.	4 ply built-up roofing.	4 ply built-up roofing.
Plumbing (8% of total cost)	2 rest rooms with 8 fixtures each, floor piping and drains to refrigerated cases.	2 rest rooms with 6 fixtures each, floor piping and drains to refrigerated cases.	2 rest rooms with 4 fixtures each, floor piping and drains to refrigerated cases.	2 rest rooms with 2 fixtures each, floor piping and drains to refrigerated cases.
Electrical (5% of total cost)	Conduit wiring, recessed 4 tube fluorescent fixtures 8' o.c., 60-80 spotlights.	Conduit wiring, 4 tube fluorescent fixtures with diffusers 8' o.c., 60-80 spotlights.	Conduit wiring, 3 tube fluorescent fixtures 8' o.c., 20 to 40 spotlights.	Conduit wiring, double tube fluorescent fixtures, 8' o.c.

Note: Use the percent of total cost to help identify the correct quality classification.

Square foot costs include the following components*:* Foundations as required for normal soil conditions. Floor, wall, and roof structures. Interior floor, wall and ceiling finishes. Exterior wall finish and roof cover. Display fronts. Interior partitions. Entry and delivery doors. A canopy and walk across the front of the building as described in the applicable building specifications. Basic lighting and electrical systems. Rough and finish plumbing. Design and engineering fees. Permits and hook-up fees. Contractor's mark-up.

Discount Houses – Masonry or Concrete

Estimating Procedure

1. Establish the structure quality class by using the information on page 78.
2. Compute the building floor area. This should include everything within the building exterior walls and all insets outside the main walls but under the main building roof.
3. Add to or subtract from the square foot cost below the appropriate amount from the Wall Height Adjustment Row (at the bottom of this page) if the wall height is more or less than 20 feet.
4. Multiply the adjusted square foot cost by the building area.
5. Deduct, if appropriate, for common walls, using the figures at the bottom of this page.
6. Multiply the total cost by the location factor listed on page 7.
7. Add the cost of heating and cooling equipment, fire sprinklers, exterior signs, yard improvements, loading docks, ramps and walk-in boxes if they are an integral part of the building. See pages 201 to 213.

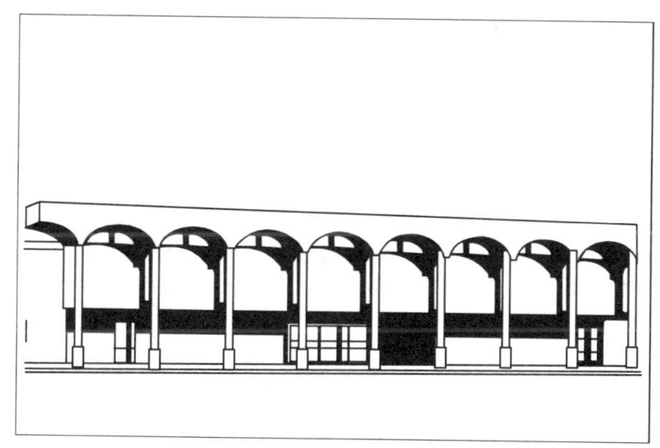

Discount House, Class 3

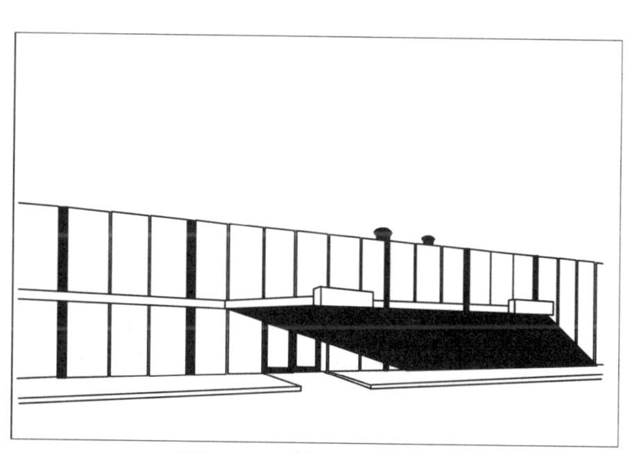

Discount House, Class 4

Square Foot Area

Quality Class	15,000	20,000	25,000	30,000	35,000	40,000	50,000	75,000	100,000	150,000	200,000
1, Best	68.43	66.09	64.51	63.32	62.38	61.63	60.48	58.68	57.61	56.29	55.50
1 & 2	64.32	62.14	60.65	59.52	58.65	57.95	56.86	55.13	54.12	52.93	52.20
2, Good	61.47	59.38	57.93	56.87	56.04	55.35	54.33	52.71	51.74	50.55	49.88
2 & 3	57.15	55.20	53.88	52.89	52.12	51.48	50.52	49.01	48.10	47.04	46.35
3, Average	53.79	51.99	50.73	49.77	49.06	48.46	47.56	46.14	45.29	44.27	43.65
3 & 4	49.18	47.50	46.35	45.48	44.82	44.28	43.44	42.16	41.38	40.46	39.88
4, Low	44.44	42.93	41.89	41.11	40.54	40.03	39.29	38.11	37.38	36.56	36.04
Wall Height Adjustment*	.70	.63	.51	.46	.43	.41	.38	.33	.31	.28	.27

***Wall Height Adjustment:** Add or subtract the amount listed in this row to the square foot of floor cost for each foot of wall height more or less than 20 feet.

Perimeter Wall Adjustment: A common wall exists when two buildings share one wall. Adjust for common walls by deducting the linear foot costs below from the total structure cost. In some structures one or more walls are not owned at all. In this case, deduct the "No Ownership" cost per linear foot of wall not owned. For common wall, deduct $174 per linear foot. For no wall ownership, deduct $348 per linear foot.

Discount Houses – Wood or Wood and Steel Frame

Quality Classification

	Class 1 Best Quality	Class 2 Good Quality	Class 3 Average Quality	Class 4 Low Quality
Foundation (15% of total cost)	Reinforced concrete.	Reinforced concrete.	Reinforced concrete.	Reinforced concrete.
Floor Structure (12% of total cost)	4" reinforced concrete on 6" rock fill.	4" reinforced concrete on 6" rock fill.	4" reinforced concrete on 6" rock fill.	4" reinforced concrete on 6" rock fill.
Wall Structure (10% of total cost)	2" x 6" - 16" o.c.	2" x 6" - 16" o.c.	2" x 6" - 16" o.c.	2" x 4" - 16" o.c.
Roof Structure (10% of total cost)	Glu-lams or steel "I" beams on steel intermediate columns, plywood sheathing.	Glu-lams or steel "I" beams on steel intermediate columns, plywood sheathing.	Glu-lams, or steel "I" beams on steel intermediate columns, plywood sheathing.	Glu-lams, or steel "I" beams on steel intermediate columns, plywood sheathing.
Floor Finish (5% of total cost)	Terrazzo, sheet vinyl or good carpet in sales area.	Resilient tile in sales area with some terrazzo, solid vinyl tile or carpet.	Resilient tile in sales area.	Inexpensive composition tile in sales area.
Interior Wall Finish (5% of total cost)	Gypsum wallboard and vinyl wall cover or interior stucco on inside of walls and on partitions.	Gypsum wallboard, texture and paint or vinyl wall cover on inside of exterior walls and on partitions.	Gypsum wallboard, texture and paint on inside of exterior walls and on partitions.	Gypsum wallboard and paint on inside of exterior walls and on partitions.
Ceiling Finish (5% of total cost)	Suspended acoustical tile, dropped ceilings over some areas.	Suspended acoustical tile or gypsum board and acoustical texture, dropped ceilings in some areas.	Ceiling tile on roof purlins, dropped ceiling over some areas.	Open.
Front (10% of total cost)	A large amount of float glass in good aluminum frames (18'-22' high for 3/4 of width), brick or stone veneer on remainder, 1 pair good automatic doors 1 per 7,000 S.F. of floor area, anodized aluminum sunshade over glass area, 8' canopy across front, 10'-12' raised walk across front.	A large amount of float glass in good aluminum frames (16'-18' high for 2/3 of width), brick or stone veneer on remainder, 1 pair good automatic doors per 10,000 S.F. of floor area, 8' canopy across front, 10' raised walk across front.	A moderate amount of float glass in average quality aluminum frames (12'-16' high extending 20' on each side of entrances), exposed aggregate on remainder, 1 pair average automatic doors per 20,000 S.F. of floor area, 6' canopy over glass areas.	Stucco or exposed aggregate with a small amount of float glass in inexpensive aluminum frames (6'-10' high for 1/2 of width), 6' canopy at entrances, 6' ground level walk across front.
Exterior Finish (7% of total cost)	Good wood siding or masonry veneer.	Wood siding, some masonry veneer.	Stucco, some masonry trim.	Stucco.
Roof Cover (5% of total cost)	5 ply built-up roofing with large rock.	5 ply built-up roofing with tar and rock.	4 ply built-up roofing.	4 ply built-up roofing.
Plumbing (10% of total cost)	2 rest rooms with 8 fixtures each, floor piping and drains to refrigerated cases.	2 rest rooms with 6 fixtures each, floor piping and drains to refrigerated cases.	2 rest rooms with 4 fixtures each, floor piping and drains to refrigerated cases.	2 rest rooms with 2 fixtures each, floor piping and drains to refrigerated cases.
Electrical (6% of total cost)	Conduit wiring, recessed 4 tube fluorescent fixtures 8' o.c., 60-80 spotlights.	Conduit wiring, 4 tube fluorescent fixtures with diffusers 8' o.c., 60-80 spotlights.	Conduit wiring, 3 tube fluorescent fixtures 8' o.c., 20 to 40 spotlights.	Conduit wiring, double tube fluorescent fixtures, 8' o.c.

Note: Use the percent of total cost to help identify the correct quality classification.

Square foot costs include the following components: Foundations as required for normal soil conditions. Floor, wall, and roof structures. Interior floor, wall and ceiling finishes. Exterior wall finish and roof cover. Display fronts. Interior partitions. Entry and delivery doors. A canopy and walk across the front of the building as described in the applicable building specifications. Basic lighting and electrical systems. Rough and finish plumbing. Design and engineering fees. Permits and hook-up fees. Contractor's mark-up.

Discount Houses – Wood or Wood and Steel Frame

Estimating Procedure

1. Establish the structure quality class by using the information on page 80.
2. Compute the building floor area. This should include everything within the building exterior walls and all insets outside the main walls but under the main building roof.
3. Add to or subtract from the square foot cost below the appropriate amount from the Wall Height Adjustment Row at the bottom of this page) if the wall height is more or less than 20 feet.
4. Multiply the adjusted square foot cost by the building area.
5. Deduct, if appropriate, for common walls, using the figures at the bottom of this page.
6. Multiply the total cost by the location factor listed on page 7.
7. Add the cost of heating and cooling equipment, fire sprinklers, exterior signs, yard improvements, loading docks, ramps and walk-in boxes if they are an integral part of the building. See pages 201 to 213.

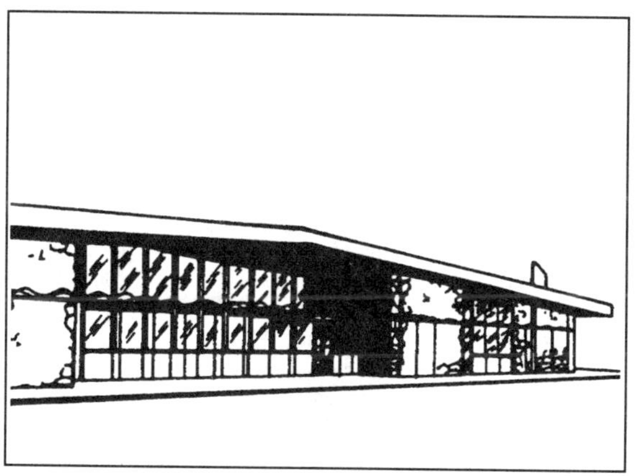

Discount House, Class 2

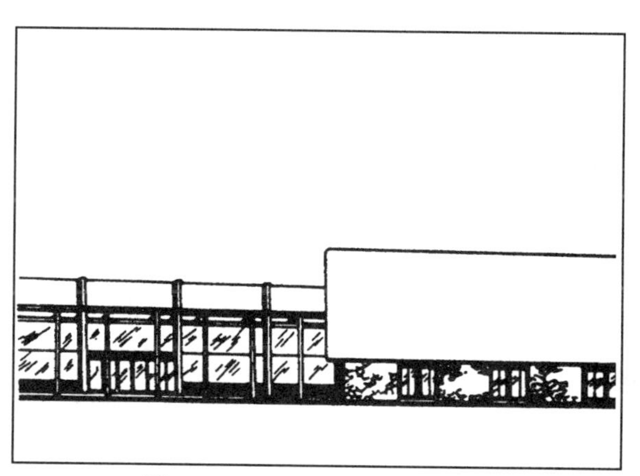

Discount House, Class 2 & 3

Square Foot Area

Quality Class	15,000	20,000	25,000	30,000	35,000	40,000	50,000	75,000	100,000	150,000	200,000
1, Best	65.05	63.44	62.34	61.54	60.89	60.38	59.60	58.38	57.64	56.77	56.25
1 & 2	61.32	59.79	58.76	58.00	57.41	56.90	56.17	55.02	54.36	53.53	53.04
2, Good	58.51	57.04	56.06	55.32	54.76	54.32	53.61	52.50	51.84	51.06	50.59
2 & 3	54.43	53.07	52.17	51.48	50.98	50.53	49.86	48.86	48.23	47.51	47.08
3, Average	51.16	49.93	49.09	48.41	47.91	47.52	46.91	45.92	45.36	44.69	44.28
3 & 4	46.77	45.61	44.83	44.26	43.76	43.43	42.85	41.97	41.46	40.83	40.47
4, Low	42.10	41.06	40.36	39.83	39.41	39.11	38.59	37.79	37.31	36.75	36.40
Wall Height Adjustment*	.39	.37	.34	.33	.31	.30	.29	.24	.22	.16	.15

***Wall Height Adjustment:** Add or subtract the amount listed in this row to the square foot of floor cost for each foot of wall height more or less than 20 feet.

Perimeter Wall Adjustment: A common wall exists when two buildings share one wall. Adjust for common walls by deducting the linear foot costs below from the total structure cost. In some structures one or more walls are not owned at all. In this case, deduct the "No Ownership" cost per linear foot of wall not owned. For common wall, deduct $128 per linear foot. For no wall ownership, deduct $256 per linear foot.

Banks and Savings Offices – Masonry or Concrete

Quality Classification

	Class 1 Best Quality	Class 2 Good Quality	Class 3 High Average Quality	Class 4 Low Average Quality	Class 5 Low Quality
Foundation (11% of total cost)	Reinforced concrete.	Reinforced concrete.	Reinforced concrete.	Reinforced concrete.	Reinforced concrete.
Floor Structure (10% of total cost)	Reinforced concrete.	Reinforced concrete.	Reinforced concrete.	Reinforced concrete.	Reinforced concrete.
Wall Structure (10% of total cost)	12" reinforced brick, 8" reinforced concrete or concrete columns with noncombustible filler walls.	12" reinforced brick, 8" reinforced concrete or concrete columns with noncombustible filler walls.	6" concrete tilt-up, 12" reinforced brick, 8" reinforced con. or concrete columns with noncombustible filler walls.	8" decorative concrete block, 6" concrete tilt-up, 8" reinforced brick or concrete columns with noncombustible filler walls.	8" concrete block, 6" reinforced concrete, 6" concrete tilt-up or 8" reinforced brick.
Roof & Cover (7% of total cost)	5 ply built-up roof, some portions copper or slate.	5 ply built-up roof, some portions heavy shake or mission tile.	4 ply built-up roof, some portions heavy shake or mission tile.	4 ply built-up roof, some portions shake or composition shingles.	4 ply built-up roof.
Exterior Wall Finish (9% of total cost)	Expensive veneers.	Ornamental brick or stone veneer.	Brick, or ornamental block veneer.	Ornamental rock imbedded in tilt-up panels or stucco with masonry trim.	Exposed aggregate or stucco.
Interior Wall Finish (7% of total cost)	Ornamental plaster painted with murals or designs. Ornamental moldings, marble and similar expensive finishes.	Ornamental plaster furred out from walls. Ornamental moldings, some hardwood panels or marble finishes.	Hard plaster furred out from walls. Ornamental cove at ceiling. Vinyl wall covering. Some hardwood veneer.	Plaster on lath with putty coat finish. Gypsum wallboard, texture and paint. Some vinyl wall cover.	Plaster on lath with putty coat finish or gypsum wallboard, texture and paint.
Glass (5% of total cost)	Tinted float glass in customized frames, (50% of exterior wall area.)	Tinted float glass in bronze frames, (50% of exterior wall area).	Tinted float glass in heavy anodized aluminum frames or custom wood frames, (50% of exterior wall area).	Moderate amount of float glass in heavy frames, (25% of exterior wall area).	Small to moderate amount of float glass in average aluminum frames, (5 to 10% of exterior wall area).
Overhang (4% of total cost)	4' closed overhang with copper gutters.	4' closed overhang with copper gutters.	4' closed overhang with painted gutters.	3' closed overhang with painted gutters.	None.
Floor Finish					
Public Area: (7% of total cost)	Terrazzo and marble.	Detailed terrazzo.	Detailed terrazzo.	Plain terrazzo.	Sheet vinyl or vinyl tile.
Officers Area: (3% of total cost)	Excellent carpet.	Very good carpet.	Very good carpet.	Good carpet.	Average quality carpet.
Work area: (2% of total cost)	Good sheet vinyl or carpet.	Average sheet vinyl, tile or carpet.	Resilient tile.	Resilient tile.	Composition tile.
Ceiling Finish (8% of total cost)	Ornamental plaster with exposed ornamental beams.	Ornamental plaster.	Suspended acoustical tile with gypsum wallboard backing or acoustical plaster.	Suspended acoustical tile with exposed grid system.	Acoustical tile on wood furring.
Lighting (5% of total cost)	Recessed panelized fluorescent lighting ,custom light fixtures or chandeliers, and many spotlights.	Recessed panelized fluorescent lighting. Many spotlights.	Recessed panelized fluorescent lighting. Many spotlights.	Recessed panelized fluorescent lighting. Some spotlights.	Nonrecessed panelized lighting.
Plumbing (12% of total cost)	2 or more rest rooms with special fixtures, good ceramic tile, marble or terrazzo wainscot walls. Hard plaster with putty coat and enamel paint. Marble toilet screens.	2 or more rest rooms with many good fixtures, good ceramic tile, marble or terrazzo wainscot walls. Hard plaster with putty coat and enamel paint. Marble toilet screens.	2 rest rooms with 4 or more fixtures, ceramic tile or terrazzo floors. Ceramic tile or terrazzo wainscot. Hard plaster walls with putty coat and enamel paint. Marble or metal toilet screens.	2 rest rooms with 4 or more fixtures each. Ceramic tile or terrazzo floor. Ceramic tile or terrazzo wainscot plaster walls with putty coat. Enamel paint. Good metal toilet partitions.	2 rest rooms with 4 fixtures each. Ceramic tile or vinyl asbestos tile floors. Plaster walls with enamel paint. Metal toilet screens.

Note: Use the percent of total cost to help identify the correct quality classification.

Square foot costs include the following components: Foundations as required for normal soil conditions. Floor, wall, and roof structures. Interior floor, wall and ceiling finishes. Exterior wall finish and roof cover. All glass and glazing. Interior partitions. Roof overhang as described above. Basic electrical systems and lighting fixtures. Rough and finish plumbing. Typical bank vault. Alarm systems. Night depository. Typical record vault and fire doors. Fire exits. Design and engineering fees. Permits and hook-up fees. Contractor's mark-up.

Banks and Savings Offices – Masonry or Concrete

Length Less Than Twice Widths

Estimating Procedure

1. Establish the structure quality class by using the information on page 82.
2. Compute the building floor area. This should include everything within the building exterior walls and all insets outside the main walls but under the main building roof.
3. Add to or subtract from the square foot cost below the appropriate amount from the Wall Height Adjustment Table on page 86 if the wall height is more or less than 16 feet for the first floor or 12 feet for higher floors.
4. Multiply the adjusted square foot cost by the building area.
5. Deduct, if appropriate, for common walls, using the figures on page 86.
6. Multiply the total cost by the location factor listed on page 7.
7. Add the cost of heating and cooling equipment, fire sprinklers, exterior signs, elevators, and yard improvements from pages 201 to 213. Add the cost of mezzanines, bank fixtures, external windows, safe deposit boxes and vault doors from page 92.

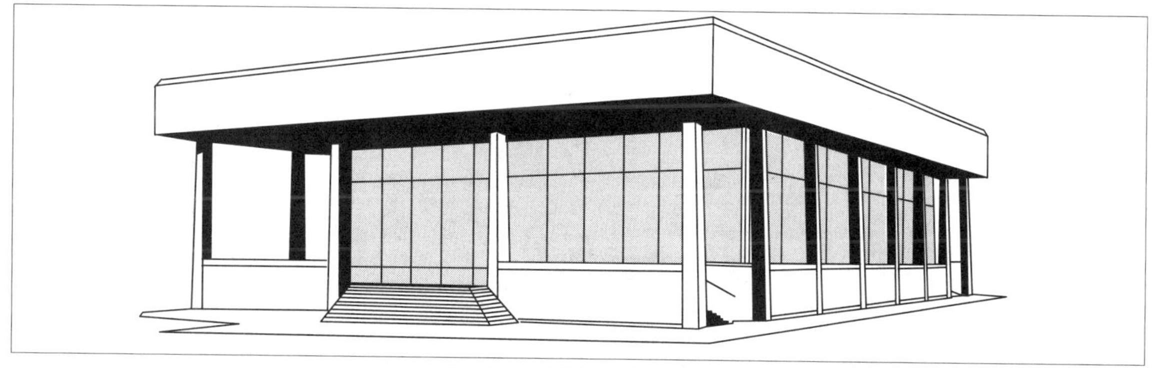

Savings Office, Class 3

First Story - Square Foot Area

Quality Class	2,500	3,000	3,500	4,000	4,500	5,000	6,000	7,500	10,000	15,000	20,000
1, Best	246.25	238.69	232.05	226.23	221.07	216.49	208.68	199.37	188.04	173.48	164.23
1 & 2	231.77	224.64	218.41	212.89	208.05	203.75	196.39	187.65	176.97	163.25	154.55
2, Very Good	220.30	213.54	207.59	202.38	197.79	193.66	186.68	178.38	168.22	155.21	146.94
2 & 3	206.20	199.82	194.29	189.41	185.12	181.27	174.74	166.94	157.44	145.27	137.51
3, Good	195.70	189.69	184.44	179.82	175.70	172.08	165.86	158.55	149.43	137.86	130.51
3 & 4	184.54	178.89	173.93	169.57	165.69	162.28	156.38	149.45	140.93	130.02	123.10
4, Average	175.71	170.30	165.55	161.40	157.75	154.47	148.90	142.25	134.15	123.80	117.17
4 & 5	162.25	157.26	152.89	149.05	145.65	142.61	137.47	131.34	123.89	114.31	108.20
5, Low	147.83	143.28	139.30	135.81	132.74	129.96	125.29	119.70	112.92	104.16	98.61

Second and Higher Stories - Square Foot Area

Quality Class	500	750	1,000	1,500	2,000	3,000	4,000	5,000	7,500	10,000	20,000
1, Best	200.75	179.84	167.49	152.96	144.32	134.16	128.13	124.06	117.69	113.91	106.79
1 & 2	189.83	170.04	158.36	144.60	136.45	126.83	121.16	117.28	111.25	107.69	100.96
2, Very Good	182.27	163.26	152.03	138.84	131.04	121.81	116.35	112.64	106.86	103.43	96.98
2 & 3	171.70	153.79	143.23	130.80	123.45	114.74	109.58	106.09	100.66	97.43	91.33
3, Good	164.35	147.23	137.11	125.20	118.17	109.83	104.90	101.57	96.36	93.25	87.42
3 & 4	155.01	138.86	129.31	118.09	111.41	103.58	98.93	95.77	90.86	87.96	82.43
4, Average	147.78	132.40	123.29	112.60	106.25	98.77	94.34	91.31	86.62	83.86	78.61
4 & 5	139.43	124.93	116.33	106.25	100.27	93.18	89.00	86.14	81.72	79.12	74.18
5, Low	130.22	116.65	108.65	99.21	93.62	87.00	83.10	80.45	76.32	73.87	69.25

Banks and Savings Offices – Masonry or Concrete

Length Between 2 and 4 Times Width

Estimating Procedure

1. Establish the structure quality class by using the information on page 82.
2. Compute the building floor area. This should include everything within the building exterior walls and all insets outside the main walls but under the main building roof.
3. Add to or subtract from the square foot cost below the appropriate amount from the Wall Height Adjustment Table on page 86 if the wall height is more or less than 16 feet for the first floor or 12 feet for higher floors.
4. Multiply the adjusted square foot cost by the building area.
5. Deduct, if appropriate, for common walls, using the figures on page 86.
6. Multiply the total cost by the location factor listed on page 7.
7. Add the cost of heating and cooling equipment, fire sprinklers, exterior signs, elevators, and yard improvements from pages 201 to 213. Add the cost of mezzanines, bank fixtures, external windows, safe deposit boxes and vault doors from page 92.

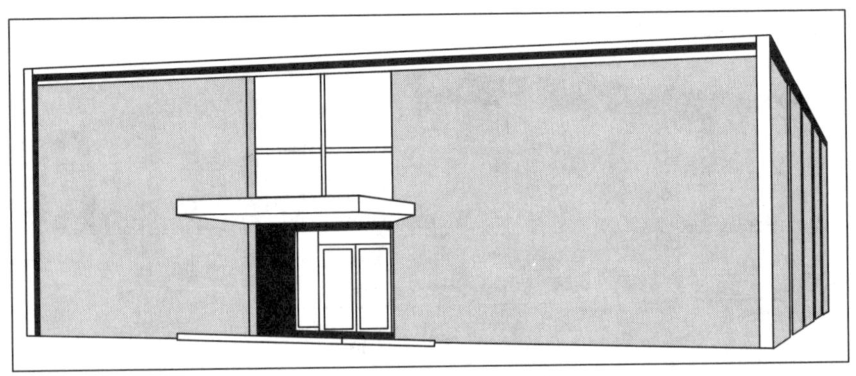

Savings Office, Class 4

First Story - Square Foot Area At 16 Foot Wall Height

Quality Class	2,500	3,000	3,500	4,000	4,500	5,000	6,000	7,500	10,000	15,000	20,000
1, Best	262.39	248.85	238.69	230.77	224.36	219.08	210.79	201.96	192.33	181.50	175.36
1 & 2	247.80	234.80	225.08	217.48	211.37	206.24	198.30	189.87	180.63	170.24	164.31
2, Very Good	236.56	223.91	214.46	207.04	201.07	196.14	188.41	180.21	171.26	161.19	155.47
2 & 3	222.18	209.86	200.71	193.60	187.87	183.19	175.86	168.11	159.79	150.52	145.33
3, Good	211.43	199.84	191.19	184.44	178.99	174.46	167.41	159.86	151.68	142.49	137.27
3 & 4	199.01	188.07	179.91	173.55	168.43	164.19	157.57	150.53	142.91	134.37	129.52
4, Average	189.13	178.69	170.90	164.85	159.96	155.95	149.72	143.08	135.90	127.94	123.45
4 & 5	174.12	164.48	157.32	151.78	147.35	143.67	137.97	131.93	125.45	118.26	114.26
5, Low	157.99	149.31	142.87	137.88	133.88	130.60	125.47	120.12	114.34	107.98	104.43

Second and Higher Stories - Square Foot Area

Quality Class	500	750	1,000	1,500	2,000	3,000	4,000	5,000	7,500	10,000	20,000
1, Best	214.78	191.93	178.26	161.93	152.21	140.63	133.70	128.96	121.61	117.18	108.83
1 & 2	203.46	181.82	168.85	153.42	144.20	133.19	126.63	122.16	115.19	111.00	103.10
2, Very Good	195.28	174.50	162.06	147.25	138.39	127.83	121.57	117.22	110.55	106.55	98.96
2 & 3	183.63	164.10	152.40	138.46	130.14	120.23	114.31	110.24	103.96	100.19	93.05
3, Good	174.52	155.98	144.87	131.62	123.71	114.27	108.66	104.80	98.81	95.24	88.46
3 & 4	165.16	147.57	137.05	124.55	117.06	108.13	102.80	99.18	93.49	90.10	83.69
4, Average	157.44	140.71	130.69	118.74	111.59	103.08	98.02	94.54	89.14	85.93	79.80
4 & 5	148.30	132.53	123.09	111.81	105.09	97.09	92.30	89.03	83.98	80.90	75.16
5, Low	138.39	123.65	114.86	104.37	98.07	90.60	86.14	83.09	78.36	75.50	70.14

Banks and Savings Offices – Masonry or Concrete

Length More Than 4 Times Width

Estimating Procedure

1. Establish the structure quality class by using the information on page 82.
2. Compute the building floor area. This should include everything within the building exterior walls and all insets outside the main walls but under the main building roof.
3. Add to or subtract from the square foot cost below the appropriate amount from the Wall Height Adjustment Table on page 86 if the wall height is more or less than 16 feet for the first floor or 12 feet for higher floors.
4. Multiply the adjusted square foot cost by the building area.
5. Deduct, if appropriate, for common walls, using the figures on page 86.
6. Multiply the total cost by the location factor listed on page 7.
7. Add the cost of heating and cooling equipment, fire sprinklers, exterior signs, elevators, and yard improvements from pages 201 to 213. Add the cost of mezzanines, bank fixtures, external windows, safe deposit boxes and vault doors from page 92.

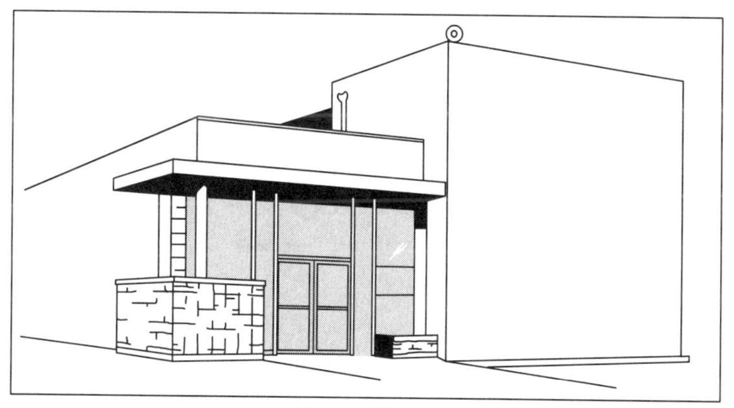

Bank, Class 5

First Story - Square Foot Area

Quality Class	2,500	3,000	3,500	4,000	4,500	5,000	6,000	7,500	10,000	15,000	20,000
1, Best	280.27	263.97	251.88	242.52	234.97	228.75	219.13	208.94	197.97	185.83	178.99
1 & 2	263.94	248.58	237.21	228.38	221.28	215.44	206.34	196.77	186.44	175.00	168.57
2, Very Good	250.85	236.27	225.44	217.03	210.31	204.75	196.11	186.99	177.20	166.31	160.20
2 & 3	235.27	221.57	211.43	203.57	197.22	192.03	183.93	175.39	166.18	155.98	150.25
3, Good	223.49	210.50	200.84	193.37	187.38	182.42	174.74	166.61	157.87	148.15	142.74
3 & 4	210.14	197.89	188.84	181.81	176.15	171.50	164.29	156.65	148.41	139.32	134.19
4, Average	199.27	187.70	179.10	172.42	167.06	162.66	155.78	148.55	140.74	132.10	127.25
4 & 5	183.36	172.69	164.80	158.66	153.72	149.67	143.35	136.68	129.52	121.57	117.11
5, Low	166.14	156.49	149.32	143.77	139.29	135.62	129.90	123.87	117.37	110.17	106.14

Second and Higher Stories - Square Foot Area

Quality Class	500	750	1,000	1,500	2,000	3,000	4,000	5,000	7,500	10,000	20,000
1, Best	237.69	210.35	194.12	174.87	163.39	149.82	141.75	136.23	127.66	122.56	112.90
1 & 2	225.01	199.13	183.76	165.53	154.67	141.83	134.17	128.95	120.85	116.02	106.89
2, Very Good	215.30	190.53	175.84	158.39	148.02	135.72	128.39	123.41	115.64	111.00	102.27
2 & 3	202.18	178.92	165.09	148.73	138.99	127.43	120.57	115.87	108.59	104.24	96.02
3, Good	192.72	170.56	157.35	141.76	132.47	121.48	114.92	110.45	103.47	99.35	91.51
3 & 4	181.19	160.33	147.95	133.31	124.57	114.21	108.06	103.82	97.31	93.41	86.07
4, Average	171.86	152.10	140.36	126.45	118.13	108.35	102.51	98.51	92.32	88.63	81.64
4 & 5	161.93	143.30	132.21	119.11	111.29	102.05	96.53	92.77	86.95	83.49	76.89
5, Low	151.03	133.66	123.32	111.11	103.81	95.21	90.04	86.56	81.09	77.87	71.75

Wall Adjustments

Wall Height Adjustment

Add or subtract the amount listed in this table to the square foot of floor cost for each foot of wall height more or less than 16 feet, if adjusting for a first floor, and 12 feet if adjusting for upper floors.

Square Foot Area

Quality Class	500	750	1,000	1,500	2,000	2,500	3,000	3,500
1, Best	9.71	7.96	6.94	5.66	4.92	4.42	4.02	3.73
2, Very Good	8.75	7.19	6.25	5.15	4.44	3.99	3.63	3.37
3, Good	7.85	6.44	5.60	4.58	3.98	3.56	3.26	3.00
4, Average	6.57	5.39	4.67	3.83	3.32	2.97	2.74	2.52
5, Low	5.20	4.27	3.71	3.04	2.64	2.38	2.14	1.98

Quality Class	4,000	4,500	5,000	6,000	7,500	10,000	15,000	20,000
1, Best	3.50	3.28	3.13	2.86	2.55	2.20	1.78	1.52
2, Very Good	3.13	2.96	2.81	2.56	2.28	1.98	1.61	1.37
3, Good	2.85	2.64	2.53	2.31	2.05	1.77	1.45	1.24
4, Average	2.37	2.19	2.11	1.91	1.72	1.49	1.20	1.05
5, Low	1.85	1.74	1.64	1.51	1.33	1.15	.92	.81

Perimeter Wall Adjustment

First Story

Class	For a Common Wall, Deduct Per L.F.	For No Wall Ownership, Deduct Per L.F.	For Lack of Exterior Finish, Deduct Per L.F.
1	$594.00	$1,188.00	$404.00
2	498.00	985.00	325.00
3	409.00	818.00	224.00
4	247.00	504.00	152.00
5	202.00	414.00	41.00

Second and Higher Stories

Class	For a Common Wall, Deduct Per L.F.	For No Wall Ownership, Deduct Per L.F.	For Lack of Exterior Finish, Deduct per L.F.
1	$336.00	$918.00	$185.00
2	281.00	544.00	140.00
3	224.00	448.00	90.00
4	190.00	381.00	62.00
5	140.00	269.00	42.00

Note: First floor costs include the cost of overhang as described on page 82. Second floor costs do not include any allowance for overhang.

Banks and Savings Offices - Wood Frame

Quality Classification

	Class 1 Best Quality	Class 2 Good Quality	Class 3 High Average Quality	Class 4 Low Average Quality	Class 5 Low Quality
Foundation (10% of total cost)	Reinforced concrete.	Reinforced concrete.	Reinforced concrete.	Reinforced concrete.	Reinforced concrete.
Floor Structure (10% of total cost)	Reinforced concrete.	Reinforced concrete.	Reinforced concrete.	Reinforced concrete.	Reinforced concrete.
Wall Structure (10% of total cost)	2" x 6" - 16" o.c. brick or concrete columns with combustible filler walls.	2" x 6" - 16" o.c. brick or concrete columns with combustible filler walls.	2" x 6" - 16" o.c.	2" x 6" - 16" o.c.	2" x 6" - 16" o.c.
Roof & Cover (10% of total cost)	Copper or slate.	Mission tile or heavy shakes.	5 ply built-up roof. Some portions heavy shake or mission tile.	4 ply built-up roof. Some portions shingle or composition shingles.	4 ply built-up roof.
Exterior Wall Finish (7% of total cost)	Expensive veneers.	Expensive ornamental veneer filler walls.	Wood siding combined with brick or stone veneers.	Good wood siding.	Good stucco or wood siding.
Interior Wall Finish (7% of total cost)	Ornamental plaster painted with murals or designs, ornamental moldings, marble and similar expensive finishes.	Ornamental plaster ornamental moldings, some hardwood panels or marble finishes.	Hard plaster, ornamental cove at ceiling, vinyl wall covering, some hardwood veneer.	Plaster on lath with putty coat finish. Gypsum wallboard, texture and paint. Some vinyl wall cover.	Plaster on lath with putty coat finish. Gypsum wallboard, texture and paint.
Glass (5% of total cost)	Tinted float glass in customized frames (50% of exterior wall area).	Tinted float glass in bronze frames (50% of exterior wall area).	Tinted float glass in heavy anodized aluminum frames or custom wood frames (50% of exterior wall area).	Moderate amount of float glass in heavy frames (25% of exterior wall area).	Small to moderate amount of float glass in average aluminum frames (5 to 10% of exterior wall area).
Overhang (4% of total cost)	4' closed overhang with copper gutters.	4' closed overhang with copper gutters.	4' closed overhang with painted gutters.	3' closed overhang with painted gutters.	None.
Floor Finish					
Public Area: (7% of total cost)	Terrazzo and marble.	Detailed terrazzo.	Detailed terrazzo.	Plain terrazzo.	Sheet vinyl or resilient tile.
Officers Area: (3% of total cost)	Excellent carpet.	Very good carpet.	Very good carpet.	Good carpet.	Average quality carpet.
Work Area: (2% of total cost)	Good sheet vinyl or carpet.	Average sheet vinyl or carpet.	Resilient tile.	Composition tile.	Composition tile.
Ceiling Finish (8% of total cost)	Ornamental plaster with exposed ornamental beams.	Ornamental plaster.	Suspended acoustical tile with gypsum wallboard backing or plain acoustical plaster.	Suspended acoustical tile with exposed grid system.	Acoustical tile on wood strips.
Lighting (5% of total cost)	Recessed panelized fluorescent light fixtures, custom light fixtures or chandeliers. Many spotlights.	Recessed panelized fluorescent lighting. Custom chandeliers. Many spotlights.	Recessed panelized fluorescent lighting. Many spotlights.	Recessed panelized fluorescent lighting. Some spotlights.	Nonrecessed panelized lighting.
Plumbing (12% of total cost)	2 or more rest rooms with special fixtures. Good ceramic tile, marble or terrazzo wainscot walls. Hard plaster with putty coat and enamel paint. Marble toilet screens. toilet screens.	2 or more rest rooms with many good fixtures. Good ceramic tile, marble or terrazzo wainscot walls. Hard plaster with putty coat and enamel paint. Marble screens.	2 rest rooms with 4 or more fixtures. Ceramic tile or terrazzo wainscot, hard plaster walls with putty coat and enamel paint. Marble or metal toilet screens.	2 rest rooms with 4 or more fixtures each. Ceramic tile or terrazzo floor. Ceramic tile or terrazzo wainscot plaster walls with putty coat. Enamel paint. Good metal toilet partitions.	2 rest rooms with 4 fixtures each. Metal toilet screens. Ceramic tile or vinyl asbestos tile floors. Plaster walls with enamel paint.

Note: Use the percent of total cost to help identify the correct quality classification.

Square foot costs include the following components: Foundations as required for normal soil conditions. Floor, wall, and roof structures. Interior floor, wall and ceiling finishes. Exterior wall finish and roof cover. All glass and glazing. Interior partitions. Roof overhang as described above. Basic electrical systems and lighting fixtures. Rough and finish plumbing. Typical bank vault. Alarm systems. Night depository. Typical record vault and fire doors. Fire exits. Design and engineering fees, permits and hook-up fees, contractor's mark-up.

Banks and Savings Offices – Wood Frame

Length Less Than Twice Width

Estimating Procedure

1. Establish the structure quality class by using the information on page 87.
2. Compute the building floor area. This should include everything within the building exterior walls and all insets outside the main walls but under the main building roof.
3. Add to or subtract from the square foot cost below the appropriate amount from the Wall Height Adjustment Table on page 91 if the wall height is more or less than 16 feet for the first floor or 12 feet for higher floors.
4. Multiply the adjusted square foot cost by the building area.
5. Deduct, if appropriate, for common walls, using the figures on page 91.
6. Multiply the total cost by the location factor listed on page 7.
7. Add the cost of heating and cooling equipment, fire sprinklers, exterior signs, elevators, and yard improvements from pages 201 to 213. Add the cost of mezzanines, bank fixtures, external windows, safe deposit boxes and vault doors from page 92.

Savings Office, Class 5

First Story - Square Foot Area

Quality Class	2,500	3,000	3,500	4,000	4,500	5,000	6,000	7,500	10,000	15,000	20,000
1, Best	242.51	229.34	219.77	212.50	206.76	202.13	195.02	187.76	180.23	172.30	168.06
1 & 2	228.85	216.42	207.40	200.54	195.15	190.73	184.06	177.20	170.09	162.58	158.59
2, Very Good	216.02	204.26	195.75	189.26	184.15	180.01	173.70	167.24	160.51	153.44	149.69
2 & 3	193.49	182.98	175.35	169.56	164.95	161.26	155.61	149.82	143.78	137.48	134.11
3, Good	190.88	180.54	173.01	167.29	162.77	159.11	153.52	147.81	141.86	135.63	132.29
3 & 4	177.86	168.20	161.18	155.85	151.63	148.22	143.04	137.72	132.17	126.36	123.27
4, Average	167.29	158.20	151.61	146.63	142.66	139.49	134.57	129.54	124.33	118.84	115.96
4 & 5	153.23	144.92	138.88	134.27	130.66	127.70	123.25	118.64	113.86	108.85	106.20
5, Low	138.58	131.04	125.58	121.43	118.16	115.50	111.46	107.29	103.00	98.46	96.03

Second and Higher Stories - Square Foot Area

Quality Class	500	750	1,000	1,500	2,000	3,000	4,000	5,000	7,500	10,000	20,000
1, Best	167.27	151.77	142.53	131.59	125.03	117.31	112.68	109.55	104.61	101.69	96.17
1 & 2	155.96	142.05	133.77	124.01	118.20	111.30	107.20	104.42	100.07	97.52	92.61
2, Very Good	147.42	134.75	127.22	118.29	112.99	106.67	102.95	100.39	96.42	94.05	89.58
2 & 3	136.21	124.97	118.30	110.42	105.74	100.18	96.87	94.62	91.11	89.04	85.07
3, Good	126.68	116.87	111.04	104.12	99.97	95.06	92.17	90.21	87.09	85.25	81.76
3 & 4	116.74	108.01	102.84	96.77	93.14	88.83	86.29	84.55	81.88	80.27	77.23
4, Average	108.06	100.56	96.07	90.77	87.62	83.87	81.61	80.11	77.74	76.34	73.66
4 & 5	99.99	93.47	89.58	84.93	82.20	78.96	77.03	75.72	73.65	72.45	70.17
5, Low	93.67	86.98	83.20	78.96	76.54	73.79	72.22	71.15	69.53	68.58	66.88

Banks and Savings Offices - Wood Frame

Length Between 2 and 4 Times Width

Estimating Procedure

1. Establish the structure quality class by using the information on page 87.
2. Compute the building floor area. This should include everything within the building exterior walls and all insets outside the main walls but under the main building roof.
3. Add to or subtract from the square foot cost below the appropriate amount from the Wall Height Adjustment Table on page 91 if the wall height is more or less than 16 feet for the first floor or 12 feet for higher floors.
4. Multiply the adjusted square foot cost by the building area.
5. Deduct, if appropriate, for common walls, using the figures on page 91.
6. Multiply the total cost by the location factor listed on page 7.
7. Add the cost of heating and cooling equipment, fire sprinklers, exterior signs, elevators, and yard improvements from pages 201 to 213. Add the cost of mezzanines, bank fixtures, external windows, safe deposit boxes and vault doors from page 92.

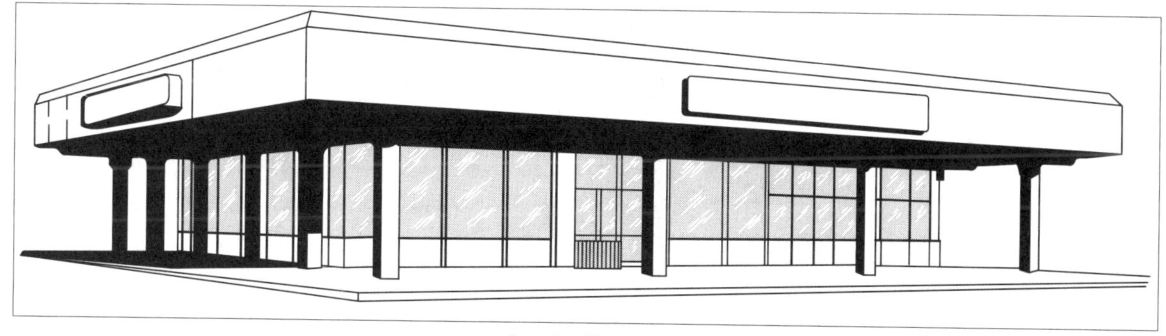

Bank, Class 4

First Story - Square Foot Area

Quality Class	2,500	3,000	3,500	4,000	4,500	5,000	6,000	7,500	10,000	15,000	20,000
1, Best	248.74	235.92	226.41	219.03	213.16	208.30	200.79	192.82	184.31	174.94	169.72
1 & 2	234.80	222.72	213.73	206.77	201.20	196.64	189.53	182.04	174.02	165.18	160.17
2, Very Good	222.73	211.24	202.74	196.11	190.84	186.52	179.78	172.67	165.06	156.63	151.94
2 & 3	207.79	197.05	189.14	183.00	178.08	173.99	167.72	161.10	153.98	146.15	141.79
3, Good	195.97	185.86	178.38	172.57	167.92	164.10	158.17	151.93	145.23	137.83	133.70
3 & 4	181.80	172.41	165.45	160.10	155.78	152.24	146.73	140.92	134.72	127.86	124.03
4, Average	170.72	161.90	155.40	150.31	146.29	142.96	137.78	132.35	126.49	120.08	116.47
4 & 5	155.89	147.85	141.88	137.27	133.55	130.53	125.82	120.85	115.48	109.62	106.36
5, Low	140.80	133.52	128.13	123.97	120.63	117.90	113.64	109.13	104.31	99.01	96.03

Second and Higher Stories - Square Foot Area

Quality Class	500	750	1,000	1,500	2,000	3,000	4,000	5,000	7,500	10,000	20,000
1, Best	168.96	154.70	146.24	136.20	130.18	123.08	118.87	115.97	111.51	108.82	103.77
1 & 2	160.02	146.52	138.49	128.97	123.31	116.59	112.58	109.84	105.62	103.07	98.30
2, Very Good	152.32	139.51	131.82	122.79	117.39	110.98	107.16	104.57	100.53	98.12	93.54
2 & 3	142.37	130.39	123.20	114.75	109.74	103.72	100.18	97.74	93.98	91.70	87.45
3, Good	134.29	122.97	116.26	108.25	103.51	97.85	94.48	92.18	88.64	86.52	82.49
3 & 4	124.87	114.34	108.07	100.67	96.26	91.01	87.85	85.73	82.40	80.42	76.69
4, Average	117.33	107.45	101.55	94.61	90.42	85.51	82.55	80.55	77.45	75.59	72.06
4 & 5	109.99	100.73	95.20	88.66	84.76	80.17	77.40	75.52	72.58	70.85	67.56
5, Low	102.50	93.87	88.71	82.62	79.00	74.70	72.12	70.36	67.67	66.03	62.97

Banks and Savings Offices – Wood Frame

Length More Than 4 Times Width

Estimating Procedure

1. Compute the building floor area. This should include everything within the building exterior walls and all insets outside the main walls but under the main building roof.
2. Add to or subtract from the square foot cost below the appropriate amount from the Wall Height Adjustment Table on page 91 if the wall height is more or less than 16 feet for the first floor or 12 feet for higher floors.
3. Multiply the adjusted square foot cost by the building area.
4. Deduct, if appropriate, for common walls, using the figures on page 91.
5. Multiply the total cost by the location factor listed on page 7.
6. Add the cost of heating and cooling equipment, fire sprinklers, exterior signs, elevators, and yard improvements from pages 201 to 213. Add the cost of mezzanines, bank fixtures, external windows, safe deposit boxes and vault doors from page 92.

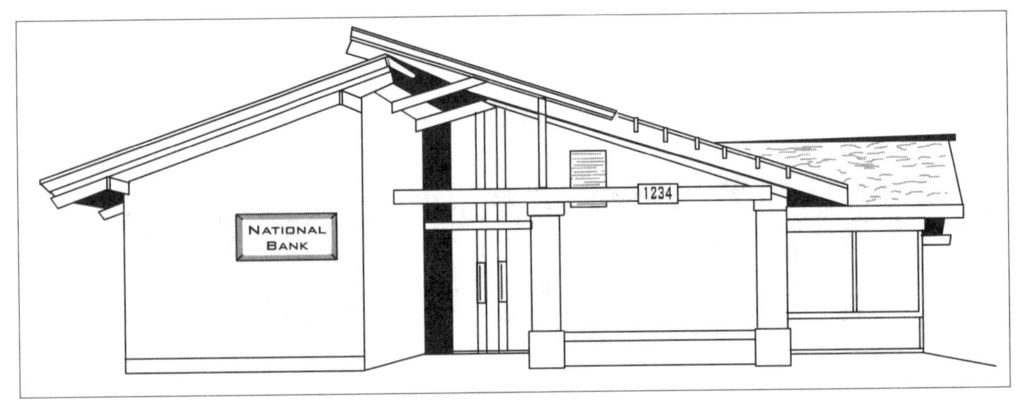

Bank, Class 4 & 5

First Story - Square Foot Area

Quality Class	2,500	3,000	3,500	4,000	4,500	5,000	6,000	7,500	10,000	15,000	20,000
1, Best	261.33	247.00	236.39	228.19	221.66	216.29	207.98	199.24	189.95	179.74	174.07
1 & 2	246.75	233.19	223.16	215.41	209.25	204.19	196.35	188.11	179.34	169.68	164.35
2, Very Good	233.88	221.05	211.54	204.19	198.35	193.56	186.11	178.30	169.97	160.83	155.76
2 & 3	218.68	206.64	197.80	190.92	185.45	180.99	174.00	166.72	158.95	150.40	145.66
3, Good	206.08	194.74	186.39	179.92	174.77	170.54	164.00	157.11	149.76	141.71	137.26
3 & 4	191.02	180.52	172.78	166.79	162.00	158.08	151.99	145.62	138.82	131.36	127.20
4, Average	178.40	168.59	161.37	155.76	151.30	147.63	141.96	136.00	129.66	122.68	118.82
4 & 5	162.28	153.37	146.77	141.69	137.63	134.28	129.12	123.72	117.94	111.61	108.07
5, Low	145.87	137.85	131.92	127.37	123.69	120.70	116.07	111.19	105.99	100.30	97.14

Second and Higher Stories - Square Foot Area

Quality Class	500	750	1,000	1,500	2,000	3,000	4,000	5,000	7,500	10,000	20,000
1, Best	194.12	173.60	161.38	146.90	138.30	128.11	122.05	117.91	111.49	107.64	100.40
1 & 2	181.03	162.44	151.41	138.31	130.53	121.31	115.82	112.08	106.26	102.80	96.27
2, Very Good	169.28	152.52	142.59	130.83	123.84	115.53	110.59	107.21	102.01	98.91	93.02
2 & 3	155.48	140.73	131.93	121.55	115.35	108.06	103.68	100.73	96.10	93.34	88.16
3, Good	143.60	130.65	122.96	113.85	108.40	101.97	98.15	95.51	91.45	89.05	84.45
3 & 4	131.55	120.16	113.36	105.33	100.59	94.91	91.55	89.25	85.69	83.57	79.55
4, Average	121.08	111.10	105.16	98.18	93.98	89.08	86.13	84.12	81.04	79.19	75.74
4 & 5	111.22	102.56	97.46	91.36	87.77	83.52	80.99	79.22	76.56	74.95	71.93
5, Low	101.30	94.00	89.65	84.51	81.47	77.89	75.75	74.30	72.03	70.67	68.15

Wall Adjustments

Wall Height Adjustment

Add or subtract the amount listed in this table to the square foot of floor cost for each foot of wall height more or less than 16 feet, if adjusting for first floor, and 12 feet if adjusting for upper floors.

Square Foot Area

Quality Class	500	750	1,000	1,500	2,000	2,500	3,000	3,500
1, Best	8.13	6.68	5.79	4.77	4.12	3.70	3.38	3.14
2, Very Good	7.33	6.04	5.24	4.29	3.74	3.36	3.05	2.85
3, Good	6.51	5.36	4.64	3.79	3.29	2.94	2.73	2.49
4, Average	4.44	3.65	3.18	2.61	2.26	2.03	1.86	1.69
5, Low	2.92	2.39	2.10	1.70	1.51	1.32	1.21	1.12

Square Foot Area

Quality Class	4,000	4,500	5,000	6,000	7,500	10,000	15,000	20,000
1, Best	3.07	2.84	2.64	2.37	2.11	1.80	1.53	1.36
2, Very Good	2.76	2.53	2.39	2.12	1.86	1.62	1.33	1.21
3, Good	2.42	2.25	2.09	1.89	1.64	1.44	1.20	1.07
4, Average	1.65	1.56	1.47	1.30	1.14	.99	.84	.75
5, Low	1.12	1.06	1.00	.88	.75	.65	.54	.48

Perimeter Wall Adjustment

First Story

Class	For a Common Wall, Deduct Per L.F.	For No Wall Ownership, Deduct Per L.F.	For Lack of Exterior Finish, Deduct Per L.F.
1	$400.00	$787.00	$388.00
2	333.00	644.00	322.00
3	266.00	544.00	221.00
4	195.00	388.00	156.00
5	90.00	183.00	54.00

Second and Higher Stories

Class	For a Common Wall, Deduct Per L.F.	For No Wall Ownership, Deduct Per L.F.	For Lack of Exterior Finish, Deduct Per L.F.
1	$244.00	$499.00	$221.00
2	200.00	388.00	156.00
3	166.00	322.00	103.00
4	103.00	206.00	72.00
5	72.00	145.00	37.00

Note: First floor costs include the cost of overhang as described on page 87. Second floor costs do not include any allowance for overhang.

Banks and Savings Offices Additional Costs

Mezzanines Without Partitions

Quality Class	1, Best	2, Good	3, High Avg.	4, Low Avg.	5, Low
S.F. Cost	$61.50	$59.70	$51.60	$47.70	$43.10

With Partitions

Quality Class	1, Best	2, Good	3, High Avg.	4, Low Avg.	5, Low
S.F. Cost	$81.00	$74.00	$65.90	$61.50	$53.80

Mezzanine costs include: Floor system, floor finish, typical stairway, lighting and structural support costs.

Fixtures, cost per square foot of floor

Shell-type counters, plastic finish, no counter screens or drawers.	$18.60 to $20.80
Counters with drawers, good hardwood, plain counter screens.	22.00 to 30.10
Counters with drawers, good hardwood, plastic countertops, average counter screens.	30.10 to 44.30
Counters with drawers, terrazzo or marble finish, marble counter tops, fancy counter screens.	44.30 to 74.30

Costs include counters, screens, and partitions. The square-foot cost should be applied only to the first floor area. Office areas used for purposes other than conducting bank business related to the immediate site should be excluded.

External Access Facilities, cost each unit

Drive-up teller, flush window	$9,724 to $13,278
Drive-up teller, projected window	11,710 to 17,670
Walk-up teller, flush window	6,137 to 7,215
Automatic teller, cash dispensing, with phone	48,459 to 58,130
Night deposit vault whole chest	10,769 to 11,815

Safe Deposit Boxes, cost per box

Box Sizes	Modular Unit	Custom Built
3" x 5"	$85.70 to $97.00	$129.00 to $140.00
5" x 5"	102.00 to 107.00	140.00 to 172.00
3" x 10"	91.30 to 117.00	156.00 to 179.00
5" x 10"	122.00 to 140.00	195.00 to 229.00
10" x 10"	172.00 to 195.00	256.00 to 323.00

Bank fixtures and safe deposit boxes are part of the structure cost only if they are fixed to the building.

Vault Doors Record Storage

Description	3' x 7', 2 hour	3' x 7', 4 hour	4' x 7', 2 hour	4' x 7', 4 hour
In Place Cost	$3,555	$3,973	$5,386	$5,753

Prices include frames, time locks, and architrave.

Department Stores - Reinforced Concrete

Quality Classification

	Class 1 Best Quality	Class 2 Good Quality	Class 3 Average Quality	Class 4 Low Quality
Foundation (17% of total cost)	Reinforced concrete.	Reinforced concrete.	Reinforced concrete.	Reinforced concrete.
Ground Floor Structure (10% of total cost)	6" reinforced concrete on 6" rock base.	6" reinforced concrete on 6" rock base.	6" reinforced concrete on 6" rock base.	4" reinforced concrete on 6" rock base.
Wall Structure (10% of total cost)	Reinforced concrete.	Reinforced concrete.	Reinforced concrete.	Reinforced concrete.
Upper Floor Structure (12% of total cost)	Reinforced concrete.	Reinforced concrete.	Reinforced concrete.	Reinforced concrete.
Roof & Cover (10% of total cost)	Reinforced concrete with 5 ply built-up roofing and insulation.	Reinforced concrete with 5 ply built-up roofing and insulation.	Reinforced concrete with 4 ply built-up roofing.	Reinforced concrete with 4 ply built-up roofing.
Floor Finish (5% of total cost)	Terrazzo and very good carpet.	Resilient tile with 50% vinyl tile, terrazzo or good carpet.	Composition tile.	Minimum grade tile.
Interior Wall Finish (5% of total cost)	Gypsum wallboard or lath and plaster finished with good paper or vinyl wall cover on hardwood veneer paneling.	Gypsum wallboard and texture or paper. Some vinyl wall cover or hardwood veneer paneling.	Interior stucco or gypsum wallboard, texture and paint.	Gypsum wallboard, texture and paint.
Ceiling Finish (5% of total cost)	Suspended good grade acoustical tile with gypsum wallboard backing.	Suspended acoustical tile with concealed grid system.	Suspended acoustical tile with exposed grid system.	Painted.
Lighting (6% of total cost)	Recessed fluorescent lighting in modular plastic panels. Many spotlights.	Continuous recessed 3 tube fluorescent strips with egg crate diffusers 8' o.c. Average number of spotlights.	Continuous 3 tube fluorescent strips with egg crate diffusers, 8' o.c. Some spotlights.	Continuous exposed 2 tube fluorescent strips, 8' o.c.
Display Fronts (6% of total cost)	Very good front as described on page 56. 15 to 25% of the first floor exterior wall is made up of display fronts.	Good quality front as described on page 56. 15 to 25% of the first floor exterior wall is made up of display fronts.	Average quality front as described on page 56. 10 to 20% of the first floor exterior wall is made up of display fronts.	Average quality flat type as described on page 56. 10 to 20% of the first floor exterior wall is made up of display fronts.
Exterior Wall Finish (8% of total cost)	Ornamental block or brick, some marble veneer.	Decorative block, some stone veneer.	Paint.	Paint.
Plumbing (6% of total cost)	6 good fixtures per 20,000 square feet of floor area. Metal toilet partitions.	6 standard fixtures per 20,000 square feet of floor area. Metal toilet partitions.	4 standard fixtures per 20,000 square feet of floor area. Metal toilet partitions.	4 standard fixtures per 20,000 square feet of floor area. Wood toilet partitions.

Note: Use the percent of total cost to help identify the correct quality classification.

Square foot costs include the following components: Foundations as required for normal soil conditions. Floor, wall and roof structures. Interior ceiling, wall and floor finishes (including carpet). Exterior wall finish and roof cover. Display fronts. Interior partitions (including perimeter wall partitions). Entry and delivery doors. Basic lighting and electrical systems. Rough and finish plumbing. Design and engineering fees. Permits and hook-up fees. Contractor's mark-up.

The in-place cost of these extra components should be added to the basic building cost to arrive at the total structure cost. See the section "Additional Costs for Commercial Structures" on page 201. Heating and air conditioning systems. Elevators and escalators. Fire sprinklers. Exterior signs. Canopies and walks. Paving and curbing. Loading docks or ramps. Miscellaneous yard improvements. Mezzanines.

Department Stores – Reinforced Concrete

First Floor

Estimating Procedure

1. Use these square foot costs to estimate the cost of retail stores designed to sell a wider variety of goods. These buildings differ from discount houses in that they have more interior partitions and more elaborate interior and exterior finishes.
2. Establish the structure quality class by using the information at the bottom of the page.
3. Compute the building floor area. This should include everything within the building exterior walls and all inset areas outside the main walls but under the main building roof.
4. Add to or subtract from the square foot cost below the appropriate amount from the Wall Height Adjustment Table shown below if the wall height is more or less than 20 feet.
5. Multiply the adjusted square foot cost by the building area.
6. Deduct, if appropriate, for common walls, using the figures at the bottom of this page.
7. Multiply the total cost by the location factor listed on page 7.
8. Add the cost of appropriate additional components from page 201: heating and air conditioning systems, elevators and escalators, fire sprinklers, exterior signs, canopies and walks, paving and curbing, loading docks and ramps, miscellaneous yard improvements, and mezzanines.
9. Add the cost of second and higher floors and basements from page 95.

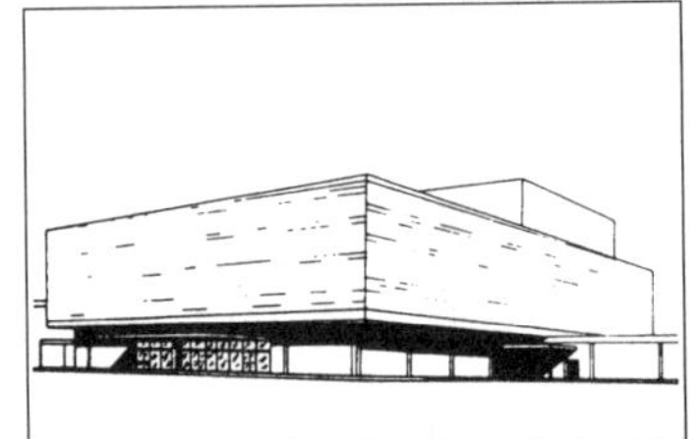

Department Store, Class 1

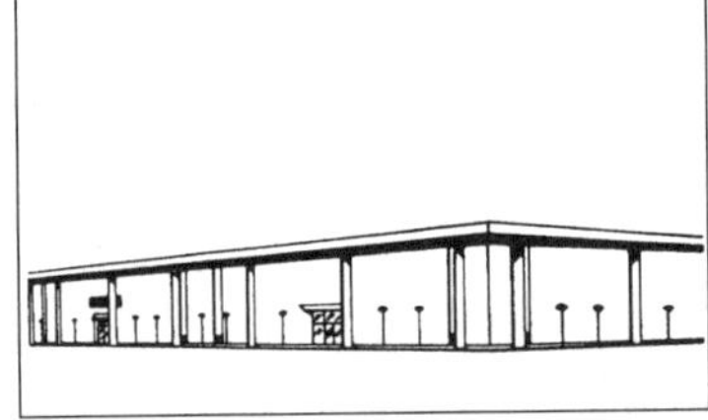

Department Store, Class 2

First Floor - Square Foot Area

Quality Class	20,000	25,000	30,000	35,000	40,000	45,000	50,000	60,000	70,000	80,000	100,000
1, Best	150.79	144.89	140.09	136.14	132.75	129.83	127.32	123.09	119.67	116.82	112.36
1 & 2	140.28	135.06	130.82	127.29	124.30	121.76	119.52	115.76	112.74	110.23	106.29
2, Good	132.10	127.48	123.72	120.61	117.99	115.72	113.74	110.45	107.77	105.60	102.13
2 & 3	122.19	118.22	115.01	112.35	110.10	108.16	106.50	103.69	101.43	99.57	96.59
3, Average	114.30	110.92	108.17	105.94	104.09	102.46	101.08	98.79	96.89	95.35	92.95
3 & 4	103.37	100.66	98.49	96.70	95.23	93.95	92.84	91.00	89.52	88.32	86.38
4, Low	93.75	91.81	90.23	88.95	87.82	86.87	86.01	84.62	83.50	82.53	81.06

Perimeter Wall Adjustment - First Floor

Class	For a Common Wall, Deduct Per L.F.	For No Wall Ownership, Deduct Per L.F.	For Lack of Exterior Finish, Deduct Per L.F.
1	$521.00	$1,042.00	$341.00
2	398.00	784.00	202.00
3	291.00	566.00	90.00
4	196.00	393.00	55.00

Wall Height Adjustment: Add or subtract the amount listed to the square foot of floor cost for each foot of first and upper story wall height more or less than 20 feet.

Square Foot Area

Quality Class	20,000	25,000	30,000	35,000	40,000	45,000	50,000	60,000	70,000	80,000	100,000
1, Best	1.56	1.30	1.17	1.09	1.00	.93	.89	.82	.80	.78	.76
2, Good	1.10	.90	.81	.75	.69	.67	.65	.59	.57	.51	.50
3, Average	.89	.76	.69	.65	.55	.51	.49	.45	.44	.43	.42
4, Low	.73	.66	.55	.47	.44	.43	.41	.39	.38	.37	.36

Department Stores – Reinforced Concrete

Estimating Procedure

1. Establish the basement and upper floor quality class. The quality class will usually be the same as the first floor of the building. Square foot costs for unfinished basements will be nearly the same regardless of the structure quality class.
2. Compute the floor area.
3. Add to or subtract from the square foot cost below the appropriate amount from the Wall Height Adjustment Table on page 94 for second and higher floors and from the bottom of this page for basements for each foot of wall height more or less than 20 feet for second and higher floors and 16 feet for basements.
4. Multiply the adjusted square foot cost from one of the 3 tables below by the floor area.
5. Deduct, if appropriate, for common or unfinished upper floor walls, using the costs in the table below titled "Second and Higher Floor Perimeter Wall Adjustments."
6. Multiply the total cost by the location factor listed on page 7.
7. Add the cost of appropriate additional components from page 201: heating and air conditioning systems, elevators and escalators, fire sprinklers, and mezzanines.
8. Add the total from this page to the total from page 94 to find the building cost.

Second and Higher Stories - Square Foot Area

Quality Class	20,000	25,000	30,000	35,000	40,000	45,000	50,000	60,000	70,000	80,000	100,000
1, Best	135.19	129.63	125.08	121.32	118.10	115.33	112.90	108.86	105.59	102.82	98.52
1 & 2	126.09	121.07	117.06	113.66	110.83	108.37	106.20	102.60	99.66	97.26	93.45
2, Good	119.00	114.52	110.91	107.93	105.37	103.20	101.29	98.13	95.57	93.44	90.09
2 & 3	110.41	106.60	103.49	100.92	98.79	96.90	95.30	92.57	90.39	88.57	85.75
3, Average	103.72	100.43	97.80	95.63	93.81	92.21	90.83	88.58	86.74	85.21	82.83
3 & 4	94.69	91.98	89.83	88.08	86.60	85.35	84.28	82.43	81.02	79.80	77.92
4, Low	85.04	82.94	81.29	79.98	78.88	77.98	77.17	75.85	74.77	73.91	72.57

Second And Higher Floor Perimeter Wall Adjustment

Class	For a Common Wall, Deduct Per L.F.	For No Wall Ownership, Deduct Per L.F.	For Lack of Exterior Finish, Deduct Per L.F.
1	$404.00	$801.00	$275.00
2	296.00	583.00	163.00
3	213.00	425.00	78.00
4	185.00	359.00	55.00

Finished Basements - Square Foot Area

Quality Class	20,000	25,000	30,000	35,000	40,000	45,000	50,000	60,000	70,000	80,000	100,000
1, Best	115.20	111.55	108.50	105.93	103.74	101.84	100.13	97.34	95.05	93.11	90.07
1 & 2	107.87	104.61	101.92	99.64	97.72	96.04	94.58	92.13	90.12	88.47	85.81
2, Good	104.14	101.10	98.61	96.49	94.71	93.18	91.82	89.53	87.71	86.18	83.73
2 & 3	97.18	94.58	92.44	90.67	89.18	87.86	86.73	84.83	83.32	82.02	80.01
3, Average	93.37	91.36	89.62	88.09	86.64	85.62	84.60	82.88	81.45	80.25	78.33
3 & 4	86.28	84.39	82.85	81.61	80.58	79.68	78.91	77.62	76.60	75.72	74.37
4, Low	78.44	76.91	75.76	74.81	74.07	73.46	72.90	72.00	71.31	70.71	69.86

Unfinished Basements

Area	20,000	25,000	30,000	35,000	40,000	45,000	50,000	60,000	70,000	80,000	100,000
Cost	53.56	52.60	51.78	51.23	50.70	50.40	50.00	49.44	49.05	48.77	48.21

Basement Wall Height Adjustment: Add or subtract the amount listed to the square foot of floor cost for each foot of basement wall height more or less than 16 feet.

Area	20,000	25,000	30,000	35,000	40,000	45,000	50,000	60,000	70,000	80,000	100,000
Finished	.74	.67	.57	.49	.47	.45	.44	.43	.42	.41	.41
Unfinished	.68	.57	.47	.45	.43	.41	.40	.39	.38	.38	.37

Department Stores – Masonry or Concrete

Quality Classification

	Class 1 Best Quality	Class 2 Good Quality	Class 3 Average Quality	Class 4 Low Quality
Foundation (20% of total cost)	Reinforced concrete.	Reinforced concrete.	Reinforced concrete.	Reinforced concrete.
Floor Structure (15% of total cost)	6" reinforced concrete on 6" rock base.	6" reinforced concrete on 6" rock base.	6" reinforced concrete on 6" rock base.	4" reinforced concrete on 6" rock base.
Wall Structure (12% of total cost)	8" reinforced decorative concrete block, 6" concrete tilt-up or 8" reinforced brick.	8" reinforced decorative concrete block, 6" concrete tilt-up or 8" reinforced brick.	8" reinforced concrete. block, 6" concrete tilt-up or 8" reinforced common brick.	8" reinforced concrete block or 6" concrete tilt-up.
Roof & Cover (12% of total cost)	Glu-lams or steel beams on steel intermediate columns. Panelized roof system, 1/2" plywood sheathing, 5 ply built-up roof with insulation.	Glu-lams or steel beams on intermediate columns. Panelized roof system, 1/2" plywood sheathing, 5 ply built-up roof with insulation.	Glu-lams on steel intermediate columns. Panelized roof system, 1/2" plywood sheathing, 4 ply built-up roof.	Glu-lams on steel intermediate columns. Panelized roof system, 1/2" plywood sheathing, 4 ply built-up roof.
Floor Finish (5% of total cost)	Terrazzo and very good carpet.	Resilient tile with 50% sheet vinyl, terrazzo or good carpet.	Composition tile.	Minimum grade tile.
Interior Wall Finish (5% of total cost)	Gypsum wallboard or lath and plaster, finished with good paper or vinyl wall covers, or hardwood veneer paneling.	Gypsum wallboard and texture or paper, some vinyl wall cover or hardwood veneer paneling.	Interior stucco or gypsum wallboard, texture and paint.	Gypsum wallboard, texture and paint.
Ceiling Finish (5% of total cost)	Suspended good grade acoustical tile with gypsum wallboard backing.	Suspended acoustical tile with concealed grid system.	Suspended acoustical tile with exposed grid system.	Painted.
Lighting (6% of total cost)	Recessed fluorescent lighting in modular plastic panels. Many spotlights.	Continuous recessed 3 tube fluorescent strips with egg crate diffusers, 8' o.c. Average number of spotlights.	Continuous 3 tube fluorescent strips with egg crate diffusers, 8' o.c. Some spotlights.	Continuous exposed 2 tube fluorescent strips, 8' o.c.
Display Fronts (6% of total cost)	Very good front as described on page 56. 15 to 25% of the first floor exterior wall is made up of display fronts.	Good quality front as described on page 56. 15 to 25% of the first floor exterior wall is made up of display fronts.	Average quality front as described on page 56. 10 to 20% of the first floor exterior wall is made up of display fronts.	Average quality flat type as described on page 56. 10 to 20% of the first floor exterior wall is made up of display fronts.
Exterior Wall Finish (8% of total cost)	Ornamental block or large rock imbedded in tilt-up panels, some stone veneer.	Exposed aggregate or decorative block. Some stone veneer.	Paint, some exposed aggregate.	Paint.
Plumbing (6% of total cost)	6 good fixtures per 20,000 square feet of floor area. Metal toilet partitions.	6 standard fixtures per 20,000 square feet of floor area. Metal toilet partitions.	4 standard fixtures per 20,000 square feet of floor area. Metal toilet partitions.	4 standard fixtures per 20,000 square feet of floor area. Wood toilet partitions.

Note: Use the percent of total cost to help identify the correct quality classification.

Square foot costs include the following components: Foundations as required for normal soil conditions. Floor, wall and roof structures. Interior ceiling, wall and floor finishes (including carpet). Exterior wall finish and roof cover. Display fronts. Interior partitions (including perimeter wall partitions). Entry and delivery doors. Basic lighting and electrical systems. Rough and finish plumbing. Design and engineering fees. Permits and hook-up fees. Contractor's mark-up.

The in-place cost of these extra components should be added to the basic building cost to arrive at the total structure cost. See the section "Additional Costs for Commercial Structures" on page 201. Heating and air conditioning systems. Elevators and escalators. Fire sprinklers. Exterior signs. Canopies and walks. Paving and curbing. Loading docks or ramps. Miscellaneous yard improvements. Mezzanines.

Department Stores – Masonry or Concrete

First Floor

Estimating Procedure

1. Use these square foot costs to estimate the cost of retail stores designed to sell a wider variety of goods. These buildings differ from discount houses in that they have more interior partitions and more elaborate interior and exterior finishes.
2. Establish the structure quality class by using the information on page 96.
3. Compute the building floor area. This should include everything within the building exterior walls and all inset areas outside the main walls but under the main building roof.
4. Add to or subtract from the square foot cost below the appropriate amount from the Wall Height Adjustment Row (near the bottom of this page) if the wall height is more or less than 16 feet.
5. Multiply the adjusted square foot cost by the building area.
6. Deduct, if appropriate, for common or unfinished walls, using the figures at the bottom of this page.
7. Multiply the total cost by the location factor listed on page 7.
8. Add the cost of appropriate additional components from page 201: heating and air conditioning systems, elevators and escalators, fire sprinklers, exterior signs, canopies and walks, paving and curbing, loading docks and ramps, miscellaneous yard improvements, and mezzanines.
9. Add the cost of second and higher floors and basements from page 98.

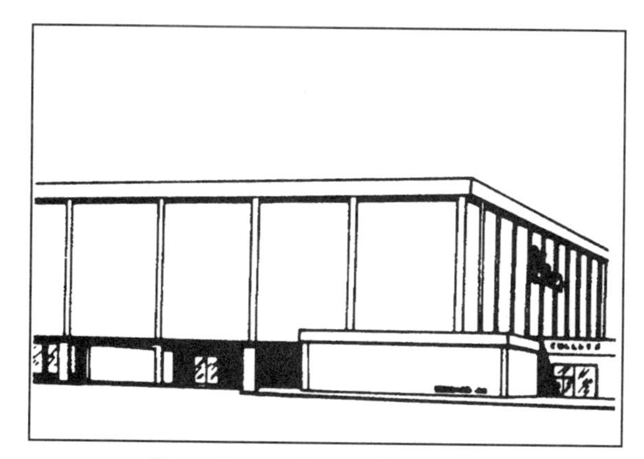

Department Store, Class 1 & 2

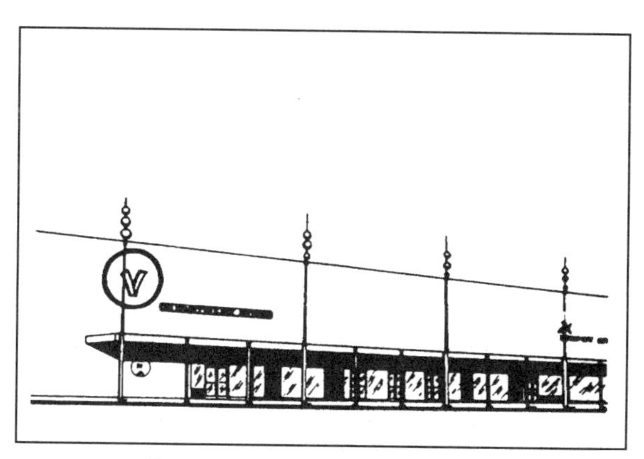

Department Store, Class 2 & 3

First Floor - Square Foot Area

Quality Class	20,000	25,000	30,000	35,000	40,000	45,000	50,000	60,000	70,000	80,000	100,000
1, Best	114.27	109.22	105.20	101.93	99.17	96.81	94.76	91.38	88.61	86.37	82.84
1 & 2	102.49	97.95	94.36	91.40	88.92	86.79	84.99	81.92	79.48	77.49	74.27
2, Good	92.24	88.18	84.95	82.30	80.05	78.16	76.50	73.76	71.55	69.72	66.89
2 & 3	82.50	78.87	75.96	73.57	71.60	69.89	68.42	65.95	64.00	62.36	59.81
3, Average	73.98	70.73	68.13	66.03	64.20	62.70	61.36	59.16	57.40	55.94	53.65
3 & 4	64.56	61.73	59.46	57.62	56.04	54.74	53.56	51.63	50.07	48.83	46.82
4, Low	55.09	52.66	50.72	49.16	47.82	46.68	45.68	44.05	42.72	41.66	39.93
Wall Height Adjustment*	.63	.49	.44	.41	.39	.38	.37	.36	.35	.35	.35

***Wall Height Adjustment:** Add or subtract the amount listed in this row to the square foot of floor cost for each foot of wall height more or less than 16 feet.

Perimeter Wall Adjustment: For common wall, deduct $170.00 per linear foot. For no wall ownership, deduct $340.00 per linear foot.

Upper Floors and Basements

Estimating Procedure

1. Establish the basement and upper floor quality class. The quality class will usually be the same as the first floor of the building. Square foot costs for unfinished basements will be nearly the same regardless of the structure quality class.
2. Compute the floor area.
3. Add to or subtract from the square foot cost below the appropriate amount from the Wall Height Adjustment Row for each foot of wall height more or less than 12 feet for second and higher floors and 12 feet for basements.
4. Multiply the adjusted square foot cost by the floor area.
5. Deduct, if appropriate, for common or unfinished upper floor walls, using the Perimeter Wall Adjustment costs listed below.
6. Multiply the total cost by the location factor listed on page 7.
7. Add the cost of the appropriate additional components from page 201: heating and air conditioning systems, elevators and escalators, fire sprinklers, and mezzanines.
8. Add the total from this page to the total from page 97 to find the building cost.

Second and Higher Floors - Square Foot Area

Quality Class	20,000	25,000	30,000	35,000	40,000	45,000	50,000	60,000	70,000	80,000	100,000
1, Best	100.04	96.37	92.98	89.97	87.30	84.92	82.80	79.16	76.15	73.60	69.48
1 & 2	88.24	85.01	82.03	79.39	77.01	74.93	73.07	69.87	67.18	64.93	61.26
2, Good	78.00	75.10	72.49	70.15	68.06	66.20	64.56	61.72	59.37	57.38	54.15
2 & 3	69.24	66.68	64.36	62.27	60.42	58.77	57.30	54.81	52.71	50.92	48.10
3, Average	61.04	58.77	56.71	54.88	53.28	51.79	50.49	48.28	46.45	44.91	42.38
3 & 4	50.28	48.43	46.72	45.19	43.89	42.68	41.62	39.78	38.28	36.99	34.92
4, Low	40.60	39.27	38.22	37.30	36.56	35.89	35.35	34.38	33.64	32.98	31.98
Wall Height Adjustment*	.45	.40	.37	.35	.34	.32	.31	.30	.30	.29	.29

***Wall Height Adjustment:** Add or subtract the amount listed in this row to the square foot of floor cost for each foot of wall height more or less than 12 feet.

Perimeter Wall Adjustment: For common wall, deduct $91.00 per linear foot. For no wall ownership, deduct $182.00 per linear foot.

Finished Basement - Square Foot Area

Quality Class	20,000	25,000	30,000	35,000	40,000	45,000	50,000	60,000	70,000	80,000	100,000
1, Best	88.89	85.31	82.43	80.05	78.05	76.37	74.85	72.37	70.35	68.68	66.08
2, Good	71.23	68.37	66.05	64.14	62.57	61.17	59.97	57.98	56.38	55.05	52.93
3, Average	57.83	55.49	53.63	52.11	50.79	49.65	48.69	47.08	45.77	44.70	42.99
4, Low	41.41	39.75	38.41	37.29	36.35	35.58	34.85	33.71	32.79	32.01	30.77

Unfinished Basements

Area	20,000	25,000	30,000	35,000	40,000	45,000	50,000	60,000	70,000	80,000	100,000
Cost	24.48	23.31	22.48	21.87	21.35	20.93	20.57	20.03	19.58	19.25	18.72

Wall Height Adjustment: Add or subtract the amount listed in this table to the square foot of floor cost for each foot of basement wall height more or less than 12 feet.

Area	20,000	25,000	30,000	35,000	40,000	45,000	50,000	60,000	70,000	80,000	100,000
Finished	.49	.46	.45	.43	.42	.42	.41	.40	.40	.39	.38
Unfinished	.47	.44	.43	.42	.41	.40	.40	.39	.38	.38	.37

Department Stores – Wood Frame

Quality Classification

	Class 1 Best Quality	Class 2 Good Quality	Class 3 Average Quality	Class 4 Low Quality
Foundation (20% of total cost)	Reinforced concrete.	Reinforced concrete.	Reinforced concrete.	Reinforced concrete.
Floor Structure (10% of total cost)	6" reinforced concrete on 6" rock base.	6" reinforced concrete on 6" rock base.	6" reinforced concrete on 6" rock base.	4" reinforced concrete on 6" rock base.
Wall Structure (9% of total cost)	2" x 6", 16" o.c.	2" x 6", 16" o.c.	2" x 6", 16" o.c.	2" x 4", 16" o.c.
Roof & Cover (12% of total cost)	Glu-lams or steel beams on steel intermediate columns. Panelized roof system, 1/2" plywood sheathing, 5 ply built-up roof with insulation.	Glu-lams or steel beams on steel intermediate columns. Panelized roof system, 1/2" plywood sheathing, 5 ply built-up roof with insulation.	Glu-lams on steel intermediate columns. Panelized roof system, 1/2" plywood sheathing, 4 ply built-up roof.	Glu-lams on steel intermediate columns. Panelized roof system, 1/2" plywood sheathing, 4 ply built-up roof.
Floor Finish (6% of total cost)	Terrazzo and very good carpet.	Resilient tile with 50% sheet vinyl, terrazzo or good carpet.	Composition tile.	Minimum grade tile.
Interior Wall Finish (5% of total cost)	Gypsum wallboard or lath and plaster finished with good paper or vinyl wall covers, or hardwood veneer paneling.	Gypsum wallboard, texture or paper, some vinyl wall cover or hardwood veneer paneling.	Interior stucco or gypsum wallboard. Texture and paint.	Gypsum wallboard. Texture and paint.
Ceiling Finish (5% of total cost)	Suspended good grade acoustical tile with gypsum wallboard backing.	Suspended acoustical tile with concealed grid system.	Suspended acoustical tile with exposed grid system.	Painted.
Lighting (8% of total cost)	Recessed fluorescent lighting in modular plastic panels. Many spotlights.	Continuous recessed 3 tube fluorescent strips with egg crate diffusers, 8' o.c. Average number of spotlights.	Continuous 3 tube fluorescent strips with egg crate diffusers, 8' o.c. Some spotlights.	Continuous exposed 2 tube fluorescent strips, 8' o.c.
Display Fronts (8% of total cost)	Very good fronts as described on page 56. 15 to 25% of the first floor exterior wall is made up of display fronts.	Good quality front as described on page 56. 15 to 25% of the first floor exterior wall is made up of display fronts.	Average quality front as described on page 56. 10 to 20% of the first floor exterior wall is made up of display fronts.	Average quality flat type as described on page 56. 10 to 20% of the first floor exterior wall is made up of display fronts.
Exterior Wall Finish (10% of total cost)	Good wood siding. Extensive stone veneer.	Wood siding. Some brick or stone veneer.	Stucco. Some brick trim.	Stucco.
Plumbing (7% of total cost)	6 good fixtures per 20,000 square feet of floor area. Metal toilet partitions.	6 standard fixtures per 20,000 square feet of floor area. Metal toilet partitions.	4 standard fixtures per 20,000 square feet of floor area. Metal toilet partitions.	4 standard fixtures per 20,000 square feet of floor area. Wood toilet partitions.

Note: Use the percent of total cost to help identify the correct quality classification.

Square foot costs include the following components: Foundations as required for normal soil conditions. Floor, wall and roof structures. Interior ceiling, wall and floor finishes (including carpet). Exterior wall finish and roof cover. Display fronts. Interior partitions (including perimeter wall partitions). Entry and delivery doors. Basic lighting and electrical systems. Rough and finish plumbing. Design and engineering fees. Permits and hook-up fees. Contractor's mark-up.

The in-place cost of these extra components should be added to the basic building cost to arrive at the total structure cost. See the section "Additional Costs for Commercial Structures" on page 201. Heating and air conditioning systems. Elevators and escalators. Fire sprinklers. Exterior signs. Canopies and walks. Paving and curbing. Loading docks or ramps. Miscellaneous yard improvements. Mezzanines.

Department Stores – Wood Frame

First Floor

Estimating Procedure

1. Use these square foot costs to estimate the cost of retail stores designed to sell a wider variety of goods. These buildings differ from discount houses in that they have more interior partitions and more elaborate interior and exterior finishes.
2. Establish the structure quality class by applying the information on page 99.
3. Compute the floor area. This should include everything within the building exterior walls and all inset areas outside the main walls but under the main building roof.
4. Add to or subtract from the square foot cost below the appropriate amount from the Wall Height Adjustment Row (near the bottom of this page) if the wall height is more or less than 16 feet.
5. Multiply the adjusted square foot cost by the building area.
6. Deduct, if appropriate, for common walls or no wall ownership, using the figures at the bottom of this page.
7. Multiply the total cost by the location factor on page 7.
8. Add the cost of appropriate additional components from page 201: heating and air conditioning systems, elevators and escalators, fire sprinklers, exterior signs, canopies and walks, paving and curbing, loading docks and ramps, miscellaneous yard improvements, and mezzanines.
9. Add the cost of second and higher floors and basements from page 101.

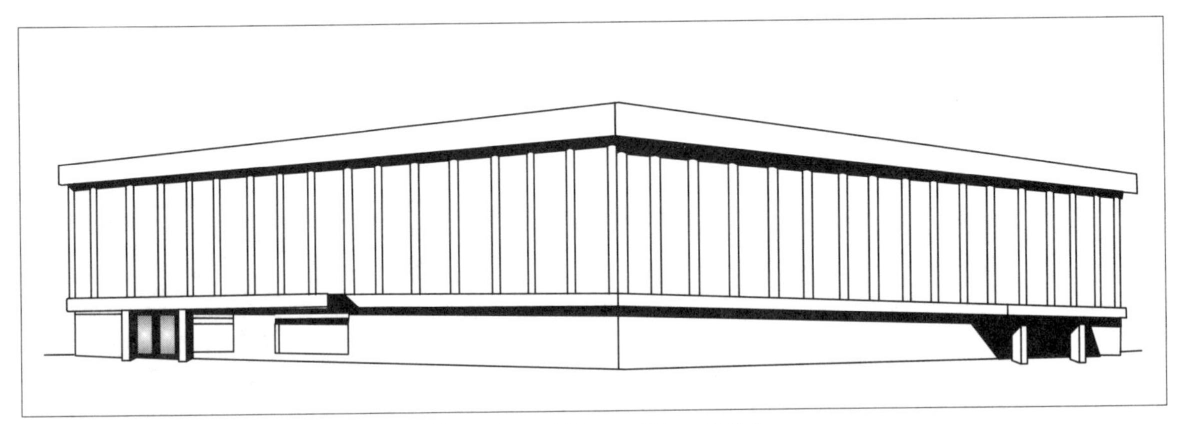

Department Store, Class 1 & 2

Square Foot Area

Quality Class	20,000	25,000	30,000	35,000	40,000	45,000	50,000	60,000	70,000	80,000	100,000
1, Best	110.73	106.23	102.59	99.59	97.06	94.86	92.94	89.76	87.17	85.03	81.64
1 & 2	99.48	95.46	92.18	89.49	87.22	85.25	83.54	80.65	78.33	76.44	73.37
2, Good	89.06	85.43	82.51	80.10	78.07	76.29	74.75	72.20	70.11	68.39	65.68
2 & 3	79.49	76.29	73.65	71.50	69.67	68.11	66.71	64.43	62.59	61.04	58.63
3, Average	71.41	68.50	66.14	64.20	62.59	61.15	59.91	57.84	56.19	54.83	52.66
3 & 4	62.00	59.49	57.45	55.77	54.34	53.12	52.03	50.25	48.82	47.62	45.71
4, Low	52.57	50.45	48.73	47.28	46.06	45.04	44.13	42.62	41.39	40.38	38.77
Wall Height Adjustment*	.31	.30	.29	.27	.26	.26	.24	.23	.22	.21	.14

***Wall Height Adjustment:** Add or subtract the amount listed in this table to the square foot of floor cost for each foot of wall height more or less than 16 feet.

Perimeter Wall Adjustment: For common wall, deduct $36.00 per linear foot. For no wall ownership, deduct $72.00 per linear foot.

Department Stores – Wood Frame

Upper Floors and Basements

Estimating Procedure

1. Establish the basement and upper floor quality class. The quality class will usually be the same as the first floor of the building. Square foot costs for unfinished basements will be nearly the same regardless of the structure quality class.
2. Compute the floor area.
3. Add to or subtract from the square foot cost below the appropriate amount from the Wall Height Adjustment Row for each foot of wall height more or less than 12 feet.
4. Multiply the adjusted square foot cost by the floor area.
5. Deduct, if appropriate, for common upper floor walls and walls not owned, using the Perimeter Wall Adjustment costs.
6. Multiply the total cost by the location factor on page 7.
7. Add the cost of appropriate additional components from page 201: heating and air conditioning systems, elevators and escalators, fire sprinklers, and mezzanines.
8. Add the total from this page to the page 100 to find the building cost.

Second and Higher Floors - Square Foot Area

Quality Class	20,000	25,000	30,000	35,000	40,000	45,000	50,000	60,000	70,000	80,000	100,000
1, Best	91.78	88.55	85.80	83.45	81.45	79.68	78.10	75.47	73.34	71.54	68.70
1 & 2	81.29	78.40	76.00	73.90	72.11	70.56	69.19	66.88	64.95	63.37	60.88
2, Good	72.14	69.62	67.44	65.62	64.03	62.64	61.41	59.35	57.68	56.26	54.00
2 & 3	64.11	61.86	59.91	58.29	56.88	55.64	54.56	52.75	51.21	49.98	48.00
3, Average	56.93	54.91	53.22	51.73	50.53	49.41	48.45	46.83	45.49	44.39	42.63
3 & 4	48.29	46.56	45.10	43.87	42.80	41.91	41.08	39.72	38.59	37.63	36.14
4, Low	39.49	38.06	36.89	35.92	35.05	34.28	33.64	32.50	31.56	30.79	29.55
Wall Height Adjustment*	.29	.26	.23	.21	.19	.16	.14	.12	.11	.09	.07

***Wall Height Adjustment:** Add or subtract the amount listed in this table to the square foot of floor cost for each foot of basement wall height more or less than 12 feet.

Perimeter Wall Adjustment: For common wall, deduct $90.00 per linear foot. For no wall ownership, deduct $180.00 per linear foot.

Finished Basements - Square Foot Area

Quality Class	20,000	25,000	30,000	35,000	40,000	45,000	50,000	60,000	70,000	80,000	100,000
1, Best	89.75	86.88	84.16	81.64	79.43	77.44	75.63	72.50	69.90	67.69	64.09
2, Good	71.96	69.67	67.45	65.47	63.69	62.06	60.63	58.14	56.04	54.26	51.39
3, Average	58.84	56.98	55.19	53.57	52.11	50.78	49.59	47.56	45.84	44.38	42.03
4, Low	40.70	39.41	38.16	37.07	36.03	35.12	34.30	32.90	31.71	30.69	29.09

Unfinished Basements

Area	20,000	25,000	30,000	35,000	40,000	45,000	50,000	60,000	70,000	80,000	100,000
Cost	22.99	22.09	21.37	20.77	20.24	19.80	19.48	18.82	18.32	17.90	17.23

Wall Height Adjustment: Add or subtract the amount listed in this table to the square foot of floor cost for each foot of basement wall height more or less than 12 feet

Area	20,000	25,000	30,000	35,000	40,000	45,000	50,000	60,000	70,000	80,000	100,000
Finished	.50	.48	.46	.45	.44	.43	.42	.40	.39	.39	.38
Unfinished	.48	.46	.44	.43	.42	.40	.40	.39	.38	.37	.36

General Office Buildings – Masonry or Concrete

Quality Classification

	Class 1 Best Quality	Class 2 Good Quality	Class 3 Average Quality	Class 4 Low Quality
Foundation (12% of total cost)	Reinforced concrete.	Reinforced concrete.	Reinforced concrete.	Reinforced concrete.
First Floor Structure (8% of total cost)	Reinforced concrete slab on grade or standard wood frame.	Reinforced concrete slab on grade or standard wood frame.	Reinforced concrete slab on grade or 4" x 6" girders with plywood sheathing.	Reinforced concrete slab on grade.
Upper Floor Structures (9% of total cost)	Standard wood frame, plywood and 1-1/2" light weight concrete sub floor.	Standard wood frame, plywood and 1-1/2" lightweight concrete sub floor.	Standard wood frame, 5/8" plywood sub floor.	Standard wood frame, 5/8" plywood sub floor.
Walls (10% of total cost)	8" decorative concrete block or 6" concrete tilt-up.	8" decorative concrete block or 6" concrete tilt-up.	8" reinforced concrete block or 8" reinforced brick or 8" clay tile.	8" reinforced concrete block or 8" clay tile.
Roof Structure (6% of total cost)	Standard wood frame, flat or low pitch.	Standard wood frame, flat or low pitch.	Standard wood frame, flat or low pitch.	Standard wood frame, flat or low pitch.
Exterior Wall Finish (8% of total cost)	Decorative block or large rock imbedded in tilt-up panels with 10 - 20% brick or stone veneer.	Decorative block or exposed aggregate and 10 - 20% brick or stone veneer.	Stucco or colored concrete block.	Painted concrete block or tile.
Windows (5% of total cost)	Average number in good aluminum frame. Fixed float glass in good frame on front side.	Average number in good aluminum frame. Some fixed float glass in front.	Average number of average aluminum sliding type.	Average number of low cost aluminum sliding type.
Roof Cover (5% of total cost)	5 ply built-up roofing on flat roofs. Heavy shake or tile on sloping roofs.	5 ply built-up roofing on flat roofs. Average shake or composition, tar and large rock on sloping roofs.	4 ply built-up roofing on flat roofs. Wood shingle or composition, tar and pea gravel on sloping roofs.	3 ply built-up roofing on flat roofs. Composition shingle on sloping roofs.
Overhang (3% of total cost)	3' closed overhang, fully guttered.	2' closed overhang, fully guttered.	None on flat roofs. 18" open on sloping roofs. Fully guttered.	None on flat roofs. 12" to 16" open on sloping roofs. Gutters over entrances.
Floor Finishes:				
Offices (3% of total cost)	Very good carpet.	Good carpet.	Average grade carpet.	Minimum grade tile.
Corridors (2% of total cost)	Solid vinyl tile or carpet.	Resilient tile.	Composition tile.	Minimum grade tile.
Bathrooms (1% of total cost)	Sheet vinyl or ceramic tile.	Sheet vinyl or ceramic tile.	Vinyl asbestos tile.	Minimum grade tile.
Interior Wall Finishes:				
Offices (6% of total cost)	Good hardwood veneer.	Hardwood veneer paneling or vinyl wall cover.	Gypsum wallboard, texture and paint.	Gypsum wallboard, texture and paint.
Corridors (4% of total cost)	Good hardwood veneer.	Gypsum wallboard and vinyl wall cover	Gypsum wallboard, texture and paint	Gypsum wallboard, texture and paint.
Bathrooms (2% of total cost)	Gypsum wallboard and enamel with ceramic tile wainscot.	Gypsum wallboard and enamel or vinyl wall covering	Gypsum wallboard and enamel.	Gypsum wallboard and texture and paint.
Ceiling Finish (4% of total cost)	Suspended "T" bar and acoustical tile.	Gypsum wallboard and acoustical tile.	Gypsum wallboard and acoustical texture.	Gypsum walboard and paint.
Plumbing (6% of total cost)	Copper tubing and top quality fixtures.	Copper tubing and good fixtures.	Copper tubing and standard fixtures.	Copper tubing and economy fixtures.
Lighting (6% of total cost)	Conduit wiring, good fixtures.	Conduit wiring, good fixtures.	Romex or conduit wiring, average fixtures.	Romex wiring, economy fixtures.

Note: Use the percent of total cost to help identify the correct quality classification.

Square foot costs include the following components: Foundations as required for normal soil conditions. Floor, wall and roof structures. Interior ceiling, wall and floor finishes. Exterior wall finish and roof cover. Interior partitions. Cabinets, doors and windows. Basic electrical systems and lighting fixtures. Rough plumbing and fixtures. Permits and fees. Contractor's mark-up.

General Office Buildings – Masonry or Concrete

Exterior Suite Entrances, Length Less Than Twice Width

Estimating Procedure

1. Use these figures to estimate general office buildings in which access to each suite is through an exterior entrance. Medical and dental offices have smaller rooms and more plumbing fixtures than general offices and should be estimated with figures from the Medical and Dental Buildings Section. See page 118.
2. Establish the building quality class by applying the information on page 102.
3. Compute the first floor area. This should include everything within the exterior walls and all insets outside the main walls but under the main roof.
4. If the first floor wall height is more or less than 10 feet, add to or subtract from the first floor square foot cost below the appropriate amount from the Wall Height Adjustment Table on page 109.
5. Multiply the adjusted square foot cost by the first floor area.
6. Deduct, if appropriate, for common walls or no wall finish, Use the figures on page 109.
7. If there are second or higher floors, compute the square foot area on each floor. Locate the appropriate square foot cost from the table at the bottom of this page. Adjust this figure for a wall height more or less than 9 feet, using the figures on page 109. Multiply the adjusted cost by the square foot area on each floor. Use the figures on page 109 to deduct for common walls or no wall finish. Add the result to the cost from step 6 above.
8. Multiply the total cost by the location factor on page 7.
9. Add the cost of heating and air conditioning systems, elevators, fire sprinklers, exterior signs, paving and curbing, miscellaneous yard improvements, covered porches and garages. See pages 201 to 213.

First Story - Square Foot Area

Quality Class	1,000	1,500	2,000	2,500	3,000	4,000	5,000	7,500	10,000	15,000	20,000
Exceptional	158.00	145.02	137.54	132.57	128.95	123.95	120.62	115.49	112.50	109.01	106.95
1, Best	145.78	133.78	126.91	122.28	118.96	114.34	111.26	106.56	103.79	100.55	98.67
1 & 2	132.33	121.45	115.22	111.02	108.01	103.83	101.02	96.75	94.26	91.31	89.61
2, Good	122.41	112.34	106.55	102.66	99.89	96.02	93.41	89.47	87.15	84.44	82.85
2 & 3	111.71	102.53	97.23	93.74	91.17	87.65	85.28	81.66	79.54	77.07	75.62
3, Average	103.32	94.81	89.90	86.65	84.32	81.04	78.84	75.50	73.55	71.27	69.93
3 & 4	93.46	85.76	81.34	78.40	76.25	73.30	71.32	68.30	66.54	64.45	63.26
4, Low	82.74	75.93	72.04	69.42	67.53	64.90	63.18	60.50	58.93	57.07	56.01

Second and Higher Stories - Square Foot Area

Quality Class	1,000	1,500	2,000	2,500	3,000	4,000	5,000	7,500	10,000	15,000	20,000
Exceptional	146.08	134.89	128.37	123.96	120.73	116.27	113.25	108.59	105.83	102.61	100.70
1, Best	134.62	124.30	118.29	114.24	111.26	107.12	104.37	100.05	97.51	94.54	92.76
1 & 2	122.98	113.56	108.07	104.37	101.64	97.88	95.30	91.40	89.08	86.35	84.76
2, Good	114.53	105.78	100.66	97.19	94.68	91.14	88.80	85.12	82.97	80.43	78.91
2 & 3	104.24	96.29	91.60	88.46	86.15	82.96	80.80	77.49	75.50	73.20	71.83
3, Average	95.57	88.25	83.05	81.08	78.99	76.07	74.05	71.02	69.21	67.13	65.86
3 & 4	85.55	79.02	75.19	72.59	70.71	68.08	66.35	63.60	61.99	60.11	58.97
4, Low	76.23	70.40	66.98	64.66	62.99	60.63	59.07	56.64	55.21	53.51	52.52

General Office Buildings – Masonry or Concrete

Exterior Suite Entrances, Length Between 2 and 4 Times Width

Estimating Procedure

1. Use these figures to estimate general office buildings in which access to each suite is through an exterior entrance. Medical and dental offices have smaller rooms and more plumbing fixtures than general offices and should be estimated with figures from the Medical and Dental Buildings Section. See page 118.
2. Establish the building quality class by applying the information on page 102.
3. Compute the first floor area. This should include everything within the exterior walls and all insets outside the main walls but under the main roof.
4. If the first floor wall height is more or less than 10 feet, add to or subtract from the first floor square foot cost below the appropriate amount from the Wall Height Adjustment Table on page 109.
5. Multiply the adjusted square foot cost by the first floor area.
6. Deduct, if appropriate, for common walls or no wall finish. Use the figures on page 109.
7. If there are second or higher floors, compute the square foot area on each floor. Locate the appropriate square foot cost from the table at the bottom of this page. Adjust this figure for a wall height more or less than 9 feet, using the figures on page 109. Multiply the adjusted cost by the square foot area on each floor. Use the figures on page 109 to deduct for common walls or no wall finish. Add the result to the cost from step 6 above.
8. Multiply the total cost by the location factor on page 7.
9. Add the cost of heating and air conditioning systems, elevators, fire sprinklers, exterior signs, paving and curbing, miscellaneous yard improvements, covered porches and garages. See pages 201 to 213.

First Story - Square Foot Area

Quality Class	1,000	1,500	2,000	2,500	3,000	4,000	5,000	7,500	10,000	15,000	20,000
Exceptional	171.65	154.98	145.55	139.36	134.89	128.81	124.79	118.67	115.13	111.06	108.68
1, Best	158.34	142.98	134.31	128.57	124.46	118.85	115.12	109.49	106.24	102.46	100.29
1 & 2	143.96	129.97	122.10	116.86	113.12	108.05	104.65	99.56	96.58	93.17	91.14
2, Good	133.06	120.14	112.86	108.04	104.59	99.85	96.73	92.01	89.27	86.11	84.28
2 & 3	121.61	109.78	103.10	98.73	95.57	91.25	88.40	84.07	81.59	78.68	76.99
3, Average	112.32	101.52	95.33	91.26	88.37	84.39	81.71	77.72	75.41	72.73	71.20
3 & 4	101.99	92.11	86.51	82.80	80.18	76.53	74.16	70.54	68.43	66.00	64.60
4, Low	90.60	81.80	76.82	73.54	71.20	67.98	65.86	62.64	60.79	58.63	57.37

Second and Higher Stories - Square Foot Area

Quality Class	1,000	1,500	2,000	2,500	3,000	4,000	5,000	7,500	10,000	15,000	20,000
Exceptional	158.53	143.50	135.03	129.47	125.44	119.97	116.35	110.86	107.70	104.06	101.90
1, Best	146.10	132.27	124.44	119.30	115.63	110.57	107.23	102.17	99.26	95.91	93.92
1 & 2	133.89	121.20	114.03	109.32	105.92	101.30	98.23	93.63	90.94	87.85	86.08
2, Good	124.50	112.67	106.06	101.64	98.50	94.21	91.36	87.05	84.56	81.69	80.01
2 & 3	113.21	102.47	96.42	92.44	89.57	85.63	83.06	79.15	76.89	74.29	72.78
3, Average	103.41	93.61	88.08	84.45	81.84	78.26	75.90	72.33	70.25	67.88	66.48
3 & 4	93.95	85.08	80.04	76.71	74.33	71.09	68.93	65.71	63.82	61.64	60.40
4, Low	83.16	75.26	69.56	67.91	65.77	62.92	61.01	58.14	56.51	54.57	53.43

General Office Buildings – Masonry or Concrete

Exterior Suite Entrances, Length More Than 4 Times Width

Estimating Procedure

1. Use these figures to estimate general office buildings in which access to each suite is through an exterior entrance. Medical and dental offices have smaller rooms and more plumbing fixtures than general offices and should be estimated with figures from the Medical and Dental Buildings Section. See page 118.
2. Establish the building quality class by applying the information on page 102.
3. Compute the first floor area. This should include everything within the exterior walls and all insets outside the main walls but under the main roof.
4. If the first floor wall height is more or less than 10 feet, add to or subtract from the first floor square foot cost below the appropriate amount from the Wall Height Adjustment Table on page 109.
5. Multiply the adjusted square foot cost by the first floor area.
6. Deduct, if appropriate, for common walls or no wall finish. Use the figures on page 109.
7. If there are second or higher floors, compute the square foot area on each floor. Locate the appropriate square foot cost from the table at the bottom of this page. Adjust this figure for a wall height more or less than 9 feet, using the figures on page 109. Multiply the adjusted cost by the square foot area on each floor. Use the figures on page 109 to deduct for common walls or no wall finish. Add the result to the cost from step 6 above.
8. Multiply the total cost by the location factor on page 7.
9. Add the cost of heating and air conditioning systems, elevators, fire sprinklers, exterior signs, paving and curbing, miscellaneous yard improvements, covered porches and garages. See pages 201 to 213.

First Story - Square Foot Area

Quality Class	1,000	1,500	2,000	2,500	3,000	4,000	5,000	7,500	10,000	15,000	20,000
Exceptional	187.54	167.30	155.91	148.44	143.06	135.77	130.94	123.62	119.44	114.58	111.76
1, Best	172.55	153.90	143.44	136.55	131.62	124.91	120.45	113.73	109.87	105.39	102.82
1 & 2	157.27	140.29	130.73	124.47	119.97	113.84	109.78	103.66	100.15	96.07	93.72
2, Good	144.97	129.29	120.51	114.74	110.59	104.92	101.21	95.57	92.30	88.55	86.37
2 & 3	132.69	118.37	110.28	105.01	101.23	96.04	92.60	87.44	84.49	81.07	79.08
3, Average	123.30	110.02	102.52	97.59	94.08	89.29	86.10	81.30	78.54	75.32	73.48
3 & 4	111.59	99.55	92.76	88.31	85.13	80.79	77.92	73.55	71.07	68.18	66.48
4, Low	99.28	88.56	82.54	78.57	75.76	71.87	69.32	65.47	63.23	60.63	59.15

Second and Higher Stories - Square Foot Area

Quality Class	1,000	1,500	2,000	2,500	3,000	4,000	5,000	7,500	10,000	15,000	20,000
Exceptional	172.21	153.79	143.46	136.67	131.84	125.20	120.82	114.26	110.45	106.08	103.53
1, Best	158.84	141.85	132.34	126.06	121.60	115.49	111.42	105.36	101.89	97.83	95.48
1 & 2	145.57	129.99	121.28	115.53	111.43	105.82	102.14	96.58	93.37	89.67	87.49
2, Good	135.85	121.35	113.17	107.82	104.01	98.80	95.30	90.14	87.14	83.67	79.28
2 & 3	124.09	110.81	103.36	98.49	94.98	90.21	87.06	82.31	79.58	76.42	74.59
3, Average	114.55	102.33	95.44	90.94	87.68	83.28	80.37	76.01	73.47	70.55	68.86
3 & 4	103.90	92.80	86.57	82.48	79.54	75.54	72.89	68.92	66.65	64.01	62.44
4, Low	92.72	82.77	77.20	73.55	70.95	67.37	65.05	61.50	59.46	57.10	55.71

Interior Suite Entrances, Length Less Than Twice Width

Estimating Procedure

1. Use these figures to estimate general office buildings in which access to each suite is through an interior corridor. Medical and dental offices have smaller rooms and more plumbing fixtures than general offices and should be estimated with figures from the Medical and Dental Buildings Section. See page 118.
2. Establish the building quality class by applying the information on page 102.
3. Compute the first floor area. This should include everything within the exterior walls and all insets outside the main walls but under the main roof.
4. If the first floor wall height is more or less than 10 feet, add to or subtract from the first floor square foot cost below the appropriate amount from the Wall Height Adjustment Table on page 109.
5. Multiply the adjusted square foot cost by the first floor area.
6. Deduct, if appropriate, for common walls or no wall finish. Use the figures on page 109.
7. If there are second or higher floors, compute the square foot area on each floor. Locate the appropriate square foot cost from the table at the bottom of this page. Adjust this figure for a wall height more or less than 9 feet, using the figures on page 109. Multiply the adjusted cost by the square foot area on each floor. Use the figures on page 109 to deduct for common walls or no wall finish. Add the result to the cost from step 6 above.
8. Multiply the total cost by the location factor on page 7.
9. Add the cost of heating and air conditioning systems, elevators, fire sprinklers, exterior signs, paving and curbing, miscellaneous yard improvements, covered porches and garages. See pages 201 to 213.

First Story - Square Foot Area

Quality Class	2,000	2,500	3,000	4,000	5,000	7,500	10,000	15,000	20,000	30,000	40,000
Exceptional	133.41	128.52	124.96	120.06	116.78	111.77	108.83	105.41	103.40	101.07	99.67
1, Best	123.03	118.51	115.23	110.71	107.67	103.04	100.37	97.21	95.37	93.19	91.92
1 & 2	112.82	108.69	105.68	101.55	98.74	94.53	92.05	89.15	87.43	85.46	84.31
2, Good	105.39	101.55	98.74	94.86	92.26	88.29	85.99	83.30	81.70	79.86	78.75
2 & 3	96.33	92.79	90.21	86.66	84.31	80.68	78.57	76.12	74.66	72.97	71.98
3, Average	89.03	85.77	83.39	80.12	77.92	74.60	72.63	70.36	69.01	67.46	66.49
3 & 4	80.40	77.48	75.33	72.39	70.37	67.35	65.60	63.55	62.34	60.90	60.09
4, Low	71.07	68.46	66.58	63.95	62.24	59.52	57.98	56.16	55.09	53.85	53.08

Second and Higher Stories - Square Foot Area

Quality Class	2,000	2,500	3,000	4,000	5,000	7,500	10,000	15,000	20,000	30,000	40,000
Exceptional	124.59	120.20	116.97	112.53	109.56	105.02	102.34	99.22	97.38	95.21	93.92
1, Best	114.93	110.87	107.93	103.81	101.09	96.85	94.40	91.53	89.82	87.82	86.64
1 & 2	105.74	102.01	99.29	95.55	93.00	89.13	86.87	84.20	82.64	80.79	79.72
2, Good	98.91	94.79	92.89	89.36	86.98	83.39	81.24	78.77	77.31	75.59	74.57
2 & 3	90.14	86.95	84.61	81.42	79.26	75.97	74.02	71.77	70.45	68.87	68.25
3, Average	82.79	79.88	77.76	74.79	72.82	69.79	68.01	65.94	64.71	63.27	62.41
3 & 4	74.56	71.90	69.99	67.32	65.55	62.83	61.21	59.37	58.26	56.96	56.20
4, Low	65.70	63.37	61.66	59.35	57.77	55.36	53.97	52.31	51.35	50.19	49.54

Interior Suite Entrances, Length Between 2 and 4 Times Width

Estimating Procedure

1. Use these figures to estimate general office buildings in which access to each suite is through an interior corridor. Medical and dental offices have smaller rooms and more plumbing fixtures than general offices and should be estimated with figures from the Medical and Dental Buildings Section. See page 118.
2. Establish the building quality class by applying the information on page 102.
3. Compute the first floor area. This should include everything within the exterior walls and all insets outside the main walls but under the main roof.
4. If the first floor wall height is more or less than 10 feet, add to or subtract from the first floor square foot cost below the appropriate amount from the Wall Height Adjustment Table on page 109.
5. Multiply the adjusted square foot cost by the first floor area.
6. Deduct, if appropriate, for common walls or no wall finish. Use the figures on page 109.
7. If there are second or higher floors, compute the square foot area on each floor. Locate the appropriate square foot cost from the table at the bottom of this page. Adjust this figure for a wall height more or less than 9 feet, using the figures on page 109. Multiply the adjusted cost by the square foot area on each floor. Use the figures on page 109 to deduct for common walls or no wall finish. Add the result to the cost from step 6 above.
8. Multiply the total cost by the location factor on page 7.
9. Add the cost of heating and air conditioning systems, elevators, fire sprinklers, exterior signs, paving and curbing, miscellaneous yard improvements, covered porches and garages. See pages 201 to 213.

General Offices, Class 1 & 2

First Story - Square Foot Area

Quality Class	2,000	2,500	3,000	4,000	5,000	7,500	10,000	15,000	20,000	30,000	40,000
Exceptional	141.42	135.36	131.01	125.04	121.06	115.03	111.55	107.49	105.11	102.32	100.70
1, Best	130.43	124.87	120.85	115.35	111.68	106.14	102.90	99.16	96.93	94.38	92.89
1 & 2	119.48	114.49	110.82	105.77	102.41	97.29	94.35	90.89	88.90	86.55	85.15
2, Good	104.83	100.45	97.33	93.04	90.19	85.85	83.32	80.38	78.67	76.68	75.47
2 & 3	102.16	97.81	94.67	90.34	87.46	83.12	80.60	77.63	75.95	73.92	72.75
3, Average	94.51	90.47	87.57	83.57	80.92	76.89	74.56	71.81	70.25	68.39	67.28
3 & 4	85.50	81.85	79.23	75.61	73.23	69.58	67.46	65.01	63.55	61.88	60.87
4, Low	75.66	72.43	70.07	66.91	64.80	61.54	59.70	57.50	56.24	54.76	53.87

Second and Higher Stories - Square Foot Area

Quality Class	2,000	2,500	3,000	4,000	5,000	7,500	10,000	15,000	20,000	30,000	40,000
Exceptional	131.55	126.12	122.15	116.79	113.18	107.74	104.56	100.88	98.73	96.20	94.71
1, Best	121.26	116.21	112.61	107.64	104.32	99.30	96.39	93.01	91.02	88.69	87.33
1 & 2	111.66	107.05	103.70	99.13	96.08	91.46	88.78	85.63	83.82	81.66	80.43
2, Good	104.83	100.45	97.33	93.04	90.19	85.85	83.32	80.38	78.67	76.68	75.47
2 & 3	95.41	91.46	88.61	84.72	82.11	78.14	75.85	73.16	71.63	69.79	68.73
3, Average	87.72	84.14	81.54	77.94	75.55	71.91	69.80	67.33	65.92	64.24	63.26
3 & 4	79.29	76.01	73.64	70.39	68.22	64.94	63.02	60.80	59.51	57.98	57.10
4, Low	70.07	67.17	65.09	62.24	60.31	57.39	55.70	53.74	52.59	51.27	50.47

General Office Buildings – Masonry or Concrete

Interior Suite Entrances, Length More Than 4 Times Width

Estimating Procedure

1. Use these figures to estimate general office buildings in which access to each suite is through an interior corridor. Medical and dental offices have smaller rooms and more plumbing fixtures than general offices and should be estimated with figures from the Medical and Dental Buildings Section. See page 118.
2. Establish the building quality class by applying the information on page 102.
3. Compute the first floor area. This should include everything within the exterior walls and all insets outside the main walls but under the main roof.
4. If the first floor wall height is more or less than 10 feet, add to or subtract from the first floor square foot cost below the appropriate amount from the Wall Height Adjustment Table on page 109.
5. Multiply the adjusted square foot cost by the first floor area.
6. Deduct, if appropriate, for common walls or no wall finish. Use the figures on page 109.
7. If there are second or higher floors, compute the square foot area on each floor. Locate the appropriate square foot cost from the table at the bottom of this page. Adjust this figure for a wall height more or less than 9 feet, using the figures on page 109. Multiply the adjusted cost by the square foot area on each floor. Use the figures on page 109 to deduct for common walls or no wall finish. Add the result to the cost from step 6 above.
8. Multiply the total cost by the location factor on page 7.
9. Add the cost of heating and air conditioning systems, elevators, fire sprinklers, exterior signs, paving and curbing, miscellaneous yard improvements, covered porches and garages. See pages 201 to 213.

First Story - Square Foot Area

Quality Class	2,000	2,500	3,000	4,000	5,000	7,500	10,000	15,000	20,000	30,000	40,000
Exceptional	151.48	143.97	138.60	131.30	126.46	119.22	115.04	110.24	107.46	104.23	102.34
1, Best	139.76	132.82	127.86	121.11	116.69	109.98	106.15	101.69	99.15	96.17	94.41
1 & 2	128.13	121.77	117.22	111.05	106.95	100.83	97.28	93.25	90.88	88.14	86.56
2, Good	119.68	113.74	109.46	103.73	99.91	94.18	90.87	87.08	84.88	82.34	80.85
2 & 3	109.39	103.94	100.05	94.76	91.30	86.05	83.05	79.58	77.57	75.25	73.89
3, Average	101.43	96.39	92.80	87.89	84.68	79.82	77.01	73.83	71.93	69.79	68.53
3 & 4	91.91	87.36	84.08	79.66	76.74	72.35	69.81	66.90	65.20	63.24	62.07
4, Low	81.62	77.58	74.68	70.73	68.16	64.22	61.99	59.40	57.91	56.16	55.17

Second and Higher Stories - Square Foot Area

Quality Class	2,000	2,500	3,000	4,000	5,000	7,500	10,000	15,000	20,000	30,000	40,000
Exceptional	139.09	132.27	127.49	121.10	117.00	111.04	107.70	103.95	101.83	99.43	98.07
1, Best	127.25	121.00	116.60	110.83	107.07	101.60	98.56	95.11	93.18	91.00	89.74
1 & 2	117.91	112.15	108.09	102.69	99.23	94.16	91.34	88.14	86.33	84.33	83.13
2, Good	111.01	105.58	101.75	96.68	93.40	88.63	85.98	82.99	81.29	79.39	78.27
2 & 3	101.04	96.08	92.59	87.99	85.03	80.67	78.26	75.53	73.97	72.25	71.27
3, Average	93.46	88.89	85.65	81.40	78.64	74.62	72.40	69.86	68.44	66.83	65.91
3 & 4	92.05	87.51	84.38	80.16	77.42	73.49	71.29	68.78	67.40	65.79	64.88
4, Low	75.02	71.34	68.77	65.34	63.11	59.90	58.10	56.09	54.94	53.65	52.90

General Office Buildings – Masonry or Concrete

Wall Height Adjustments

The square foot costs for general offices are based on the wall heights of 10 feet for first floors and 9 feet for higher floors. The main or first floor height is the distance from the bottom of the floor slab or joists to the top of the roof slab or ceiling joists. Second and higher floors are measured from the top of the floor slab or floor joists to the top of the roof slab or ceiling joists. Add or subtract the amount listed in this table to the square foot of floor cost for each foot of wall height more or less than 10 feet, if adjusting for a first floor, and 9 feet, if adjusting for upper floors.

Square Foot Area

Quality Class	1,000	1,500	2,000	3,000	4,000	5,000	7,500	10,000	15,000	20,000	40,000
1, Best	4.23	3.31	2.81	2.23	1.89	1.69	1.34	1.16	.95	.80	.59
2, Good	3.68	2.87	2.42	1.92	1.66	1.46	1.18	1.01	.81	.71	.49
3, Average	3.24	2.54	2.14	1.72	1.46	1.28	1.01	.87	.73	.62	.43
4, Low	2.87	2.22	1.91	1.52	1.29	1.14	.90	.80	.66	.51	.39

Perimeter Wall Adjustment

A common wall exists when two buildings share one wall. Adjust for common walls by deducting the linear foot cost below from the total structure cost. In some structures one or more walls are not owned at all. In this case, deduct the "No Ownership" cost per linear foot of wall not owned. Where a perimeter wall remains unfinished, deduct the "lack of exterior finish" cost.

First Story

Class	For a Common Wall, Deduct Per L.F.	For No Wall Ownership, Deduct Per L.F.	For Lack of Exterior Finish, Deduct Per L.F.
1	$218.00	$437.00	$169.00
2	179.00	359.00	107.00
3	135.00	263.00	68.00
4	104.00	208.00	39.00

Second and Higher Stories

Class	For a Common Wall, Deduct Per L.F.	For No Wall Ownership, Deduct Per L.F.	For Lack of Exterior Finish, Deduct Per L.F.
1	$208.00	$414.00	$169.00
2	179.00	348.00	107.00
3	135.00	263.00	68.00
4	129.00	252.00	39.00

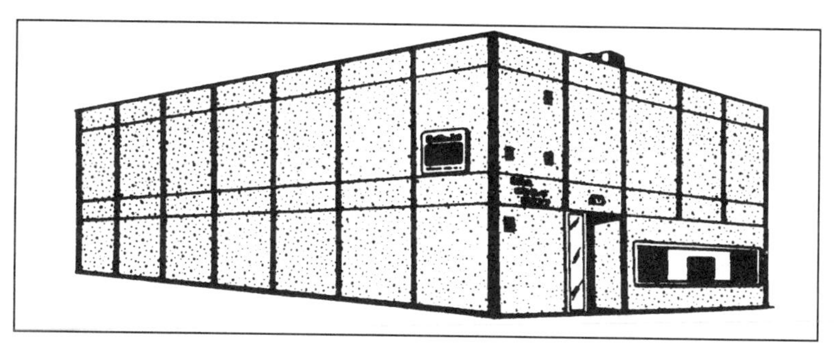

General Office, Class 3

General Office Buildings – Wood Frame

Quality Classification

	Class 1 Best Quality	Class 2 Good Quality	Class 3 Average Quality	Class 4 Low Quality
Foundation (12% of total cost)	Reinforced concrete.	Reinforced concrete.	Reinforced concrete.	Reinforced concrete.
First Floor Structure (8% of total cost)	Reinforced concrete slab on grade or standard wood frame.	Reinforced concrete slab on grade or standard wood frame.	Reinforced concrete slab on grade or 4" x 6" girders with 2" T&G subfloor.	Reinforced concrete slab on grade.
Upper Floor Structure (10% of total cost)	Standard wood frame. Plywood and 1-1/2" light-weight concrete subfloor.	Standard wood frame. Plywood and 1-1/2" light- weight concrete subfloor.	Standard wood frame. 5/8" plywood subfloor.	Standard wood frame. 5/8" plywood subfloor.
Walls (9% of total cost)	Standard wood frame.	Standard wood frame.	Standard wood frame.	Standard wood frame.
Roof Structure (6% of total cost)	Standard wood frame, flat or low pitch.	Standard wood frame, flat or low pitch.	Standard wood frame, flat or low pitch.	Standard wood frame, flat or low pitch.
Exterior Wall Finish (8% of total cost)	Good wood siding with 10 to 20% brick or stone veneer.	Average wood siding or stucco and 10 to 20% brick or stone veneer.	Stucco with some wood trim or cheap wood siding.	Stucco.
Windows (5% of total cost)	Average number in good aluminum frame. Fixed float glass in good frame on front side.	Average number in good aluminum frame. Some fixed float glass in front.	Average number of average aluminum sliding type.	Average number of low cost aluminum sliding type.
Roof Cover (5% of total cost)	5 ply built-up roofing on flat roofs. Heavy shake or tile on sloping roofs.	5 ply built-up roofing on flat roofs. Average shake or composition, tar and large rock on sloping roofs.	4 ply built-up roofing on flat roofs. Wood shingle or composition, tar and pea gravel on sloping roofs.	3 ply built-up roofing on flat roofs. Composition shingle on sloping roofs.
Overhang (3% of total cost)	3' closed overhang, fully guttered.	2' closed overhang, fully guttered.	None on flat roofs. 18" on sloping roofs, fully guttered.	None on flat roofs. 12" to 16" open on sloping roofs, gutters over entrances.
Floor Finishes: Offices (3% of total cost)	Very good carpet.	Good carpet.	Average grade carpet.	Minimum grade tile.
Corridors (2% of total cost)	Solid vinyl tile or carpet.	Resilient tile.	Composition tile.	Minimum grade tile.
Bathrooms (1% of total cost)	Sheet vinyl or ceramic tile.	Sheet vinyl or ceramic tile.	Composition tile.	Minimum grade tile.
Interior Wall Finish: Offices (6% of total cost)	Good hardwood veneer paneling.	Hardwood veneer paneling or vinyl wall cover.	Gypsum wallboard, texture and paint.	Gypsum wallboard, texture and paint.
Corridors (4% of total cost)	Good hardwood veneer paneling.	Gypsum wallboard and vinyl wall cover.	Gypsum wallboard, texture and paint.	Gypsum wallboard, texture and paint.
Bathrooms (2% of total cost)	Gypsum wallboard and enamel with ceramic tile wainscot.	Gypsum wallboard and enamel or vinyl wall covering.	Gypsum wallboard and enamel.	Gypsum wallboard, texture and paint.
Ceiling Finish (4% of total cost)	Suspended "T" bar and acoustical tile.	Gypsum wallboard and acoustical tile.	Gypsum wallboard and acoustical texture.	Gypsum wallboard and paint.
Plumbing (6% of total cost)	Copper tubing, top quality fixtures.	Copper tubing, good fixtures.	Copper tubing, standard fixtures.	Copper tubing, economy fixtures.
Lighting (6% of total cost)	Conduit wiring, good fixtures.	Conduit wiring, good fixtures.	Romex or conduit wiring, average fixtures.	Romex wiring economy fixtures.

Note: Use the percent of total cost to help identify the correct quality classification.

Square foot costs include the following components: Foundations as required for normal soil conditions. Floor, wall and roof structures. Interior floor, wall and ceiling finishes. Exterior wall finish and roof cover. Interior partitions. Cabinets, doors and windows. Basic electrical systems and lighting fixtures. Rough plumbing and fixtures. Permits and fees. Contractors' mark-up.

General Office Buildings - Wood Frame

Exterior Suite Entrances, Length Less Than Twice Width

Estimating Procedure

1. Use these figures to estimate general office buildings in which access to each suite is through an exterior entrance. Medical and dental offices have smaller rooms and more plumbing fixtures than general offices and should be estimated with figures from the Medical and Dental Buildings Section. See page 118.
2. Establish the building quality class by applying the information on page 110.
3. Compute the first floor area. This should include everything within the exterior walls and all insets outside the main walls but under the main roof.
4. If the first floor wall height is more or less than 10 feet, add to or subtract from the first floor square foot cost below the appropriate amount from the Wall Height Adjustment Table on page 117.
5. Multiply the adjusted square foot cost by the first floor area.
6. Deduct, if appropriate, for common walls or no wall finish. Use the figures on page 117.
7. If there are second or higher floors, compute the square foot area on each floor. Locate the appropriate square foot cost from the table at the bottom of this page. Adjust this figure for a wall height more or less than 9 feet, using the figures on page 117. Multiply the adjusted cost by the square foot area on each floor. Use the figures on page 117 to deduct for common walls or no wall finish. Add the result to the cost from step 6 above.
8. Multiply the total cost by the location factor on page 7.
9. Add the cost of heating and air conditioning systems, elevators, fire sprinklers, exterior signs, paving and curbing, miscellaneous yard improvements, covered porches and garages. See pages 201 to 213.

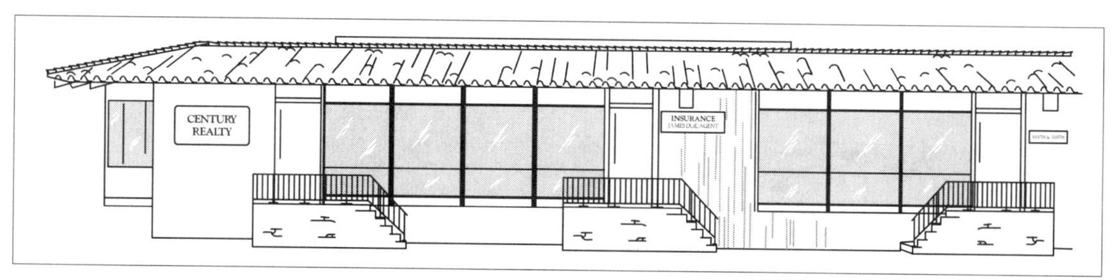

General Offices, Class 1 & 2

First Story - Square Foot Area

Quality Class	1,000	1,500	2,000	2,500	3,000	4,000	5,000	7,500	10,000	15,000	20,000
Exceptional	131.50	124.35	120.26	117.51	115.52	112.79	110.97	108.19	106.58	104.66	103.56
1, Best	121.35	114.72	110.95	108.43	106.59	104.08	102.40	99.84	98.32	96.56	95.52
1 & 2	109.91	103.89	100.48	98.21	96.52	94.26	92.73	90.42	89.05	87.48	86.54
2, Good	100.13	94.69	91.57	89.45	88.00	85.88	84.50	82.39	81.13	79.70	78.86
2 & 3	91.38	86.42	83.56	81.65	80.27	78.38	77.09	75.20	74.06	72.74	71.95
3, Average	83.61	79.05	76.43	74.70	73.44	71.67	70.53	68.76	67.73	66.52	65.83
3 & 4	74.72	70.65	68.29	66.76	65.61	64.09	63.05	61.46	60.55	59.45	58.81
4, Low	65.38	61.86	59.80	58.42	57.46	56.08	55.18	53.81	52.99	52.07	51.49

Second and Higher Stories - Square Foot Area

Quality Class	1,000	1,500	2,000	2,500	3,000	4,000	5,000	7,500	10,000	15,000	20,000
Exceptional	114.90	110.30	107.66	105.91	104.64	102.91	101.72	99.96	98.93	97.71	97.00
1, Best	105.86	101.59	99.16	97.59	96.42	94.80	93.71	92.08	91.13	89.99	89.38
1 & 2	96.61	92.77	90.55	89.08	88.01	86.54	85.56	84.07	83.18	82.18	81.57
2, Good	88.25	84.73	82.71	81.37	80.38	79.04	78.13	76.82	76.00	75.03	74.51
2 & 3	79.61	76.43	74.62	73.38	72.51	71.32	70.51	69.27	68.54	67.70	67.22
3, Average	71.56	68.70	67.09	65.98	65.20	64.12	63.39	62.26	61.64	60.88	60.45
3 & 4	63.32	60.78	59.34	58.38	57.67	56.73	56.07	55.10	54.51	53.87	53.48
4, Low	55.11	52.92	51.65	50.80	50.19	49.36	48.79	47.93	47.44	46.87	46.53

General Office Buildings – Wood Frame

Exterior Suite Entrances, Length Between 2 and 4 Times Width

Estimating Procedure

1. Use these figures to estimate general office buildings in which access to each suite is through an exterior entrance. Medical and dental offices have smaller rooms and more plumbing fixtures than general offices and should be estimated with figures from the Medical and Dental Buildings Section. See page 118.
2. Establish the building quality class by applying the information on page 110.
3. Compute the first floor area. This should include everything within the exterior walls and all insets outside the main walls but under the main roof.
4. If the first floor wall height is more or less than 10 feet, add to or subtract from the first floor square foot cost below the appropriate amount from the Wall Height Adjustment Table on page 117.
5. Multiply the adjusted square foot cost by the first floor area.
6. Deduct, if appropriate, for common walls or no wall finish. Use the figures on page 117.
7. If there are second or higher floors, compute the square foot area on each floor. Locate the appropriate square foot cost from the table at the bottom of this page. Adjust this figure for a wall height more or less than 9 feet, using the figures on page 117. Multiply the adjusted cost by the square foot area on each floor. Use the figures on page 117 to deduct for common walls or no wall finish. Add the result to the cost from step 6 above.
8. Multiply the total cost by the location factor on page 7.
9. Add the cost of heating and air conditioning systems, elevators, fire sprinklers, exterior signs, paving and curbing, miscellaneous yard improvements, covered porches and garages. See pages 201 to 213.

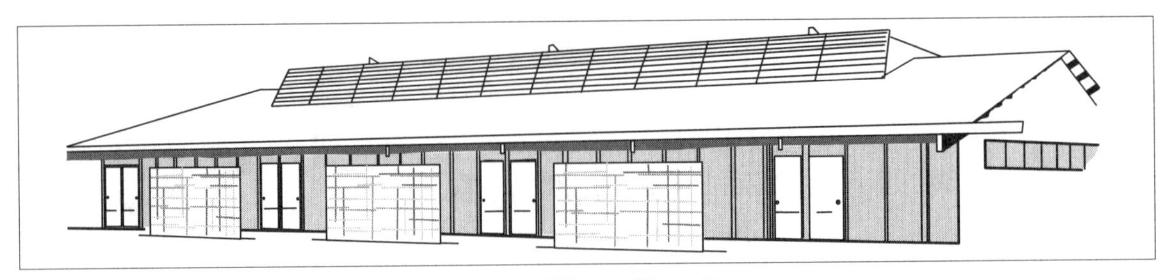

General Offices, Class 3

First Story - Square Foot Area

Quality Class	1,000	1,500	2,000	2,500	3,000	4,000	5,000	7,500	10,000	15,000	20,000
Exceptional	138.94	129.84	124.68	121.31	118.85	115.52	113.34	110.01	108.07	105.85	104.54
1, Best	128.14	119.76	115.01	111.90	109.64	106.57	104.55	101.46	99.67	97.65	96.44
1 & 2	115.97	108.38	104.09	101.25	99.24	96.45	94.62	91.84	90.22	88.33	87.26
2, Good	105.71	98.79	94.86	92.27	90.44	87.90	86.23	83.69	82.22	80.55	79.54
2 & 3	96.56	90.25	86.67	84.31	82.64	80.34	78.78	76.49	75.14	73.60	72.68
3, Average	88.36	82.55	79.29	77.13	75.60	73.49	72.06	69.93	68.72	67.31	66.48
3 & 4	79.11	73.90	70.99	69.06	67.66	65.78	64.53	62.62	61.53	60.25	59.52
4, Low	69.36	64.77	62.21	60.56	59.33	57.66	56.56	54.90	53.94	52.81	52.17

Second and Higher Stories - Square Foot Area

Quality Class	1,000	1,500	2,000	2,500	3,000	4,000	5,000	7,500	10,000	15,000	20,000
Exceptional	119.62	113.57	110.21	108.02	106.49	104.37	102.98	100.92	99.76	98.37	97.58
1, Best	110.35	104.74	101.68	99.64	98.21	96.27	94.98	93.12	92.02	90.78	90.04
1 & 2	100.32	95.35	92.51	90.69	89.39	87.63	86.48	84.74	83.73	82.61	81.93
2, Good	91.53	86.89	84.31	82.65	81.48	79.87	78.78	77.22	76.33	75.28	74.68
2 & 3	83.06	78.86	76.53	75.00	73.93	72.47	71.50	70.07	69.26	68.32	67.77
3, Average	74.81	71.02	68.93	67.56	66.58	65.26	64.41	63.11	62.39	61.56	61.06
3 & 4	66.00	62.69	60.83	59.63	58.77	57.62	56.85	55.71	55.07	54.29	53.88
4, Low	57.80	54.88	53.24	52.19	51.44	50.44	49.76	48.76	48.20	47.53	47.18

General Office Buildings - Wood Frame

Exterior Suite Entrances, Length More Than 4 Times Width

Estimating Procedure

1. Use these figures to estimate general office buildings in which access to each suite is through an exterior entrance. Medical and dental offices have smaller rooms and more plumbing fixtures than general offices and should be estimated with figures from the Medical and Dental Buildings Section. See page 118.
2. Establish the building quality class by applying the information on page 110.
3. Compute the first floor area. This should include everything within the exterior walls and all insets outside the main walls but under the main roof.
4. If the first floor wall height is more or less than 10 feet, add to or subtract from the first floor square foot cost below the appropriate amount from the Wall Height Adjustment Table on page 117.
5. Multiply the adjusted square foot cost by the first floor area.
6. Deduct, if appropriate, for common walls or no wall finish. Use the figures on page 117.
7. If there are second or higher floors, compute the square foot area on each floor. Locate the appropriate square foot cost from the table at the bottom of this page. Adjust this figure for a wall height more or less than 9 feet, using the figures on page 117. Multiply the adjusted cost by the square foot area on each floor. Use the figures on page 117 to deduct for common walls or no wall finish. Add the result to the cost from step 6 above.
8. Multiply the total cost by the location factor on page 7.
9. Add the cost of heating and air conditioning systems, elevators, fire sprinklers, exterior signs, paving and curbing, miscellaneous yard improvements, covered porches and garages. See pages 201 to 213.

General Offices, Class 2

First Story - Square Foot Area

Quality Class	1,000	1,500	2,000	2,500	3,000	4,000	5,000	7,500	10,000	15,000	20,000
Exceptional	147.80	136.51	130.18	126.03	123.04	119.02	116.32	112.33	110.01	107.37	105.80
1, Best	136.35	125.93	120.07	116.25	113.50	109.78	107.31	103.61	101.46	99.02	97.60
1 & 2	123.50	114.07	108.76	105.30	102.82	99.43	97.20	93.84	91.92	89.69	88.36
2, Good	112.43	103.83	98.99	95.83	93.56	90.48	88.50	85.41	83.66	81.63	80.44
2 & 3	102.80	94.93	90.55	87.64	85.58	82.75	80.91	78.10	76.51	74.66	73.59
3, Average	94.28	87.07	83.02	80.38	78.50	75.91	74.18	71.61	70.18	68.46	67.48
3 & 4	84.39	77.95	74.32	71.95	70.26	67.97	66.42	64.15	62.80	61.32	60.40
4, Low	74.08	68.41	65.25	63.16	61.65	59.64	58.28	56.30	55.12	53.81	53.01

Second and Higher Stories - Square Foot Area

Quality Class	1,000	1,500	2,000	2,500	3,000	4,000	5,000	7,500	10,000	15,000	20,000
Exceptional	124.79	117.62	113.59	110.95	109.06	106.50	104.78	102.19	100.76	99.04	98.03
1, Best	115.11	108.51	104.81	102.38	100.61	98.24	96.67	94.31	92.93	91.38	90.45
1 & 2	104.88	98.85	95.46	93.25	91.65	89.48	88.07	85.89	84.65	83.22	82.40
2, Good	96.04	90.49	87.41	85.38	83.93	81.94	80.63	78.66	77.51	76.19	75.44
2 & 3	86.86	81.86	79.06	77.22	75.91	74.11	72.90	71.13	70.10	68.92	68.23
3, Average	78.38	73.90	71.40	69.71	68.51	66.89	65.84	64.23	63.30	62.23	61.62
3 & 4	69.87	65.87	63.62	62.15	61.06	59.64	58.69	57.25	56.42	55.47	54.89
4, Low	60.79	57.28	55.35	54.04	53.15	51.87	51.04	49.80	49.05	48.24	47.77

General Office Buildings – Wood Frame

Interior Suite Entrances, Length Less Than Twice Width

Estimating Procedure

1. Use these figures to estimate general office buildings in which access to each suite is through an interior corridor. Medical and dental offices have smaller rooms and more plumbing fixtures than general offices and should be estimated with figures from the Medical and Dental Buildings Section. See page 118.
2. Establish the building quality class by applying the information on page 110.
3. Compute the first floor area. This should include everything within the exterior walls and all insets outside the main walls but under the main roof.
4. If the first floor wall height is more or less than 10 feet, add to or subtract from the first floor square foot cost below the appropriate amount from the Wall Height Adjustment Table on page 117.
5. Multiply the adjusted square foot cost by the first floor area.
6. Deduct, if appropriate, for common walls or no wall finish. Use the figures on page 117.
7. If there are second or higher floors, compute the square foot area on each floor. Locate the appropriate square foot cost from the table at the bottom of this page. Adjust this figure for a wall height more or less than 9 feet, using the figures on page 117. Multiply the adjusted cost by the square foot area on each floor. Use the figures on page 117 to deduct for common walls or no wall finish. Add the result to the cost from step 6 above.
8. Multiply the total cost by the location factor on page 7.
9. Add the cost of heating and air conditioning systems, elevators, fire sprinklers, exterior signs, paving and curbing, miscellaneous yard improvements, covered porches and garages. See page 201.

General Offices, Class 3 & 4

First Story - Square Foot Area

Quality Class	2,000	2,500	3,000	4,000	5,000	7,500	10,000	15,000	20,000	30,000	40,000
Exceptional	116.03	113.38	111.48	108.81	107.02	104.30	102.70	100.81	99.72	98.38	97.61
1, Best	106.91	104.47	102.70	100.27	98.61	96.07	94.63	92.89	91.87	90.68	89.95
1 & 2	102.63	95.65	94.02	91.76	90.27	87.99	86.62	85.05	84.10	83.02	82.35
2, Good	90.35	88.29	86.81	84.74	83.33	81.20	79.97	78.52	77.66	76.63	76.04
2 & 3	82.48	80.59	79.25	77.37	76.07	74.13	73.00	71.66	70.87	69.94	69.39
3, Average	75.44	73.73	72.48	70.75	69.57	67.80	66.78	65.55	64.81	63.98	63.51
3 & 4	67.63	66.10	64.96	63.40	62.36	60.79	59.83	58.76	58.12	57.33	56.89
4, Low	59.18	57.81	56.85	55.48	54.58	53.18	52.36	51.42	50.85	50.17	49.80

Second and Higher Stories - Square Foot Area

Quality Class	2,000	2,500	3,000	4,000	5,000	7,500	10,000	15,000	20,000	30,000	40,000
Exceptional	103.54	101.86	100.61	98.92	97.77	96.02	95.01	93.78	93.13	92.26	91.80
1, Best	95.50	93.98	92.81	91.25	90.21	88.60	87.64	86.55	85.88	85.13	84.69
1 & 2	87.96	86.52	85.45	84.03	83.07	81.57	80.71	79.68	79.09	78.37	77.98
2, Good	80.98	79.64	78.68	77.35	76.46	75.05	74.29	73.35	72.81	72.17	71.78
2 & 3	75.37	74.18	71.07	69.89	69.08	67.83	67.14	66.30	65.79	65.21	64.87
3, Average	66.03	64.94	64.18	63.08	62.33	61.23	60.58	59.81	59.35	58.83	58.53
3 & 4	58.74	57.78	57.09	56.12	55.48	54.48	53.90	53.20	52.81	52.34	52.09
4, Low	51.01	50.16	49.54	48.73	48.16	47.30	46.78	46.20	45.88	45.46	45.23

General Office Buildings - Wood Frame

Interior Suite Entrances, Length Between 2 and 4 Times Width

Estimating Procedure

1. Use these figures to estimate general office buildings in which access to each suite is through an interior corridor. Medical and dental offices have smaller rooms and more plumbing fixtures than general offices and should be estimated with figures from the Medical and Dental Buildings Section. See page 118.
2. Establish the building quality class by applying the information on page 110.
3. Compute the first floor area. This should include everything within the exterior walls and all insets outside the main walls but under the main roof.
4. If the first floor wall height is more or less than 10 feet, add to or subtract from the first floor square foot cost below the appropriate amount from the Wall Height Adjustment Table on page 117.
5. Multiply the adjusted square foot cost by the first floor area.
6. Deduct, if appropriate, for common walls or no wall finish. Use the figures on page 117.
7. If there are second or higher floors, compute the square foot area on each floor. Locate the appropriate square foot cost from the table at the bottom of this page. Adjust this figure for a wall height more or less than 9 feet, using the figures on page 117. Multiply the adjusted cost by the square foot area on each floor. Use the figures on page 117 to deduct for common walls or no wall finish. Add the result to the cost from step 6 above.
8. Multiply the total cost by the location factor on page 7.
9. Add the cost of heating and air conditioning systems, elevators, fire sprinklers, exterior signs, paving and curbing, miscellaneous yard improvements, covered porches and garages. See page 201.

First Story - Square Foot Area

Quality Class	2,000	2,500	3,000	4,000	5,000	7,500	10,000	15,000	20,000	30,000	40,000
Exceptional	120.12	117.01	114.75	111.61	109.49	106.20	104.26	101.98	100.61	99.03	98.07
1, Best	111.57	108.71	106.60	103.67	101.71	98.66	96.85	94.75	93.49	91.98	91.11
1 & 2	101.31	98.71	96.83	94.14	92.37	89.58	87.95	86.01	84.85	83.52	82.72
2, Good	93.48	91.09	89.33	86.86	85.20	82.65	81.13	79.37	78.29	77.05	76.33
2 & 3	85.41	83.21	81.62	79.38	77.85	75.52	74.13	72.51	71.54	70.41	69.75
3, Average	78.21	76.19	74.74	72.69	71.29	69.13	67.86	66.40	65.51	64.45	63.86
3 & 4	70.16	68.31	67.02	65.17	63.91	62.00	60.88	59.54	58.76	57.82	57.26
4, Low	61.40	59.82	58.66	57.05	55.97	54.28	53.27	52.14	51.44	50.60	50.13

Second and Higher Stories - Square Foot Area

Quality Class	2,000	2,500	3,000	4,000	5,000	7,500	10,000	15,000	20,000	30,000	40,000
Exceptional	106.20	104.08	102.55	100.49	99.11	97.05	95.86	94.45	93.66	92.71	92.16
1, Best	98.01	96.02	94.64	92.72	91.45	89.54	88.42	87.17	86.40	85.54	85.02
1 & 2	90.06	88.27	87.01	85.26	84.08	82.30	81.29	80.12	79.43	78.64	78.18
2, Good	83.10	81.44	80.26	78.61	77.55	75.92	74.98	73.93	73.25	72.54	72.10
2 & 3	75.10	73.61	72.52	71.05	70.10	68.62	67.77	66.79	66.22	65.56	65.19
3, Average	67.86	66.52	65.55	64.24	63.36	62.03	61.25	60.38	59.86	59.25	58.91
3 & 4	60.47	59.28	58.41	57.20	56.44	55.26	54.59	53.79	53.34	52.76	52.50
4, Low	52.32	51.32	50.55	49.52	48.87	47.82	47.26	46.55	46.16	45.69	45.43

General Office Buildings – Wood Frame

Interior Suite Entrances, Length More Than 4 Times Width

Estimating Procedure

1. Use these figures to estimate general office buildings in which access to each suite is through an interior corridor. Medical and dental offices have smaller rooms and more plumbing fixtures than general offices and should be estimated with figures from the Medical and Dental Buildings Section. See page 118.
2. Establish the building quality class by applying the information on page 110.
3. Compute the first floor area. This should include everything within the exterior walls and all insets outside the main walls but under the main roof.
4. If the first floor wall height is more or less than 10 feet, add to or subtract from the first floor square foot cost below the appropriate amount from the Wall Height Adjustment Table on page 117.
5. Multiply the adjusted square foot cost by the first floor area.
6. Deduct, if appropriate, for common walls or no wall finish. Use the figures on page 117.
7. If there are second or higher floors, compute the square foot area on each floor. Locate the appropriate square foot cost from the table at the bottom of this page. Adjust this figure for a wall height more or less than 9 feet, using the figures on page 117. Multiply the adjusted cost by the square foot area on each floor. Use the figures on page 117 to deduct for common walls or no wall finish. Add the result to the cost from step 6 above.
8. Multiply the total cost by the location factor on page 7.
9. Add the cost of heating and air conditioning systems, elevators, fire sprinklers, exterior signs, paving and curbing, miscellaneous yard improvements, covered porches and garages. See page 201.

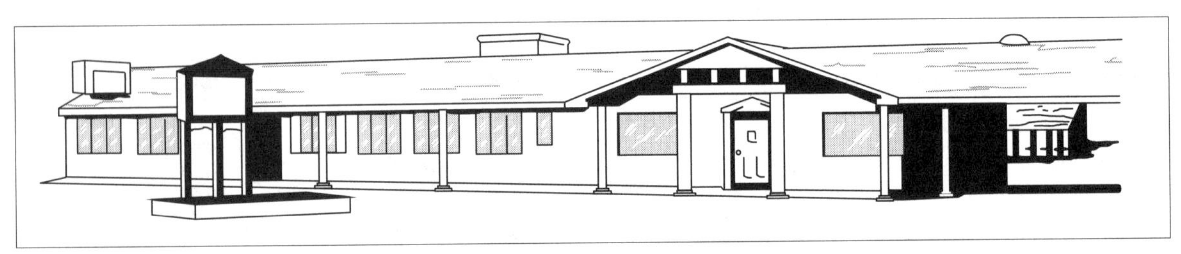

General Offices, Class 2

First Story - Square Foot Area

Quality Class	2,000	2,500	3,000	4,000	5,000	7,500	10,000	15,000	20,000	30,000	40,000
Exceptional	125.90	121.83	118.87	114.92	112.27	108.34	106.06	103.45	101.93	100.14	99.10
1, Best	116.12	112.36	109.65	105.99	103.58	99.85	97.81	95.41	94.02	92.37	91.42
1 & 2	106.16	102.74	100.27	96.92	94.68	91.35	89.42	87.35	85.92	84.46	83.58
2, Good	97.90	94.73	92.45	89.36	87.30	84.21	82.47	80.41	79.25	77.86	77.05
2 & 3	89.53	86.60	84.51	81.73	79.82	77.01	75.39	73.59	72.48	71.22	70.50
3, Average	82.08	79.44	77.52	74.94	73.21	70.64	69.14	67.44	66.45	65.28	64.65
3 & 4	73.58	71.19	69.44	67.15	65.60	63.32	61.97	60.44	59.55	58.51	57.91
4, Low	64.58	62.50	60.98	58.94	57.60	55.55	54.41	53.05	52.26	51.37	50.85

Second and Higher Stories - Square Foot Area

Quality Class	2,000	2,500	3,000	4,000	5,000	7,500	10,000	15,000	20,000	30,000	40,000
Exceptional	109.31	106.73	104.89	102.37	100.68	98.20	96.74	95.08	94.12	92.96	92.32
1, Best	101.09	98.71	96.99	94.66	93.13	90.81	89.46	87.91	87.03	85.99	85.37
1 & 2	93.12	90.91	89.34	87.17	85.75	83.63	82.40	80.98	80.16	79.20	78.64
2, Good	85.79	83.73	82.29	80.30	79.00	77.04	75.89	74.59	73.84	72.95	72.42
2 & 3	77.63	75.79	74.47	72.69	71.50	69.75	68.71	67.51	66.81	66.01	65.57
3, Average	70.26	68.61	67.40	65.81	64.72	63.10	62.17	61.09	60.50	59.76	59.32
3 & 4	62.67	61.15	60.12	58.68	57.70	56.29	55.45	54.49	53.94	53.30	52.93
4, Low	54.51	53.23	52.31	51.05	50.22	48.97	48.26	47.41	46.94	46.37	46.04

General Office Buildings - Wood Frame

Wall Height Adjustment

The square foot costs for general offices are based on the wall heights of 10 feet for first floors and 9 feet for higher floors. Add to or subtract from the amount listed in this table the square foot of floor cost for each foot of wall height more or less than 10 feet, if adjusting for a first floor, and 9 feet, if adjusting for upper floors.

Square Foot Area

Quality Class	1,000	1,500	2,000	3,000	4,000	5,000	7,500	10,000	15,000	20,000	40,000
1, Best	1.49	1.26	1.10	.93	.79	.72	.56	.47	.38	.33	.14
2, Good	1.30	1.09	.96	.81	.72	.58	.49	.44	.34	.29	.11
3, Average	1.13	.96	.86	.70	.58	.53	.44	.38	.31	.26	.11
4, Low	1.05	.86	.77	.63	.52	.48	.40	.35	.27	.22	.09

Perimeter Wall Adjustment

A common wall exists when two buildings share one wall. Adjust for common walls by deducting the linear foot cost below from the total structure cost. In some structures one or more walls are not owned at all. In this case, deduct the "No Ownership" cost per linear foot of wall not owned. Where a perimeter wall remains unfinished, deduct the "Lack of Exterior Finish" cost.

First Story

Class	For a Common Wall, Deduct Per L.F.	For No Wall Ownership, Deduct Per L.F.	For Lack of Exterior Finish, Deduct Per L.F.
1	$246.00	$498.00	$218.00
2	213.00	414.00	190.00
3	197.00	381.00	152.00
4	146.00	280.00	112.00

Second and Higher Stories

Class	For a Common Wall, Deduct Per L.F.	For No Wall Ownership, Deduct Per L.F.	For Lack of Exterior Finish, Deduct Per L.F.
1	$168.00	$320.00	$218.00
2	140.00	265.00	190.00
3	134.00	252.00	152.00
4	124.00	241.00	112.00

Medical-Dental Buildings – Masonry or Concrete

	Class 1 Best Quality	Class 2 Good Quality	Class 3 Average Quality	Class 4 Low Quality
First Floor Structure (10% of total cost)	Reinforced concrete slab on grade or standard wood frame.	Reinforced concrete slab on grade or standard wood frame.	Reinforced concrete slab on grade.	Reinforced concrete slab on grade.
Upper Floor Structure (8% of total cost)	Standard wood frame, plywood and 1-1/2" lightweight concrete subfloor.	Standard wood frame, plywood and 1-1/2" lightweight concrete subfloor.	Standard wood frame. 5/8" plywood subfloor.	Standard wood frame 5/8" plywood subfloor.
Walls (9% of total cost)	8" decorative concrete block or 6" concrete tilt-up.	8" decorative concrete block or 6" concrete tilt-up.	8" reinforced concrete block or 8" reinforced brick.	8" reinforced concrete block or clay tile.
Roof Structure (6% of total cost)	Standard wood frame, flat or low pitch.	Standard wood frame, flat or low pitch.	Standard wood frame, flat or low pitch.	Standard wood frame, flat or low pitch.
Exterior Wall Finish (8% of total cost)	Decorative block or large rock imbedded in tilt-up panels with 10 to 20% brick or stone veneer.	Decorative block or exposed aggregate and 10 to 20% brick or stone veneer.	Stucco or colored concrete block.	Painted.
Windows (5% of total cost)	Average number in good aluminum frame. Fixed float glass in good frame on front.	Average number in good aluminum frame. Some fixed float glass in front.	Average number of average aluminum sliding type.	Average number of low cost aluminum sliding type.
Roof Cover (5% of total cost)	5 ply built-up roofing on flat roofs. Heavy shake or tile on on sloping roofs.	5 ply built-up roofing on flat roofs. Average shake or composition, tar and large rock on sloping roofs.	4 ply built-up roofing on flat roofs. Wood shingle or composition, tar and pea gravel on sloping roofs.	3 ply built-up roofing on flat roofs. Composition shingle on sloping roofs.
Overhang (3% of total cost)	3' closed overhang, fully guttered.	2' closed overhang, fully guttered.	None on flat roofs. 18" open on sloping roofs, fully guttered.	None on flat roofs. 12" to 16" open on sloping roofs, gutters over entrances.
Business Offices (3% of total cost)	Good hardwood veneer paneling. Solid vinyl or carpet.	Hardwood paneling or vinyl wall cover. Resilient tile or carpet.	Gypsum wallboard, texture and paint. composition tile.	Gypsum wallboard, texture and paint. Minimum grade tile.
Corridors (6% of total cost)	Good hardwood veneer paneling. Solid vinyl or carpet.	Gypsum wallboard and vinyl wall cover. Resilient tile.	Gypsum wallboard, texture and paint. Composition tile.	Gypsum wallboard, texture and paint. Minimum grade tile.
Waiting Rooms (7% of total cost)	Good hardwood veneer paneling. Carpet.	Hardwood paneling. Carpet.	Gypsum wallboard, some paneling. Resilient tile or carpet.	Gypsum wallboard, texture and paint. Minimum grade tile.
Private Offices (2% of total cost)	Good hardwood veneer paneling. Carpet.	Textured wall cover and hardwood paneling. Carpet.	Gypsum wallboard and paper, wood paneling. Resilient tile or carpet.	Gypsum wallboard, texture and paint. Minimum grade tile.
Treatment Rooms (5% of total cost)	Gypsum wallboard and vinyl wall covering. Sheet vinyl or carpet.	Gypsum wallboard and enamel. Sheet vinyl.	Gypsum wallboard and enamel. Resilient tile.	Gypsum wallboard, texture and paint. Minimum grade tile.
Bathrooms (3% of total cost)	Gypsum wallboard and enamel with ceramic tile wainscot. Sheet vinyl or ceramic tile.	Gypsum wallboard and enamel or vinyl wall covering. Sheet vinyl or ceramic tile.	Gypsum wallboard and enamel. Resilient tile.	Gypsum wallboard, texture and paint. Minimum grade tile.
Ceiling Finish (4% of total cost)	Suspended "T" bar and acoustical tile.	Gypsum wallboard and acoustical tile.	Gypsum wallboard and acoustical tile.	Gypsum wallboard and paint.
Utilities				
Plumbing (6% of total cost)	Copper tubing, good economy fixtures.	Copper tubing, good fixtures.	Copper tubing, average fixtures.	Copper tubing, fixtures.
Lighting (6% of total cost)	Conduit wiring, good fixtures.	Conduit wiring, good fixtures.	Romex or conduit wiring, average fixtures.	Romex wiring, economy fixtures.
Cabinets (4% of total cost)	Formica faced with formica tops.	Good grade of hardwood with formica tops.	Average amount of painted wood or low grade hardwood with formica top.	Minimum amount of painted wood with formica top.

Note: Use the percent of total cost to help identify the correct quality classification.

Square foot costs include the following components: Foundations as required for normal soil conditions. Floor, wall and roof structures. Interior floor, wall and ceiling finishes. Exterior wall finish and roof cover. Interior partitions. Cabinets, doors and windows. Basic electrical systems and lighting fixtures. Rough plumbing and fixtures. Permits and fees. Contractors' mark-up. In addition to the above components, costs for buildings with more than 10,000 feet include the cost of lead shielding for typical x-ray rooms.

Medical-Dental Buildings – Masonry or Concrete

Exterior Suite Entrances, Length Less Than Twice Width

Estimating Procedure

1. Use these figures to estimate medical, dental, psychiatric, optometry and similar professional buildings in which access to each office suite is through an exterior entrance. Buildings in this section have more plumbing fixtures per square foot of floor and smaller room sizes than general office buildings. Note also that buildings with more than 10,000 square feet are assumed to have lead shielded x-ray rooms.
2. Establish the building quality class by applying the information on page 118.
3. Compute the first floor area. This should include everything within the exterior walls and all insets outside the main walls but under the main roof.
4. If the first floor wall height is more or less than 10 feet, add to or subtract from the first floor square foot cost below the appropriate amount from the Wall Height Adjustment Table on page 125.
5. Multiply the adjusted square foot cost by the first floor area.
6. Deduct, if appropriate, for common walls or no wall finish. Use the figures on page 125.
7. If there are second or higher floors, compute the square foot area on each floor. Locate the appropriate square foot cost from the table at the bottom of this page. Adjust this figure for a wall height more or less than 9 feet, using the figures on page 125. Multiply the adjusted cost by the square foot area on each floor. Use the figures on page 125 to deduct for common walls or no wall finish. Add the result to the cost from step 6 above.
8. Multiply the total cost by the location factor listed on page 7.
9. Add the cost of heating and air conditioning systems, elevators, fire sprinklers, exterior signs, paving and curbing, miscellaneous yard improvements, covered porches and garages. See page 201.

Medical-Dental Building, Class 2

First Story - Square Foot Area

Quality Class	1,000	1,500	2,000	2,500	3,000	4,000	5,000	7,500	10,000	15,000	20,000
Exceptional	177.87	167.38	161.27	157.20	154.19	150.06	147.25	143.00	140.47	137.50	135.77
1, Best	164.64	154.91	149.27	145.47	142.72	138.90	136.30	132.36	130.01	127.27	125.67
1 & 2	156.10	146.85	141.53	137.94	135.32	131.69	129.24	125.49	123.27	120.68	119.14
2, Good	149.08	140.29	135.18	131.76	129.24	125.78	123.45	119.85	117.75	115.27	113.79
2 & 3	137.82	129.66	124.94	121.76	119.45	116.25	114.11	110.78	108.81	106.53	105.18
3, Average	129.23	121.71	117.25	114.30	112.12	108.00	107.07	103.97	102.13	99.98	98.72
3 & 4	121.82	114.64	110.47	107.65	105.63	102.79	100.85	97.93	96.21	94.19	92.98
4, Low	113.46	106.79	102.89	100.29	98.35	95.74	93.96	91.23	89.62	87.72	86.59

Second and Higher Stories - Square Foot Area

Quality Class	1,000	1,500	2,000	2,500	3,000	4,000	5,000	7,500	10,000	15,000	20,000
Exceptional	159.25	148.60	142.84	139.15	136.59	133.15	130.91	127.70	125.87	123.83	122.66
1, Best	155.49	145.10	139.50	135.89	133.38	130.04	127.85	124.67	122.90	120.93	119.77
1 & 2	147.74	137.85	132.53	129.11	126.72	123.53	121.45	118.46	116.77	114.87	113.80
2, Good	142.57	133.05	127.89	124.61	122.30	119.21	117.24	114.32	112.67	110.86	109.82
2 & 3	131.65	122.86	118.09	115.03	112.89	110.10	108.24	105.55	104.06	102.38	101.39
3, Average	123.58	115.33	110.86	107.99	106.04	103.34	101.62	99.09	97.68	96.09	95.21
3 & 4	116.37	108.59	104.39	101.68	99.81	97.31	95.66	93.29	91.79	90.47	89.62
4, Low	108.47	101.24	97.29	94.80	93.05	90.69	89.18	86.97	85.72	84.37	83.57

Exterior Suite Entrances, Length Between 2 and 4 Times Width

Estimating Procedure

1. Use these figures to estimate medical, dental, psychiatric, optometry and similar professional buildings in which access to each office suite is through an exterior entrance. Buildings in this section have more plumbing fixtures per square foot of floor and smaller room sizes than general office buildings. Note also that buildings with more than 10,000 square feet are assumed to have lead shielded x-ray rooms.
2. Establish the building quality class by applying the information on page 118.
3. Compute the first floor area. This should include everything within the exterior walls and all insets outside the main walls but under the main roof.
4. If the first floor wall height is more or less than 10 feet, add to or subtract from the first floor square foot cost below the appropriate amount from the Wall Height Adjustment Table on page 125.
5. Multiply the adjusted square foot cost by the first floor area.
6. Deduct, if appropriate, for common walls or no wall finish. Use the figures on page 125.
7. If there are second or higher floors, compute the square foot area on each floor. Locate the appropriate square foot cost from the table at the bottom of this page. Adjust this figure for a wall height more or less than 9 feet, using the figures on page 125. Multiply the adjusted cost by the square foot area on each floor. Use the figures on page 125 to deduct for common walls or no wall finish. Add the result to the cost from step 6 above.
8. Multiply the total cost by the location factor listed on page 7.
9. Add the cost of heating and air conditioning systems, elevators, fire sprinklers, exterior signs, paving and curbing, miscellaneous yard improvements, covered porches and garages. See page 201.

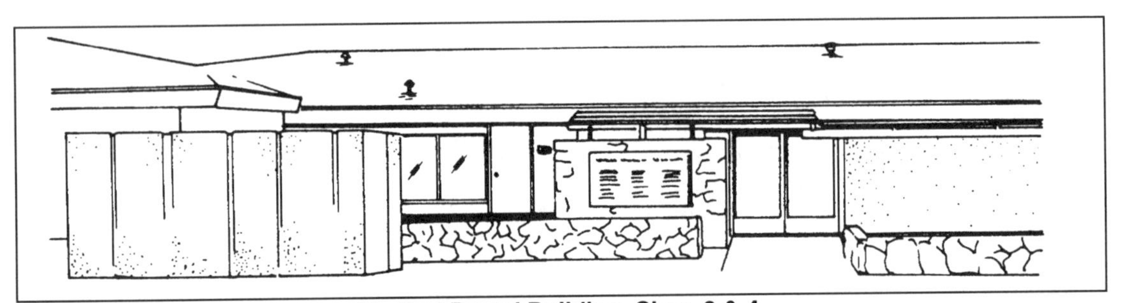

Medical Dental Building, Class 3 & 4

First Story - Square Foot Area

Quality Class	1,000	1,500	2,000	2,500	3,000	4,000	5,000	7,500	10,000	15,000	20,000
Exceptional	191.74	177.53	169.55	164.26	160.48	155.28	151.87	146.70	143.68	140.22	138.18
1, Best	177.77	164.62	157.19	152.29	148.79	143.99	140.79	135.99	133.23	130.01	128.13
1 & 2	167.18	154.83	147.86	143.25	139.96	135.44	132.44	127.94	125.31	122.27	120.54
2, Good	159.53	147.75	141.06	136.69	133.53	129.23	126.40	122.08	119.58	116.70	115.01
2 & 3	147.44	136.55	130.36	126.33	123.41	119.42	116.79	112.81	110.52	107.82	106.27
3, Average	138.44	128.19	122.41	118.61	115.87	112.14	109.68	105.90	103.74	101.23	99.78
3 & 4	130.33	120.68	115.23	111.64	109.08	105.58	103.22	99.71	97.67	95.31	93.94
4, Low	121.37	112.38	107.32	104.02	101.59	98.32	96.12	92.84	90.98	88.75	87.46

Second and Higher Stories - Square Foot Area

Quality Class	1,000	1,500	2,000	2,500	3,000	4,000	5,000	7,500	10,000	15,000	20,000
Exceptional	176.72	164.29	157.29	152.66	149.33	144.79	141.80	137.27	134.61	131.57	129.78
1, Best	163.84	152.45	145.82	141.50	138.44	134.20	131.46	127.24	124.78	121.94	120.31
1 & 2	155.61	144.66	138.46	134.42	131.50	127.48	124.83	120.83	118.52	115.84	114.25
2, Good	150.11	139.52	133.59	129.66	126.84	122.97	120.43	116.57	114.32	111.76	110.22
2 & 3	138.56	128.84	123.34	119.71	117.11	113.57	111.18	107.63	105.56	103.18	101.77
3, Average	130.07	120.93	115.76	112.36	109.90	106.58	104.39	101.01	99.08	96.85	95.55
3 & 4	122.52	113.93	109.07	105.87	103.57	100.40	98.32	95.17	93.36	91.21	89.98
4, Low	114.29	106.26	101.71	98.74	96.58	93.65	91.71	88.75	87.03	85.10	83.96

Medical-Dental Buildings – Masonry or Concrete

Exterior Suite Entrances, Length More Than 4 Times Width

Estimating Procedure

1. Use these figures to estimate medical, dental, psychiatric, optometry and similar professional buildings in which access to each office suite is through an exterior entrance. Buildings in this section have more plumbing fixtures per square foot of floor and smaller room sizes than general office buildings. Note also that buildings with more than 10,000 square feet are assumed to have lead shielded x-ray rooms.
2. Establish the building quality class by applying the information on page 118.
3. Compute the first floor area. This should include everything within the exterior walls and all insets outside the main walls but under the main roof.
4. If the first floor wall height is more or less than 10 feet, add to or subtract from the first floor square foot cost below the appropriate amount from the Wall Height Adjustment Table on page 125.
5. Multiply the adjusted square foot cost by the first floor area.
6. Deduct, if appropriate, for common walls or no wall finish. Use the figures on page 125.
7. If there are second or higher floors, compute the square foot area on each floor. Locate the appropriate square foot cost from the table at the bottom of this page. Adjust this figure for a wall height more or less than 9 feet, using the figures on page 125. Multiply the adjusted cost by the square foot area on each floor. Use the figures on page 125 to deduct for common walls or no wall finish. Add the result to the cost from step 6 above.
8. Multiply the total cost by the location factor listed on page 7.
9. Add the cost of heating and air conditioning systems, elevators, fire sprinklers, exterior signs, paving and curbing, miscellaneous yard improvements, covered porches and garages. See page 201.

Medical-Dental Building, Class 2 & 3

First Story - Square Foot Area

Quality Class	1,000	1,500	2,000	2,500	3,000	4,000	5,000	7,500	10,000	15,000	20,000
Exceptional	206.66	188.75	178.76	172.22	167.54	161.22	157.05	150.81	147.20	143.10	140.70
1, Best	191.58	175.00	165.71	159.66	155.34	149.46	145.60	139.79	136.48	132.66	130.43
1 & 2	179.89	164.28	155.56	149.88	145.85	140.30	136.66	131.24	128.11	124.55	122.44
2, Good	172.14	157.21	148.89	143.44	139.57	134.29	130.82	125.60	122.59	119.18	117.21
2 & 3	158.41	144.68	137.02	131.99	128.43	123.55	120.39	115.58	112.83	109.69	107.82
3, Average	148.56	135.67	128.47	123.80	120.43	115.88	112.88	108.38	105.81	102.83	101.13
3 & 4	139.83	127.68	120.94	116.50	113.37	109.08	106.26	102.41	99.57	96.80	95.18
4, Low	130.20	118.89	112.60	108.48	105.54	101.57	98.90	94.98	92.72	90.12	88.60

Second and Higher Stories - Square Foot Area

Quality Class	1,000	1,500	2,000	2,500	3,000	4,000	5,000	7,500	10,000	15,000	20,000
Exceptional	189.01	173.76	165.15	159.51	155.43	149.88	146.17	140.63	137.38	133.66	131.50
1, Best	174.50	160.42	152.49	147.24	143.50	138.36	134.94	129.82	126.84	123.41	121.41
1 & 2	166.32	152.92	145.35	140.37	136.77	131.89	128.65	123.73	120.92	117.63	115.70
2, Good	160.40	147.47	140.16	135.33	131.90	127.19	124.07	119.33	116.59	113.43	111.60
2 & 3	148.12	136.18	129.43	124.97	121.78	117.41	114.52	110.18	107.65	104.74	103.02
3, Average	138.98	127.79	121.44	117.30	114.31	110.22	107.50	103.40	101.03	98.29	96.70
3 & 4	131.12	120.51	114.57	110.63	107.81	103.95	101.41	97.53	95.30	92.71	91.19
4, Low	122.10	112.25	106.69	103.03	100.40	96.80	94.40	90.84	88.75	86.36	84.92

Medical-Dental Buildings – Masonry or Concrete

Interior Suite Entrances, Length Less Than Twice Width

Estimating Procedure

1. Use these figures to estimate medical, dental, psychiatric, optometry and similar professional buildings in which access to each office suite is through an interior corridor. Buildings in this section have more plumbing fixtures per square foot of floor and smaller room sizes than general office buildings. Note also that buildings with more than 10,000 square feet are assumed to have lead shielded x-ray rooms.
2. Establish the building quality class by applying the information on page 118.
3. Compute the first floor area. This should include everything within the exterior walls and all insets outside the main walls but under the main roof.
4. If the first floor wall height is more or less than 10 feet, add to or subtract from the first floor square foot cost below the appropriate amount from the Wall Height Adjustment Table on page 125.
5. Multiply the adjusted square foot cost by the first floor area.
6. Deduct, if appropriate, for common walls or no wall finish. Use the figures on page 125.
7. If there are second or higher floors, compute the square foot area on each floor. Locate the appropriate square foot cost from the table at the bottom of this page. Adjust this figure for a wall height more or less than 9 feet, using the figures on page 125. Multiply the adjusted cost by the square foot area on each floor. Use the figures on page 125 to deduct for common walls or no wall finish. Add the result to the cost from step 6 above.
8. Multiply the total cost by the location factor listed on page 7.
9. Add the cost of heating and air conditioning systems, elevators, fire sprinklers, exterior signs, paving and curbing, miscellaneous yard improvements, covered porches and garages. See page 201.

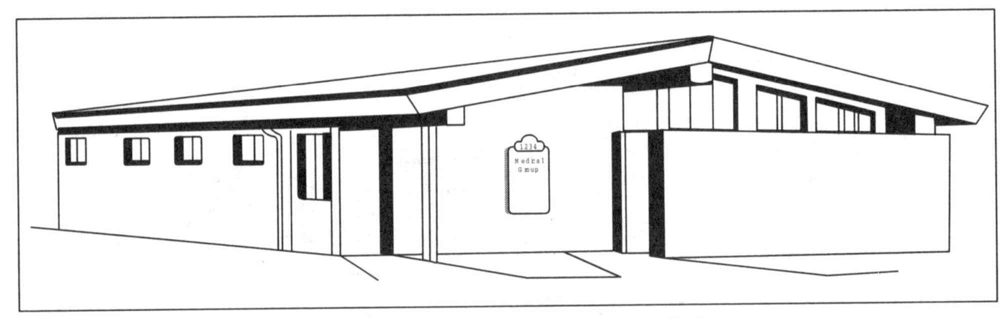

Medical-Dental Building, Class 3 & 4

First Story - Square Foot Area

Quality Class	2,000	2,500	3,000	4,000	5,000	7,500	10,000	15,000	20,000	30,000	40,000
Exceptional	150.36	146.29	143.44	139.94	136.54	132.38	129.93	127.08	125.42	123.48	122.32
1, Best	143.27	139.42	136.59	132.74	130.12	126.17	123.85	121.11	119.55	117.68	116.57
1 & 2	135.18	131.52	128.83	125.19	122.77	118.99	116.81	114.24	112.76	111.00	109.98
2, Good	129.36	125.86	123.32	119.84	117.47	113.88	111.80	109.36	107.93	106.23	105.24
2 & 3	119.90	116.67	114.30	111.06	108.89	105.56	103.65	101.35	100.03	98.45	97.54
3, Average	111.87	108.83	106.63	103.63	101.59	98.50	96.67	94.56	93.32	91.87	91.00
3 & 4	105.87	103.00	100.93	98.07	96.12	93.21	91.50	89.50	88.32	86.94	86.13
4, Low	99.25	96.54	94.60	91.91	90.12	87.37	85.76	83.92	82.78	81.51	80.74

Second and Higher Stories - Square Foot Area

Quality Class	2,000	2,500	3,000	4,000	5,000	7,500	10,000	15,000	20,000	30,000	40,000
Exceptional	142.15	138.51	135.88	132.26	129.82	126.11	123.94	121.43	119.96	118.21	117.19
1, Best	135.56	132.09	129.60	126.14	123.83	120.28	118.23	115.85	114.41	112.74	111.80
1 & 2	128.66	125.38	122.97	119.72	117.53	114.15	112.22	109.95	108.59	107.00	106.08
2, Good	123.95	120.80	118.51	115.33	113.20	110.00	108.11	105.90	104.60	103.10	102.19
2 & 3	114.60	111.66	109.54	106.65	104.66	101.69	99.97	97.91	96.72	95.31	94.49
3, Average	107.58	104.85	102.84	100.08	98.25	95.46	93.83	91.93	90.80	89.47	88.70
3 & 4	101.20	98.59	96.74	94.17	92.46	89.83	88.24	86.51	85.42	84.17	83.46
4, Low	95.29	92.84	91.08	88.65	87.01	84.54	83.12	81.43	80.43	79.25	78.57

Medical-Dental Buildings - Masonry or Concrete

Interior Suite Entrances, Length Between 2 and 4 Times Width

Estimating Procedure

1. Use these figures to estimate medical, dental, psychiatric, optometry and similar professional buildings in which access to each office suite is through an interior corridor. Buildings in this section have more plumbing fixtures per square foot of floor and smaller room sizes than general office buildings. Note also that buildings with more than 10,000 square feet are assumed to have lead shielded x-ray rooms.
2. Establish the building quality class by applying the information on page 118.
3. Compute the first floor area. This should include everything within the exterior walls and all insets outside the main walls but under the main roof.
4. If the first floor wall height is more or less than 10 feet, add to or subtract from the first floor square foot cost below the appropriate amount from the Wall Height Adjustment Table on page 125.
5. Multiply the adjusted square foot cost by the first floor area.
6. Deduct, if appropriate, for common walls or no wall finish. Use the figures on page 125.
7. If there are second or higher floors, compute the square foot area on each floor. Locate the appropriate square foot cost from the table at the bottom of this page. Adjust this figure for a wall height more or less than 9 feet, using the figures on page 125. Multiply the adjusted cost by the square foot area on each floor. Use the figures on page 125 to deduct for common walls or no wall finish. Add the result to the cost from step 6 above.
8. Multiply the total cost by the location factor listed on page 7.
9. Add the cost of heating and air conditioning systems, elevators, fire sprinklers, exterior signs, paving and curbing, miscellaneous yard improvements, covered porches and garages. See page 201.

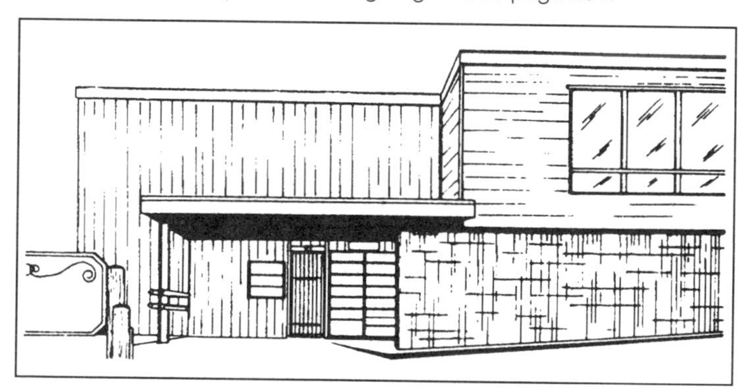

Medical-Dental Building, Class 3

First Story - Square Foot Area

Quality Class	2,000	2,500	3,000	4,000	5,000	7,500	10,000	15,000	20,000	30,000	40,000
Exceptional	157.39	152.38	148.75	143.79	140.49	135.48	132.58	129.19	127.20	124.90	123.52
1, Best	150.02	145.25	141.80	137.06	133.92	129.13	126.36	123.12	121.23	119.01	117.73
1 & 2	141.38	136.90	133.65	129.17	126.20	121.70	119.08	116.06	114.25	112.21	110.97
2, Good	135.18	130.90	127.80	123.52	120.69	116.39	113.88	110.97	109.28	107.28	106.13
2 & 3	125.21	121.23	118.33	114.42	111.79	107.76	105.45	102.78	101.21	99.35	98.28
3, Average	116.73	113.01	110.33	106.63	104.19	100.48	98.30	95.79	94.33	92.62	91.59
3 & 4	110.58	107.05	104.54	101.03	98.69	95.20	93.14	90.73	89.37	87.73	86.81
4, Low	103.49	100.20	97.82	94.55	92.38	89.10	87.21	84.92	83.65	82.12	81.22

Second and Higher Stories - Square Foot Area

Quality Class	2,000	2,500	3,000	4,000	5,000	7,500	10,000	15,000	20,000	30,000	40,000
Exceptional	148.36	143.87	140.68	136.24	133.29	128.83	126.24	123.23	121.45	119.44	118.19
1, Best	141.40	137.12	134.06	129.85	127.01	122.81	120.29	117.44	115.74	113.82	112.65
1 & 2	134.14	130.10	127.18	123.18	120.54	116.46	114.13	111.42	109.81	107.97	106.88
2, Good	129.26	125.38	122.54	118.68	116.14	112.27	110.02	107.39	105.83	104.05	102.99
2 & 3	119.34	115.74	113.16	109.60	107.23	103.67	101.56	99.13	97.73	96.04	95.11
3, Average	111.95	108.61	106.13	102.81	100.59	97.21	95.24	92.98	91.67	90.11	89.18
3 & 4	106.01	102.81	100.49	97.35	95.24	92.05	90.21	88.06	86.78	85.29	84.44
4, Low	99.34	96.35	94.20	91.24	89.28	86.27	84.51	82.53	81.34	79.97	79.15

Medical-Dental Buildings – Masonry or Concrete

Interior Suite Entrances, Length More Than 4 Times Width

Estimating Procedure

1. Establish the building quality class by applying the information on page 118.
3. Compute the first floor area. This should include everything within the exterior walls and all insets outside the main walls but under the main roof.
4. If the first floor wall height is more or less than 10 feet, add to or subtract from the first floor square foot cost below the appropriate amount from the Wall Height Adjustment Table on page 125.
5. Multiply the adjusted square foot cost by the first floor area.
6. Deduct, if appropriate, for common walls or no wall finish. Use the figures on page 125.
7. If there are second or higher floors, compute the square foot area on each floor. Locate the appropriate square foot cost from the table at the bottom of this page. Adjust this figure for a wall height more or less than 9 feet, using the figures on page 125. Multiply the adjusted cost by the square foot area on each floor. Use the figures on page 125 to deduct for common walls or no wall finish. Add the result to the cost from step 6 above.
8. Multiply the total cost by the location factor listed on page 7.
9. Add the cost of heating and air conditioning systems, elevators, fire sprinklers, exterior signs, paving and curbing, miscellaneous yard improvements, covered porches and garages. See page 201.

Medical-Dental Building, Class 2 & 3

First Story - Square Foot Area

Quality Class	2,000	2,500	3,000	4,000	5,000	7,500	10,000	15,000	20,000	30,000	40,000
Exceptional	166.42	160.13	155.63	149.51	145.48	139.43	135.93	131.90	129.56	126.87	125.30
1, Best	158.62	152.62	148.35	142.51	138.65	132.87	129.55	125.72	123.48	120.93	119.38
1 & 2	149.36	143.74	139.69	134.19	130.57	125.14	122.02	118.41	116.32	113.87	112.44
2, Good	142.72	137.34	133.48	128.23	124.76	119.55	116.57	113.11	111.13	108.78	107.43
2 & 3	132.03	127.03	123.48	118.62	115.42	110.60	107.84	104.64	102.80	100.68	99.38
3, Average	124.13	119.46	116.10	111.53	108.49	104.00	101.38	98.39	96.65	94.61	93.45
3 & 4	117.44	113.01	109.81	105.54	102.67	98.37	95.93	93.09	91.45	89.51	88.43
4, Low	109.99	105.82	102.87	98.83	96.16	92.15	89.82	87.21	85.63	83.85	82.78

Second and Higher Stories - Square Foot Area

Quality Class	2,000	2,500	3,000	4,000	5,000	7,500	10,000	15,000	20,000	30,000	40,000
Exceptional	156.05	150.46	146.44	141.01	137.41	132.03	128.93	125.39	123.32	120.94	119.55
1, Best	148.82	143.47	139.64	134.45	131.05	125.89	122.94	119.57	117.58	115.30	113.95
1 & 2	141.11	136.04	132.43	127.51	124.26	119.38	116.59	113.37	111.52	109.34	108.07
2, Good	135.93	131.06	127.56	122.84	119.71	115.02	112.31	109.22	107.42	105.32	104.11
2 & 3	125.37	120.90	117.65	113.30	110.41	106.11	103.61	100.76	99.09	97.15	96.03
3, Average	117.76	113.54	110.53	106.42	103.71	99.66	97.33	94.61	93.06	91.26	90.19
3 & 4	111.49	107.48	104.60	100.74	98.15	94.34	92.11	89.57	88.10	86.37	85.38
4, Low	104.47	100.75	98.02	94.38	91.98	88.42	86.32	83.96	82.55	80.94	80.00

Medical-Dental Buildings - Masonry or Concrete

Wall Height Adjustment

The square foot costs for medical-dental buildings are based on the wall heights of 10 feet for first floors and 9 feet for higher floors. The main or first floor height is the distance from the bottom of the floor slab or joists to the top of the roof slab or ceiling joists. Second and higher floors are measured from the top of the floor slab or floor joists to the top of the roof slab or ceiling joists. Add or subtract the amount listed in this table to the square foot of floor cost for each foot of wall height more or less than 10 feet, if adjusting for a first floor, and 9 feet, if adjusting for upper floors.

Square Foot Area

Quality Class	1,000	1,500	2,000	3,000	4,000	5,000	7,500	10,000	15,000	20,000	40,000
1, Best	4.50	3.51	2.96	2.32	1.96	1.75	1.38	1.19	.97	.80	.58
2, Good	3.88	3.05	2.55	2.03	1.72	1.54	1.21	1.03	.82	.72	.48
3, Average	3.39	2.65	2.21	1.78	1.51	1.33	1.09	.89	.74	.65	.43
4, Low	2.98	2.37	1.97	1.55	1.32	1.16	.92	.81	.66	.51	.38

Perimeter Wall Adjustment

A common wall exists when two buildings share one wall. Adjust for common walls by deducting the linear foot costs below from the total structure cost. In some structures, one or more walls are not owned at all. In this case, deduct the "No Ownership" cost per linear foot of wall not owned. If a wall has no exterior finish, deduct the "Lack of Exterior Finish" cost.

First Story

Class	For a Common Wall, Deduct Per L.F.	For No Wall Ownership, Deduct Per L.F.	For Lack of Exterior Finish, Deduct Per L.F.
1	$257.00	$521.00	$196.00
2	202.00	353.00	129.00
3	146.00	291.00	68.00
4	109.00	224.00	44.00

Second and Higher Stories

Class	For a Common Wall, Deduct Per L.F.	For No Wall Ownership, Deduct Per L.F.	For Lack of Exterior Finish, Deduct Per L.F.
1	$241.00	$471.00	$196.00
2	196.00	381.00	129.00
3	146.00	291.00	68.00
4	129.00	257.00	44.00

Medical-Dental Buildings – Wood Frame

	Class 1 Best Quality	Class 2 Good Quality	Class 3 Average Quality	Class 4 Low Quality
Foundation (9% of total cost)	Reinforced concrete.	Reinforced concrete.	Reinforced concrete.	Reinforced concrete.
First Floor Structure (4% of total cost)	Reinforced concrete slab on grade or standard wood frame.	Reinforced concrete slab on grade or standard wood frame.	Reinforced concrete slab on grade or 4" x 6" girders with 2" T&G subfloor.	Reinforced concrete slab on grade.
Upper Floor Structure (6% of total cost)	Standard wood frame, plywood and 1-1/2" lightweight concrete subfloor.	Standard wood frame, plywood and 1-1/2" lightweight concrete subfloor.	Standard wood frame, 5/8" plywood subfloor.	Standard wood frame, 5/8" plywood subfloor.
Walls (9% of total cost)	Standard wood frame.	Standard wood frame.	Standard wood frame.	Standard wood frame.
Roof Structure (6% of total cost)	Standard wood frame, flat or low pitch.	Standard wood frame, flat or low pitch.	Standard wood frame, flat or low pitch.	Standard wood frame, flat or low pitch.
Exterior Finishes:				
Walls (8% of total cost)	Good wood siding with 10 to 20% brick or stone veneer.	Average wood siding or stucco and 10 to 20% brick or stone veneer.	Stucco with some wood trim or cheap wood siding.	Stucco.
Windows (5% of total cost)	Average number in good aluminum frame. Fixed float glass in good frame on front.	Average number in good aluminum frame. Some fixed float glass in front.	Average amount of average aluminum sliding type.	Average number of low cost aluminum sliding type.
Roof Cover (5% of total cost)	5 ply built-up roofing on flat roofs. Heavy shake or tile on sloping roofs.	5 ply built-up roofing on flat roofs. Avg. shake or composition, tar and large rock on sloping roofs.	4 ply built-up roofing on flat roofs. Wood shingle or composition, tar and pea gravel on sloping roofs.	3 ply built-up roofing on flat roofs. Composition shingle on sloping roofs.
Overhang (3% of total cost)	3' sealed overhang, fully guttered.	2' sealed overhang, fully guttered.	None on flat roofs. 18" unsealed on sloping roofs, fully guttered.	None on flat roofs. 12" to 16" unsealed on sloping roofs, gutters over entrances.
Floor Finishes:				
Business Offices (3% of total cost)	Solid vinyl tile or carpet.	Resilient tile or carpet.	Composition tile.	Minimum grade tile.
Corridors (2% of total cost)	Solid vinyl tile or carpet.	Resilient tile or carpet.	Composition tile.	Minimum grade tile.
Waiting Rooms (1% of total cost)	Carpet.	Carpet.	Composition tile or carpet.	Minimum grade tile.
Private Offices (2% of total cost)	Carpet.	Carpet.	Composition tile or carpet.	Minimum grade tile.
Treatment Room (2% of total cost)	Sheet vinyl or carpet.	Sheet vinyl.	Composition tile.	Minimum grade tile.
Bathrooms (1% of total cost)	Sheet vinyl or ceramic tile.	Sheet vinyl or ceramic tile.	Composition tile.	Minimum grade tile.
Interior Wall Finishes:				
Business Offices (3% of total cost)	Good hardwood veneer paneling.	Hardwood paneling or vinyl wall cover.	Gypsum wallboard, texture and paint.	Gypsum wallboard, texture and paint.
Corridors (2% of total cost)	Good hardwood veneer paneling.	Gypsum wallboard and vinyl wall cover.	Gypsum wallboard, texture and paint.	Gypsum wallboard, texture and paint.
Waiting Room (1% of total cost)	Good hardwood veneer paneling.	Hardwood paneling.	Gypsum wallboard and paper, some wood paneling.	Gypsum wallboard, texture and paint.
Treatment Room (3% of total cost)	Gypsum wallboard and vinyl wall covering.	Gypsum wallboard and enamel.	Gypsum wallboard and enamel.	Gypsum wallboard, texture and paint.
Bathrooms (2% of total cost)	Gypsum wallboard and enamel with ceramic tile wainscot.	Gypsum wallboard and enamel or vinyl wall covering.	Gypsum wallboard and enamel.	Gypsum wallboard, texture and paint.
Ceiling Finish (4% of total cost)	Suspended "T" bar and acoustical tile.	Gypsum wallboard and acoustical tile.	Gypsum wallboard and acoustical tile.	Gypsum wallboard and paint.
Plumbing (6% of total cost)	Copper tubing, good fixtures.	Copper tubing, good fixtures.	Copper tubing, average fixtures.	Copper tubing, economy fixtures.
Lighting (6% of total cost)	Conduit wiring, good fixtures.	Conduit wiring, good fixtures.	Romex or conduit wiring, average fixtures.	Romex wiring, economy fixtures.
Cabinets (7% of total cost)	Formica faced with formica tops.	Good grade of hardwood with formica tops.	Average amount of painted wood or low grade hardwood with formica top.	Minimum amount of painted wood with formica top.

Note: Use the percent of total cost to help identify the correct quality classification.

Square foot costs include the following components: Foundations as required for normal soil conditions. Floor, wall and roof structures. Interior floor, wall and ceiling finishes. Exterior wall finish and roof cover. Interior partitions. Cabinets, doors and windows. Basic electrical systems and lighting fixtures. Rough plumbing and fixtures. Permits and fees. Contractors' mark-up. In addition to the above components, costs for buildings with more than 10,000 feet include the cost of lead shielding for typical x-ray rooms.

Medical-Dental Buildings - Wood Frame

Exterior Suite Entrances, Length Less Than Twice Width

Estimating Procedure

1. Use these figures to estimate medical, dental, psychiatric, optometry and similar professional buildings in which access to each office suite is through an exterior entrance. Buildings in this section have more plumbing fixtures per square foot of floor and smaller room sizes than general office buildings. Note also that buildings with more than 10,000 square feet are assumed to have lead shielded x-ray rooms.
2. Establish the building quality class by applying the information on page 126.
3. Compute the first floor area. This should include everything within the exterior walls and all insets outside the main walls but under the main roof.
4. If the first floor wall height is more or less than 10 feet, add to or subtract from the first floor square foot cost below the appropriate amount from the Wall Height Adjustment Table on page 133.
5. Multiply the adjusted square foot cost by the first floor area.
6. Deduct, if appropriate, for common walls or no wall finish. Use the figures on page 133.
7. If there are second or higher floors, compute the square foot area on each floor. Locate the appropriate square foot cost from the table at the bottom of this page. Adjust this figure for a wall height more or less than 9 feet, using the figures on page 133. Multiply the adjusted cost by the square foot area on each floor. Use the figures on page 133 to deduct for common walls or no wall finish. Add the result to the cost from step 6 above.
8. Multiply the total cost by the location factor listed on page 7.
9. Add the cost of heating and air conditioning systems, elevators, fire sprinklers, exterior signs, paving and curbing, miscellaneous yard improvements, covered porches and garages. See page 201.

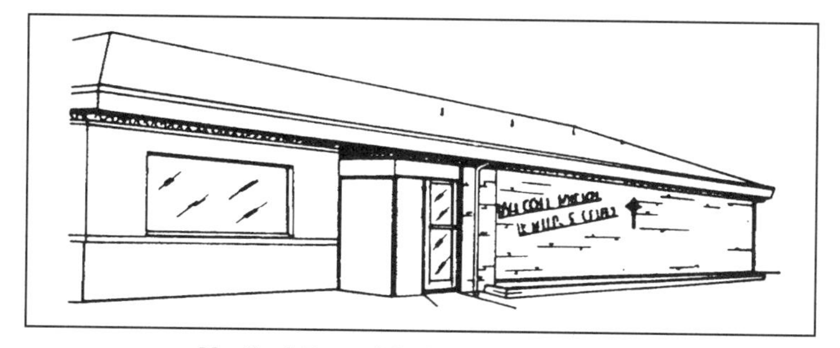

Medical-Dental Building, Class 2 & 3

First Story - Square Foot Area

Quality Class	1,000	1,500	2,000	2,500	3,000	4,000	5,000	7,500	10,000	15,000	20,000
Exceptional	158.00	151.87	148.36	146.00	144.32	142.01	140.47	138.13	136.77	135.16	134.24
1, Best	145.05	139.39	136.21	134.05	132.53	130.39	128.99	126.83	125.55	124.11	123.24
1 & 2	136.75	131.44	128.42	126.39	124.95	122.94	121.59	119.56	118.35	116.98	116.16
2, Good	129.37	124.33	121.46	119.55	118.20	116.28	115.02	113.09	111.99	110.67	109.89
2 & 3	120.13	115.44	112.77	111.02	109.74	107.95	106.81	104.99	103.99	102.74	102.04
3, Average	112.44	108.08	105.56	103.89	102.73	101.07	99.97	98.31	97.33	96.20	95.52
3 & 4	105.10	101.02	98.66	97.12	96.01	94.44	93.47	91.89	90.97	89.91	89.27
4, Low	98.13	94.32	92.17	90.71	89.67	88.25	87.27	85.83	84.98	83.97	83.37

Second and Higher Stories - Square Foot Area

Quality Class	1,000	1,500	2,000	2,500	3,000	4,000	5,000	7,500	10,000	15,000	20,000
Exceptional	140.85	137.29	135.21	133.79	132.76	131.31	130.34	128.82	127.92	126.87	126.25
1 Best	129.49	126.20	124.29	123.01	122.06	120.74	119.84	118.43	117.62	116.63	116.06
1 & 2	121.11	118.07	116.28	115.07	114.17	112.92	112.08	110.80	110.01	109.09	108.56
2, Good	117.97	115.01	113.29	112.09	111.22	110.02	109.21	107.93	107.12	106.28	105.75
2 & 3	109.02	106.25	104.65	103.57	102.74	101.62	100.87	99.71	99.00	98.21	97.71
3, Average	101.45	99.25	97.38	96.34	95.61	94.58	93.87	92.77	92.13	91.40	90.93
3 & 4	94.81	92.41	91.02	90.06	89.37	88.39	87.73	86.73	86.12	85.39	85.00
4, Low	88.68	86.45	85.15	84.25	83.60	82.70	82.05	81.12	80.57	79.90	79.48

Medical-Dental Buildings – Wood Frame

Exterior Suite Entrances, Length Between 2 and 4 Times Width

Estimating Procedure

1. Use these figures to estimate medical, dental, psychiatric, optometry and similar professional buildings in which access to each office suite is through an exterior entrance. Buildings in this section have more plumbing fixtures per square foot of floor and smaller room sizes than general office buildings. Note also that buildings with more than 10,000 square feet are assumed to have lead shielded x-ray rooms.
2. Establish the building quality class by applying the information on page 126.
3. Compute the first floor area. This should include everything within the exterior walls and all insets outside the main walls but under the main roof.
4. If the first floor wall height is more or less than 10 feet, add to or subtract from the first floor square foot cost below the appropriate amount from the Wall Height Adjustment Table on page 133.
5. Multiply the adjusted square foot cost by the first floor area.
6. Deduct, if appropriate, for common walls or no wall finish. Use the figures on page 133.
7. If there are second or higher floors, compute the square foot area on each floor. Locate the appropriate square foot cost from the table at the bottom of this page. Adjust this figure for a wall height more or less than 9 feet, using the figures on page 133. Multiply the adjusted cost by the square foot area on each floor. Use the figures on page 133 to deduct for common walls or no wall finish. Add the result to the cost from step 6 above.
8. Multiply the total cost by the location factor listed on page 7.
9. Add the cost of heating and air conditioning systems, elevators, fire sprinklers, exterior signs, paving and curbing, miscellaneous yard improvements, covered porches and garages. See page 201.

Medical-Dental Building, Class 1 & 2

First Story - Square Foot Area

Quality Class	1,000	1,500	2,000	2,500	3,000	4,000	5,000	7,500	10,000	15,000	20,000
Exceptional	161.34	153.71	149.43	146.61	144.62	141.81	140.00	137.21	135.61	133.74	132.65
1, Best	148.09	141.11	137.18	134.59	132.73	130.19	128.53	125.97	124.51	122.79	121.80
1 & 2	139.42	132.87	129.16	126.71	124.99	122.58	120.99	118.62	117.20	115.64	115.29
2, Good	131.68	125.63	122.09	119.80	118.17	115.88	114.40	112.12	110.82	109.29	108.43
2 & 3	122.35	116.61	113.32	111.20	109.65	107.54	106.17	104.08	102.84	101.46	100.63
3, Average	114.46	109.10	106.06	104.07	102.60	100.63	99.34	97.38	96.24	94.88	94.16
3 & 4	106.92	101.90	99.06	97.20	95.87	94.02	92.79	90.96	89.91	88.68	87.96
4, Low	99.86	95.16	92.49	90.75	89.48	87.80	86.65	84.92	83.94	82.78	82.10

Second and Higher Stories - Square Foot Area

Quality Class	1,000	1,500	2,000	2,500	3,000	4,000	5,000	7,500	10,000	15,000	20,000
Exceptional	142.50	137.97	135.37	133.64	132.38	130.65	129.49	127.70	126.68	125.48	124.78
1, Best	130.84	126.65	124.27	122.67	121.54	119.93	118.87	117.24	116.31	115.20	114.56
1 & 2	124.31	120.37	118.10	116.57	115.49	113.99	112.98	111.41	110.53	109.49	108.85
2, Good	119.32	115.52	113.35	111.91	110.84	109.38	108.44	106.96	106.09	105.05	104.48
2 & 3	110.35	106.83	104.85	103.48	102.49	101.15	100.28	98.88	98.09	97.14	96.60
3, Average	102.23	99.00	97.12	95.89	94.99	93.74	92.93	91.63	90.88	90.04	89.52
3 & 4	96.10	93.08	91.28	90.09	89.25	88.10	87.30	86.12	85.45	84.63	84.15
4, Low	89.19	86.36	84.75	83.64	82.86	81.77	81.04	79.95	79.32	78.54	78.09

Exterior Suite Entrances, Length More Than 4 Times Width

Estimating Procedure

1. Use these figures to estimate medical, dental, psychiatric, optometry and similar professional buildings in which access to each office suite is through an exterior entrance. Buildings in this section have more plumbing fixtures per square foot of floor and smaller room sizes than general office buildings. Note also that buildings with more than 10,000 square feet are assumed to have lead shielded x-ray rooms.
2. Establish the building quality class by applying the information on page 126.
3. Compute the first floor area. This should include everything within the exterior walls and all insets outside the main walls but under the main roof.
4. If the first floor wall height is more or less than 10 feet, add to or subtract from the first floor square foot cost below the appropriate amount from the Wall Height Adjustment Table on page 133.
5. Multiply the adjusted square foot cost by the first floor area.
6. Deduct, if appropriate, for common walls or no wall finish. Use the figures on page 133.
7. If there are second or higher floors, compute the square foot area on each floor. Locate the appropriate square foot cost from the table at the bottom of this page. Adjust this figure for a wall height more or less than 9 feet, using the figures on page 133. Multiply the adjusted cost by the square foot area on each floor. Use the figures on page 133 to deduct for common walls or no wall finish. Add the result to the cost from step 6 above.
8. Multiply the total cost by the location factor listed on page 7.
9. Add the cost of heating and air conditioning systems, elevators, fire sprinklers, exterior signs, paving and curbing, miscellaneous yard improvements, covered porches and garages. See page 201.

Medical-Dental Building, Class 1

First Story - Square Foot Area

Quality Class	1,000	1,500	2,000	2,500	3,000	4,000	5,000	7,500	10,000	15,000	20,000
Exceptional	168.66	159.31	154.08	150.64	148.19	144.83	142.65	139.29	137.38	135.16	133.89
1, Best	155.18	146.58	141.78	138.62	136.35	133.26	131.25	128.19	126.42	124.35	123.18
1 & 2	145.93	137.85	133.30	130.34	128.20	125.33	123.41	120.53	118.85	116.95	115.84
2, Good	137.77	130.16	125.88	123.08	121.05	118.32	116.54	113.80	112.21	110.43	109.37
2 & 3	127.85	120.74	116.79	114.18	112.34	109.78	108.12	105.57	104.11	102.45	101.48
3, Average	119.54	112.97	109.20	106.78	105.04	102.67	101.08	98.73	97.38	95.80	94.88
3 & 4	111.60	105.44	101.98	99.71	98.05	95.85	94.37	92.19	90.89	89.45	88.62
4, Low	104.17	98.40	95.18	93.05	91.52	89.44	88.09	86.02	84.83	83.51	82.70

Second and Higher Stories - Square Foot Area

Quality Class	1,000	1,500	2,000	2,500	3,000	4,000	5,000	7,500	10,000	15,000	20,000
Exceptional	147.32	141.49	138.20	136.07	134.55	132.46	131.05	128.99	127.78	126.40	125.57
1, Best	135.30	129.96	126.97	125.00	123.59	121.67	120.40	118.48	117.38	116.09	115.36
1 & 2	128.42	123.33	120.49	118.64	117.31	115.45	114.25	112.44	111.38	110.18	109.50
2, Good	122.84	117.97	115.27	113.45	112.17	110.44	109.28	107.54	106.54	105.41	104.70
2 & 3	113.53	109.04	106.51	104.88	103.67	102.08	101.02	99.40	98.44	97.42	96.78
3, Average	105.53	101.39	99.05	97.51	96.41	94.91	93.91	92.41	91.55	90.58	89.98
3 & 4	98.74	94.83	92.68	91.23	90.20	88.79	87.85	86.47	85.66	84.75	84.19
4, Low	92.20	88.59	86.49	85.19	84.23	82.91	82.03	80.75	79.97	79.11	78.61

Medical-Dental Buildings – Wood Frame

Interior Suite Entrances, Length Less Than Twice Width

Estimating Procedure

1. Use these figures to estimate medical, dental, psychiatric, optometry and similar professional buildings in which access to each office suite is through an interior corridor. Buildings in this section have more plumbing fixures per square foot or floor and smaller room sizes than general office buildings. Note also that buildings with more than 10,000 square feet are assumed to have lead shielded x-ray rooms.
2. Establish the building quality class by applying the information on page 126.
3. Compute the first floor area. This should include everything within the exterior walls and all insets outside the main walls but under the main roof.
4. If the first floor wall height is more or less than 10 feet, add or subtract from the first floor square foot cost below the appropriate amount from the Wall Height Adjustment Table on page 133.
5. Multiply the adjusted square foot cost by the first floor area.
6. Deduct, if appropriate, for common walls or no wall finish. Use the figures on page 133.
7. If there are second or higher floors, compute the square foot area on each floor. Locate the appropriate square foot cost from the table at the
bottom of this page. Adjust this figure for a wall height more or less than 9 feet, using the figures on page 133. Multiply the adjusted cost by the square foot area on each floor. Use the figures on page 133 to deduct for common walls or no wall finish. Add the result to the cost from step 6 above.
8. Multiply the total cost by the location factor listed on page 7.
9. Add the cost of heating and air conditioning systems, elevators, fire sprinklers, exterior signs, paving and curbing, miscellaneous yard improvements, covered porches and garages. See page 201.

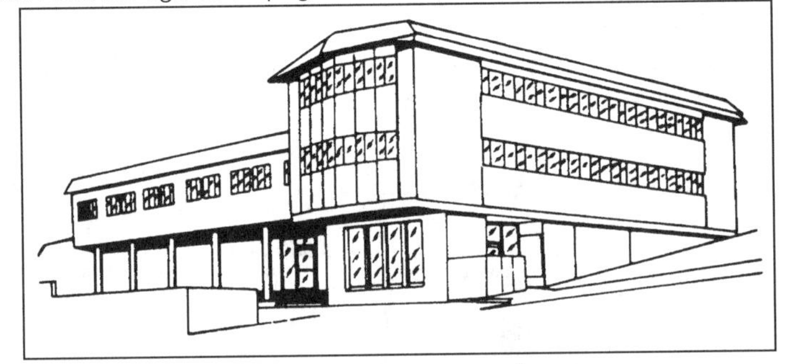

Medical-Dental Building, Class 3

First Story - Square Foot Area

Quality Class	2,000	2,500	3,000	4,000	5,000	7,500	10,000	15,000	20,000	30,000	40,000
Exceptional	133.76	131.58	129.99	127.80	126.36	124.14	122.84	121.31	120.41	119.34	118.76
1, Best	126.65	124.58	123.06	121.01	119.61	117.51	116.29	114.86	114.01	113.01	112.43
1 & 2	119.62	117.68	116.25	114.31	113.00	110.99	109.86	108.48	107.66	106.71	106.20
2, Good	113.48	111.64	110.29	108.44	107.18	105.30	104.19	102.94	102.15	101.29	100.75
2 & 3	105.41	103.67	102.44	100.73	99.55	97.80	96.77	95.52	94.88	94.05	93.59
3, Average	98.85	97.25	96.05	94.43	93.36	91.72	90.76	89.64	88.98	88.24	87.77
3 & 4	92.97	91.44	90.35	88.81	87.82	86.27	85.34	84.31	83.67	82.95	82.52
4, Low	87.44	86.01	84.98	83.55	82.58	81.13	80.29	79.32	78.68	78.04	77.63

Second and Higher Stories - Square Foot Area

Quality Class	2,000	2,500	3,000	4,000	5,000	7,500	10,000	15,000	20,000	30,000	40,000
Exceptional	122.64	121.31	120.35	118.99	118.09	116.68	115.86	114.92	114.35	113.71	113.32
1, Best	116.30	115.02	114.12	112.81	111.97	110.66	109.87	108.98	108.45	107.81	107.43
1 & 2	110.42	109.21	108.31	107.11	106.29	105.03	104.32	103.47	102.95	102.37	101.65
2, Good	105.45	104.28	103.45	102.29	101.50	100.31	99.64	98.79	98.32	97.75	97.41
2 & 3	97.43	96.34	95.58	94.51	93.77	92.70	92.04	91.29	90.84	90.32	90.00
3, Average	90.63	89.64	88.90	87.90	87.24	86.23	85.61	84.92	84.50	84.00	83.73
3 & 4	85.19	84.27	83.60	82.75	82.02	81.07	80.49	79.83	79.43	78.98	78.71
4, Low	80.02	79.14	78.51	77.63	77.04	76.13	75.61	74.99	74.60	74.19	73.93

Medical-Dental Buildings - Wood Frame

Interior Suite Entrances, Length Between 2 and 4 Times Width

Estimating Procedure

1. Use these figures to estimate medical, dental, psychiatric, optometry and similar professional buildings in which access to each office suite is through an interior corridor. Buildings in this section have more plumbing fixtures per square foot of floor and smaller room sizes than general office buildings. Note also that buildings with more than 10,000 square feet are assumed to have lead shielded x-ray rooms.
2. Establish the building quality class by applying the information on page 126.
3. Compute the first floor area. This should include everything within the exterior walls and all insets outside the main walls but under the main roof.
4. If the first floor wall height is more or less than 10 feet, add to or subtract from the first floor square foot cost below the appropriate amount from the Wall Height Adjustment Table on page 133.
5. Multiply the adjusted square foot cost by the first floor area.
6. Deduct, if appropriate, for common walls or no wall finish. Use the figures on page 133.
7. If there are second or higher floors, compute the square foot area on each floor. Locate the appropriate square foot cost from the table at the bottom of this page. Adjust this figure for a wall height more or less than 9 feet, using the figures on page 133. Multiply the adjusted cost by the square foot area on each floor. Use the figures on page 133 to deduct for common walls or no wall finish. Add the result to the cost from step 6 above.
8. Multiply the total cost by the location factor listed on page 7.
9. Add the cost of heating and air conditioning systems, elevators, fire sprinklers, exterior signs, paving and curbing, miscellaneous yard improvements, covered porches and garages. See page 201.

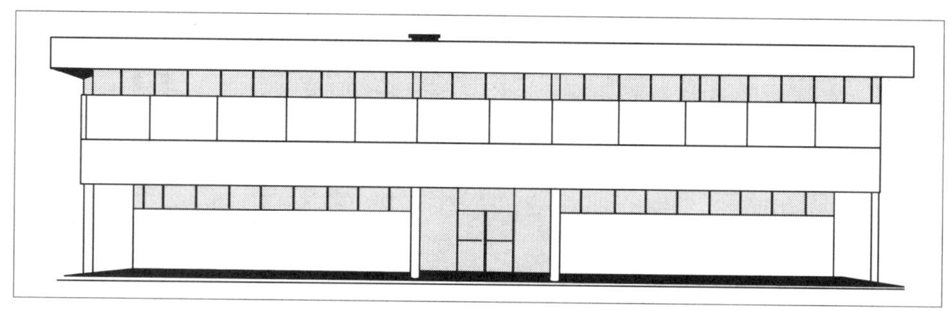

Medical-Dental Building, Class 1 & 2

First Story - Square Foot Area

Quality Class	2,000	2,500	3,000	4,000	5,000	7,500	10,000	15,000	20,000	30,000	40,000
Exceptional	137.11	134.46	132.54	129.92	128.19	125.52	124.00	122.18	121.18	119.92	119.23
1, Best	130.28	127.73	125.92	123.44	121.77	119.26	117.77	116.08	115.11	113.92	113.25
1 & 2	122.86	120.46	118.75	116.40	114.83	112.48	111.08	109.49	108.55	107.47	106.78
2, Good	116.43	114.19	112.56	110.34	108.83	106.59	105.29	103.77	102.90	101.85	101.25
2 & 3	108.19	106.09	104.58	102.50	101.12	99.03	97.83	96.41	95.58	94.62	94.07
3, Average	101.37	99.39	97.97	96.02	94.76	92.77	91.63	90.31	89.56	88.65	88.11
3 & 4	95.65	93.80	92.45	90.62	89.42	87.56	86.48	85.24	84.49	83.65	83.16
4, Low	89.62	87.86	86.63	84.87	83.74	82.02	81.01	79.86	79.16	78.38	77.89

Second and Higher Stories - Square Foot Area

Quality Class	2,000	2,500	3,000	4,000	5,000	7,500	10,000	15,000	20,000	30,000	40,000
Exceptional	125.31	123.47	122.15	120.40	119.25	117.52	116.57	115.45	114.83	114.09	113.64
1, Best	119.41	117.66	116.42	114.73	113.63	112.02	111.08	110.01	109.41	108.69	108.30
1 & 2	112.79	111.13	109.95	108.36	107.34	105.80	104.92	103.93	103.34	102.69	102.32
2, Good	107.66	106.09	104.96	103.46	102.46	101.00	100.14	99.20	98.66	98.03	97.67
2 & 3	99.85	98.38	97.36	95.94	95.02	93.67	92.87	91.99	91.49	90.89	90.57
3, Average	92.57	91.19	90.25	88.93	88.09	86.82	86.10	85.29	84.81	84.27	83.96
3 & 4	87.44	86.14	85.23	84.00	83.20	82.01	81.32	80.57	80.10	79.59	79.28
4, Low	81.74	80.55	79.69	78.54	77.80	76.65	76.05	75.31	74.93	74.45	74.14

Medical-Dental Buildings – Wood Frame

Interior Suite Entrances, Length More Than 4 Times Width

Estimating Procedure

1. Use these figures to estimate medical, dental, psychiatric, optometry and similar professional buildings in which access to each office suite is through an interior corridor. Buildings in this section have more plumbing fixtures per square foot of floor and smaller room sizes than general office buildings. Note also that buildings with more than 10,000 square feet are assumed to have lead shielded x-ray rooms.
2. Establish the building quality class by applying the information on page 126.
3. Compute the first floor area. This should include everything within the exterior walls and all insets outside the main walls but under the main roof.
4. If the first floor wall height is more or less than 10 feet, add to or subtract from the first floor square foot cost below the appropriate amount from the Wall Height Adjustment Table on page 133.
5. Multiply the adjusted square foot cost by the first floor area.
6. Deduct, if appropriate, for common walls or no wall finish. Use the figures on page 133.
7. If there are second or higher floors, compute the square foot area on each floor. Locate the appropriate square foot cost from the table at the bottom of this page. Adjust this figure for a wall height more or less than 9 feet, using the figures on page 133. Multiply the adjusted cost by the square foot area on each floor. Use the figures on page 133 to deduct for common walls or no wall finish. Add the result to the cost from step 6 above.
8. Multiply the total cost by the location factor listed on page 7.
9. Add the cost of heating and air conditioning systems, elevators, fire sprinklers, exterior signs, paving and curbing, miscellaneous yard improvements, covered porches and garages. See page 201.

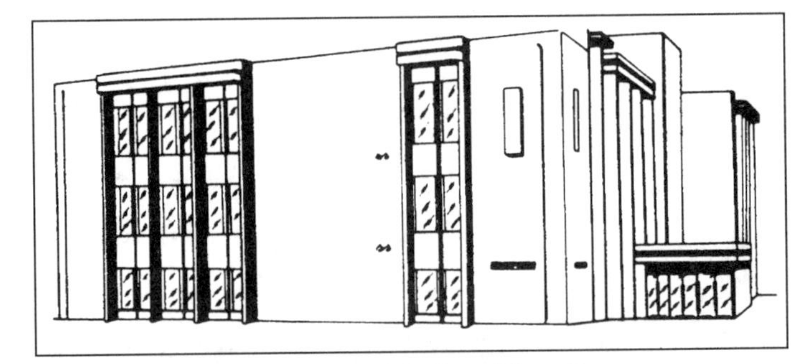

Medical-Dental Building, Class 2

First Story - Square Foot Area

Quality Class	2,000	2,500	3,000	4,000	5,000	7,500	10,000	15,000	20,000	30,000	40,000
Exceptional	142.44	139.27	136.97	133.71	131.50	128.09	126.07	123.71	122.25	120.58	119.60
1, Best	135.06	132.07	129.87	126.80	124.70	121.45	119.54	117.26	115.92	114.33	113.38
1 & 2	127.28	124.47	122.39	119.49	117.50	114.47	112.63	110.53	109.25	107.76	106.88
2, Good	120.50	117.85	115.87	113.13	111.26	108.39	106.66	104.65	103.45	102.02	101.17
2 & 3	112.47	109.97	108.16	105.55	103.84	101.15	99.52	97.67	96.54	95.20	94.44
3, Average	104.83	102.51	100.80	98.41	96.80	94.29	92.78	91.06	89.96	88.74	88.02
3 & 4	101.63	99.41	97.72	95.44	93.84	91.42	89.97	88.27	87.25	86.07	85.33
4, Low	92.52	90.48	88.98	86.87	85.46	83.22	81.93	80.38	79.43	78.35	77.69

Second and Higher Stories - Square Foot Area

Quality Class	2,000	2,500	3,000	4,000	5,000	7,500	10,000	15,000	20,000	30,000	40,000
Exceptional	127.89	125.82	124.33	122.30	120.96	118.91	117.78	116.43	115.68	114.73	114.23
1, Best	121.27	119.31	117.90	115.97	114.69	112.79	111.67	110.42	109.65	108.81	108.28
1 & 2	115.09	113.20	111.87	110.04	108.82	107.02	105.96	104.76	104.06	103.22	102.75
2, Good	109.83	108.07	106.78	105.03	103.87	102.15	101.14	99.99	99.31	98.55	98.09
2 & 3	102.01	100.35	99.16	97.54	96.48	94.87	93.91	92.86	92.23	91.53	91.11
3, Average	94.43	92.93	91.82	90.30	89.32	87.82	86.99	85.99	85.40	84.74	84.33
3 & 4	89.18	87.72	86.69	85.27	84.34	82.94	82.10	81.19	80.51	80.01	79.63
4, Low	83.32	81.99	81.04	79.69	78.82	77.52	76.77	75.87	75.36	74.78	74.44

Medical-Dental Buildings - Wood Frame

Wall Height Adjustment

The square foot costs for medical and dental buildings are based on the wall heights of 10 feet for first floors and 9 feet for higher floors. The first floor height is the distance from the bottom of the floor slab or joists to the top of the roof slab or ceiling joists. Second and higher floors are measured from the top of the floor slab or floor joists to the top of the roof slab or ceiling joists. Add or subtract the amount listed in this table to the square foot of floor cost for each foot of wall height more or less than 10 feet, if adjusting for first floor, and 9 feet, if adjusting for upper floors.

Square Foot Area

Quality Class	1,000	1,500	2,000	3,000	4,000	5,000	7,500	10,000	15,000	20,000	40,000
1, Best	1.49	1.18	1.01	.80	.70	.60	.48	.43	.36	.33	.22
2, Good	1.31	1.00	.87	.71	.58	.52	.43	.38	.32	.29	.21
3, Average	1.15	.88	.75	.61	.51	.46	.39	.34	.30	.24	.12
4, Low	1.02	.79	.67	.52	.46	.43	.34	.31	.24	.21	.12

Perimeter Wall Adjustment

A common wall exists when two buildings share one wall. Adjust for common walls by deducting the linear foot costs below from the total structure cost. In some structures, one or more walls are not owned at all. In this case, deduct the "No Ownership" cost per linear foot of wall not owned. If a wall has no exterior finish, deduct the "Lack of Exterior Finish" cost.

First Story

Class	For a Common Wall, Deduct Per L.F.	For No Wall Ownership, Deduct Per L.F.	For Lack of Exterior Finish, Deduct Per L.F.
1	$190.00	$369.00	$162.00
2	152.00	295.00	128.00
3	123.00	240.00	89.00
4	95.00	196.00	67.00

Second and Higher Stories

Class	For a Common Wall, Deduct Per L.F.	For No Wall Ownership, Deduct Per L.F.	For Lack of Exterior Finish, Deduct Per L.F.
1	$128.00	$245.00	$162.00
2	95.00	190.00	128.00
3	89.00	184.00	89.00
4	84.00	172.00	67.00

Convalescent Hospitals – Masonry or Concrete

Quality Classification

	Class 1 Best Quality	Class 2 Good Quality	Class 3 Average Quality	Class 4 Low Quality
Foundation (18% of total cost)	Reinforced concrete.	Reinforced concrete.	Reinforced concrete.	Reinforced concrete.
Floor Structure (10% of total cost)	Reinforced concrete.	Reinforced concrete.	Reinforced concrete.	Reinforced concrete.
Roof & Cover (7% of total cost)	Standard wood frame and composition roofing with large colored rock or tile. 4' to 6' sealed overhang.	Standard wood frame and composition roofing with large colored rock or asbestos shingles. 3' to 4' sealed overhang.	Standard wood frame and composition roofing small colored rock. 4' exposed or 3' sealed overhang.	Standard wood frame and composition roofing pea gravel or composition shingles. 2' exposed overhang.
Exterior Finish				
Walls (12% of total cost)	8" colored concrete block.	8" colored concrete block.	8" colored concrete block.	8" colored concrete block.
Front (5% of total cost)	Brick or stone veneer and float glass in aluminum frame.	Stone or brick veneer. Small amount of float glass in aluminum frame.	Rustic siding or some brick or stone veneer.	Stucco and small amount siding.
Windows (5% of total cost)	Good aluminum sliding glass doors or anodized aluminum casement windows.	Good aluminum sliding glass doors with aluminum jalousie window for vent.	Average aluminum sliding type. 6' sliding glass door in bedrooms.	Low cost aluminum sliding type.
Doors (5% of total cost)	Anodized aluminum entry doors.	Aluminum and glass entry doors.	Good hardwood slab door with side lights or aluminum and glass entry door.	Wood slab door.
Interior Finishes				
Floors (5% of total cost)	Resilient tile. Good carpet in entry and day room. Quarry tile or epoxy type floor in kitchen.	Resilient tile. Good carpet in day room. May have coved sheet vinyl in kitchen.	Composition tile. Day room may have average grade carpet.	Low cost tile. Composition tile in kitchen. Concrete in storage areas.
Walls (10% of total cost)	Gypsum wallboard and enamel. Large amount of plastic wall cover. Some wood paneling or glass walls in entry and day rooms. 4' ceramic tile wainscot in kitchen. Gypsum wallboard and paint in storage areas.	Gypsum wallboard and enamel. Gypsum wallboard and some plastic wall cover or wood paneling in entry, bedrooms, dining room, day room, and kitchen. Gypsum wallboard and paint in storage areas.	Gypsum wallboard and enamel. Some plastic wall cover or wood paneling in entry and day room. Gypsum wallboard and paint in storage areas.	Gypsum wallboard and texture. Gypsum wallboard and enamel in kitchen. Unfinished gypsum wallboard in storage areas.
Ceilings (8% of total cost)	Acoustical tile. Washable acoustical tile in kitchen. Gypsum wallboard and paint in storage areas.	Gypsum wallboard and enamel. Acoustical tile in corridor. Open beam ceilings with wood decking in entry and day room. Gypsum wallboard and paint in storage areas.	Gypsum wallboard & enamel or acoustical texture. Open beams with wood decking. Gypsum wallboard and paint in storage areas.	Gypsum wallboard and texture. Gypsum wallboard and acoustical texture in entry. Gypsum wallboard and enamel in kitchen. Unfinished gypsum in storage areas.
Bath Finish Detail				
Main Bath & Showers (10% of total cost)	Ceramic tile floors. Ceramic tile walls. Gypsum wallboard and enamel ceilings.	Ceramic tile floors. 6' ceramic tile wainscot. Gypsum wallboard and enamel walls and ceilings.	Ceramic tile floors. 4' ceramic tile wainscot. Gypsum wallboard and enamel walls and ceilings.	Resilient tile floors. Gypsum wallboard and enamel walls and ceilings.
Toilet Rooms (5% of total cost)	Sheet vinyl or ceramic tile floors. 4' ceramic tile wainscot. Gypsum wallboard and enamel walls and ceilings.	Sheet vinyl floors. 4' plastic or ceramic tile wainscot.	Resilient tile or linoleum floors. 4' plastic or marlite wainscot.	Composition tile floors. Gypsum wallboard and enamel walls and ceilings.

Note: Use the percent of total cost to help identify the correct quality classification.

Square foot costs include the following components: Foundations as required for normal soil conditions. Floor, wall and roof structures. Interior floor, wall and ceiling finishes. Exterior wall finish and roofing. Interior partitions. Cabinets, doors and windows. Basic electrical systems and all plumbing. Permits and fees. Contractors' mark-up. Cost of nurses' built-in station desks and cabinets. Nurses' call system. Emergency lighting system. Cubicle curtain tracks. Kitchen range hood.

The in-place cost of these extra components should be added to the basic building cost to arrive at the total structure cost. See the section "Additional Costs for Commercial Buildings" on page 201. Heating and air conditioning systems. Elevators. Fire sprinklers. Dumbwaiters. Fireplaces. *Square foot costs do not include the following items:* Drapes. Cubicle curtains. Incinerators. Laundry or kitchen equipment, other than the range hood.

Convalescent Hospitals – Masonry or Concrete

Estimating Procedure

1. Use these figures to estimate institutions such as nursing or rest homes which have facilities limited to bed care for people unable to care for themselves.
2. Establish the building quality class by applying the information on page 134.
3. Compute the floor area. This should include everything within the exterior walls and all insets outside the main walls but under the main roof. Areas such as storage rooms which have a substantially inferior finish or any area with a slightly interior finish and which does not perform a function related to the hospital operation should be estimated at 1/2 to 3/4 of the square foot costs listed.
4. Estimate second stories and basements at 100% of the first floor cost if they are of the same general quality. A storage area of the same or slightly inferior quality located in the basement, second floor or separate building should be estimated at 100% of the building square foot cost.
5. Multiply the appropriate square foot cost below by the floor area.
6. Multiply the total cost by the location factor on page 7.
7. Add the cost of heating and air conditioning systems, fire sprinklers, drapes, therapy facilities, cubicle curtains, incinerators, elevators, dumbwaiters, fireplaces, laundry equipment and kitchen equipment other than a range hood. See the section beginning on page 201.

Convalescent Hospital, Class 2

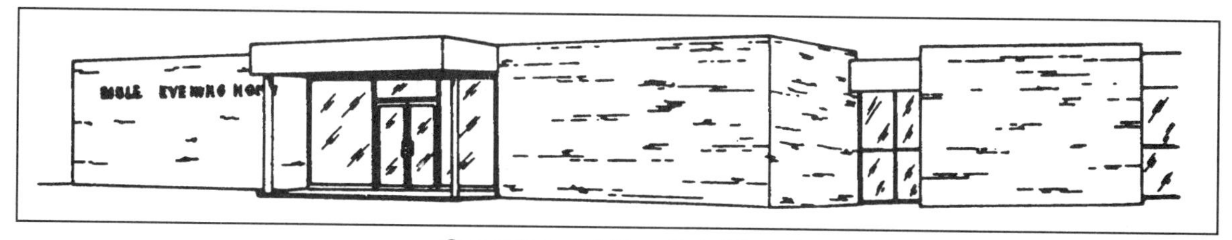

Convalescent Hospital, Class 2

Square Foot Area

Quality Class	5,000	6,000	7,000	8,000	9,000	10,000	12,500	15,000	17,500	20,000	25,000	30,000
Exceptional	136.40	132.12	129.04	126.73	124.96	123.53	120.99	119.30	118.10	117.20	115.99	115.06
1, Best	128.77	124.68	121.81	119.62	117.94	116.59	114.22	112.62	111.49	110.63	109.46	108.70
1 & 2	121.38	117.54	114.81	112.73	111.20	109.92	107.67	106.17	105.11	104.28	103.18	102.46
2, Good	114.50	110.88	108.31	106.37	104.89	103.70	101.58	100.16	99.18	98.40	97.36	96.67
2 & 3	105.73	102.39	100.01	98.24	96.86	95.74	93.78	92.51	91.53	90.86	89.88	89.27
3, Average	99.73	96.59	94.36	92.68	91.39	90.34	88.46	87.26	86.37	85.72	84.80	84.21
3 & 4	91.38	88.49	86.47	84.88	83.71	82.77	81.08	79.93	79.12	78.52	77.68	77.15
4, Low	83.64	80.98	79.10	77.70	76.60	75.74	74.18	73.14	72.42	71.82	71.10	70.59

Convalescent Hospitals - Wood Frame

Quality Classification

	Class 1 Best Quality	Class 2 Good Quality	Class 3 Average Quality	Class 4 Low Quality
Foundation (15% of total cost)	Reinforced concrete.	Reinforced concrete.	Reinforced concrete.	Reinforced concrete.
Floor Structure (7% of total cost)	Reinforced concrete.	Reinforced concrete.	Reinforced concrete.	Reinforced concrete.
Roof & Cover (7% of total cost)	Standard wood frame and composition roofing with large colored rock or tile. 4' to 6' sealed overhang.	Standard wood frame and composition roofing with large colored rock or asbestos shingles. 3' to 4' sealed overhang.	Standard wood frame and composition roofing small colored rock. 4' exposed or 3' sealed overhang.	Standard wood frame and composition with pea gravel or composition shingles. 2' exposed overhang.
Exterior Finish				
Walls (6% of total cost)	Wood siding or stucco with extensive trim.	Stucco or wood siding.	Stucco or good plywood siding.	Stucco.
Front (5% of total cost)	Brick or stone veneer and float glass in aluminum frame.	Stone or brick veneer. Small amount of float glass in aluminum frame.	Rustic siding or some brick or stone veneer.	Stucco and small amount siding.
Windows (5% of total cost)	Good aluminum sliding glass doors or anodized aluminum casement windows.	Good aluminum sliding glass doors with aluminum jalousie window for vent.	Average aluminum sliding type. 6' sliding glass door in bedrooms.	Low cost aluminum sliding type.
Doors (5% of total cost)	Anodized aluminum entry doors.	Aluminum and glass entry doors.	Good hardwood slab door with side lights or aluminum and glass entry door.	Wood slab door.
Interior Finishes				
Floors (5% of total cost)	Resilient tile. Good carpet in entry and day room. Quarry tile or epoxy type floor in kitchen.	Resilient tile. Good carpet in day room. May have coved sheet vinyl in kitchen.	Composition tile. Day room may have average grade carpet.	Low cost tile. Composition tile in kitchen. Concrete in storage areas.
Walls (10% of total cost)	Gypsum wallboard and enamel. Large amount of plastic wall cover. Some wood paneling or glass walls in entry and day rooms. 4' ceramic tile wainscot in kitchen. Gypsum wallboard and paint in storage areas.	Gypsum wallboard and enamel. Gypsum wallboard and some plastic wall cover or wood paneling in entry, bedrooms, dining room, day room, and kitchen. Gypsum wallboard and paint in storage areas.	Gypsum wallboard and enamel. Some plastic wall cover or wood paneling in entry and day room. Gypsum wallboard and paint in storage areas.	Gypsum wallboard and texture. Gypsum wallboard and enamel in kitchen. Unfinished gypsum wallboard in storage areas.
Ceilings (8% of total cost)	Acoustical tile. Washable acoustical tile in kitchen. Gypsum wallboard and paint in storage areas.	Gypsum wallboard and enamel. Acoustical tile in corridor. Open beam ceilings with wood decking in entry and day room. Gypsum wallboard and paint in storage areas.	Gypsum wallboard & enamel or acoustical texture. Open beams with wood decking. Gypsum wallboard and paint in storage areas.	Gypsum wallboard and texture. Gypsum wallboard and acoustical texture in entry. Gypsum wallboard and enamel in kitchen. Unfinished gypsum in storage areas.
Bath Finish Detail				
Main Bath & Showers (10% of total cost)	Ceramic tile floors. Ceramic tile walls. Gypsum wallboard and enamel ceilings.	Ceramic tile floors. 6' ceramic tile wainscot. Gypsum wallboard and enamel walls and ceilings.	Ceramic tile floors. 4' ceramic tile wainscot. Gypsum wallboard and enamel walls and ceilings.	Resilient tile floors. Gypsum wallboard and enamel walls and ceilings.
Toilet Rooms (5% of total cost)	Vinyl or ceramic tile floors. 4' ceramic tile wainscot. Gypsum wallboard and enamel walls and ceilings.	Vinyl floors. 4' plastic or ceramic tile wainscot.	Resilient tile or linoleum floors. 4' plastic wainscot.	Composition tile floors. Gypsum wallboard and enamel walls and ceilings.

Note: Use the percent of total cost to help identify the correct quality classification.

Square foot costs include the following components: Foundations as required for normal soil conditions. Floor, wall and roof structures. Interior floor, wall and ceiling finishes. Exterior wall finish and roofing. Interior partitions. Cabinets, doors and windows. Basic electrical systems and all plumbing. Permits and fees. Contractors' mark-up. Cost of nurses' built-in station desks and cabinets. Nurses' call system. Emergency lighting system. Cubicle curtain tracks. Kitchen range hood.

The in-place cost of these extra components should be added to the basic building cost to arrive at the total structure cost. See the section "Additional Costs for Commercial Buildings" on page 201. Heating and air conditioning systems. Elevators. Fire sprinklers. Dumbwaiters. Fireplaces. *Square foot costs do not include the following items:* Drapes. Cubicle curtains. Incinerators. Laundry or kitchen equipment, other than the range hood.

Convalescent Hospitals - Wood Frame

Estimating Procedure

1. Use these figures to estimate institutions such as nursing or rest homes which have facilities limited to bed care for people unable to care for themselves.
2. Establish the building quality class by applying the information on page 136.
3. Compute the floor area. This should include everything within the exterior walls and all insets outside the main walls but under the main roof. Areas such as storage rooms which have a substantially inferior finish or any area with a slightly interior finish and which does not perform a function related to the hospital operation should be estimated at 1/2 to 3/4 of the square foot costs listed.
4. Estimate second stories and basements at 100% of the first floor cost if they are of the same general quality. A storage area of the same or slightly inferior quality located in a basement, second floor or separate building should be estimated at 100% of the building square foot cost.
5. Multiply the appropriate square foot cost below by the floor area.
6. Multiply the total cost by the location factor on page 7.
7. Add the cost of heating and air conditioning systems, fire sprinklers, drapes, therapy facilities, cubicle curtains, incinerators, elevators, dumbwaiters, fireplaces, laundry equipment and kitchen equipment other than a range hood. See the section beginning on page 201.

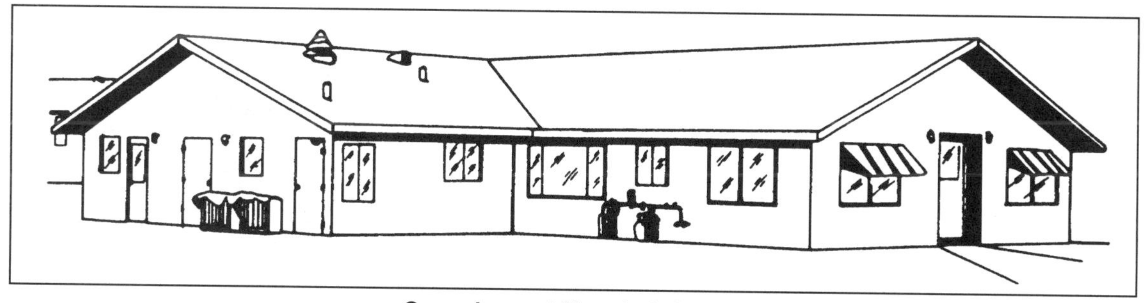

Convalescent Hospital, Class 4

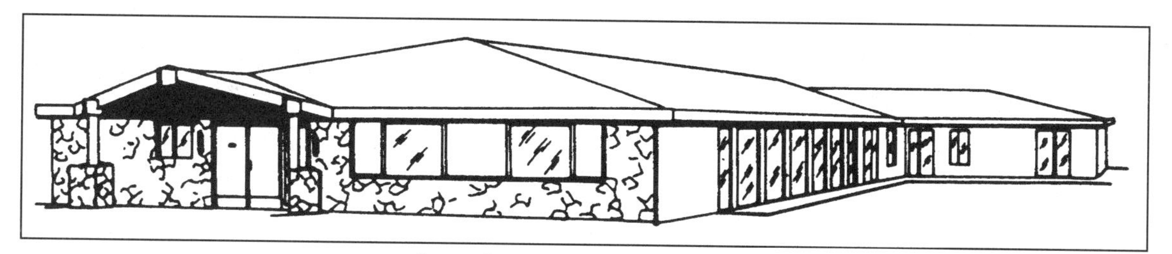

Convalescent Hospital, Class 3

Square Foot Area

Quality Class	5,000	6,000	7,000	8,000	9,000	10,000	12,500	15,000	17,500	20,000	25,000	30,000
Exceptional	120.71	116.92	114.18	112.07	110.43	109.13	106.72	105.12	103.91	103.02	101.73	100.86
1, Best	114.18	110.57	107.99	106.02	104.47	103.22	100.94	99.43	98.32	97.44	96.24	95.42
1 & 2	107.32	103.93	101.47	99.65	98.18	97.02	94.88	93.42	92.40	91.59	90.45	89.58
2, Good	101.50	98.31	95.99	94.25	92.85	91.76	89.75	88.38	87.37	86.62	85.56	84.80
2 & 3	94.81	91.81	89.66	88.01	86.71	85.71	83.80	82.55	81.62	80.91	79.90	79.21
3, Average	88.42	85.68	83.66	82.11	80.94	79.98	78.22	77.02	76.13	75.51	74.56	73.93
3 & 4	79.58	77.09	75.26	73.89	72.82	71.94	70.37	69.29	68.51	67.93	67.07	66.49
4, Low	74.52	72.19	70.45	69.17	68.20	67.39	65.91	64.89	64.14	63.61	62.81	62.26

Funeral Homes

Quality Classification

	Class 1 Best Quality	Class 2 Good Quality	Class 3 High Average Quality	Class 4 Low Average Quality	Class 5 Low Quality
Foundation (15% of total cost)	Reinforced concrete.	Reinforced concrete.	Reinforced concrete.	Reinforced concrete.	Reinforced concrete.
Floor Structure (7% of total cost)	Standard wood frame or reinforced concrete.	Standard wood frame or reinforced concrete.	Standard wood frame or reinforced concrete.	Standard wood frame or reinforced concrete.	Standard wood frame or reinforced concrete.
Wall Structure (12% of total cost)	Standard wood frame.	Standard wood frame.	Standard wood frame.	Standard wood frame.	Standard wood frame.
Roof & Cover (8% of total cost)	Standard wood frame with slate and 3' sealed overhang.	Standard wood frame with bar tile and 3' sealed overhang.	Standard wood frame with heavy shake and 3' sealed overhang.	Standard wood frame with composition roofing and large colored rock, asbestos shingles or medium shake, 3' exposed overhang.	Standard wood frame with compositon roofing and small colored rock or good composition shingles, 2' exposed overhang.
Exterior Finish					
Walls (5% of total cost)	Brick or stone veneer.	Good wood siding, brick or stone veneer.	Wood siding or stucco with extensive wood trim.	Stucco or wood siding.	Stucco.
Front (5% of total cost)	Brick or stone veneer. Some tinted float glass in bronze frames.	Brick or stone veneer. Some tinted float glass in anodized aluminum frames.	Brick or stone veneer. Some tinted float glass in anodized aluminum frames.	Stone or brick veneer. Some tinted crystal or float glass in wood frame.	Wood siding or stucco with stone brick veneer. Some colored veneer windows with wood frame (bottle glass type).
Windows (5% of total cost)	Good anodized aluminum windows. Extensive stained glass in chapel.	Good aluminum sliding type. Stained glass in chapel.	Good aluminum sliding type. Leaded glass in chapel.	Good aluminum sliding type. Mock leaded glass in chapel.	Average aluminum sliding type. Colored plastic in chapel.
Doors (5% of total cost)	Elaborate hardwood entry doors.	Custom-built hardwood entry doors.	Custom hardwood or anodized aluminum and glass.	Good hardwood or anodized aluminum and glass doors.	Good hardwood slab or aluminum and glass entry doors.
Interior Finish					
Floors (6% of total cost)	Excellent carpet. Good tile in kitchen.	Excellent carpet. Good carpet in living quarters. Good tile in kitchen.	Good carpet. Good tile in kitchen.	Good carpet. Average carpet in living quarters. Tile in kitchen.	Average carpet. Composition tile in hallways and living quarters.
Walls (12% of total cost)	Hardwood paneling, extensive brick or stone work in lobby and chapel. Gypsum wallboard and texture in living quarters.	Good hardwood veneer with brick or stone work in lobby and chapel. Gypsum wallboard and texture in living quarters.	Gypsum wallboard and embossed wallpaper or fabric textured wallcover. Hardwood in chapel. Gypsum wallboard and texture in living quarters.	Gypsum wallboard and wallpaper. Hardwood veneer in chapel.	Gypsum wallboard and texture. Some hardwood veneer in chapel.
Ceilings (8% of total cost)	Acoustical plaster. Exposed ornamental beams with ornamental hardwood decking or ornamental acoustical plaster in chapel.	Acoustical plaster. Exposed ornamental beams with ornamental wood decking or acoustical plaster in chapel.	Gypsum wallboard and heavy acoustical texture. Exposed architectural glu-lams w/ deck in chapel. Gypsum wallboard and texture in living quarters.	Gypsum wallboard and acoustical texture. Exposed wood beam with wood deck in chapel.	Gypsum wallboard and acoustical texture. Exposed beam with wood deck in chapel.
Special Lighting					
Fixtures (3% of total cost)	Two expensive chandeliers in lobby (4' to 6' diameter). Recessed and spotlighting in chapel.	Two expensive chandeliers in lobby (4' to 6' diameter). Recessed and spotlighting in chapel.	Two good chandeliers in lobby (4' diameter). Recessed and spotlighting in chapel.	Two average chandeliers in lobby (3' diameter).	One average chandelier in lobby.
Bath Detail (9% of total cost)	Ceramic tile floors and walls. Gypsum wallboard and enamel ceilings. Excellent fixtures.	Ceramic tile floors. 4' ceramic tile wainscot. Gypsum wallboard and enamel ceilings. Excellent fixtures.	Ceramic tile floors. Gypsum wallboard and enamel walls and ceilings. Good fixtures.	Good tile floors. Gypsum wallboard and enamel walls and ceilings. Average fixtures.	Tile floors. Gypsum wallboard and enamel walls and ceilings. Average fixtures.

Note: Use the percent of total cost to help identify the correct quality classification.

Square foot costs include the following components: Foundations as required for normal soil conditions. Floor, wall and roof structures. Interior floor, wall and ceiling finishes. Exterior wall finish and roof cover. Interior partitions. Cabinets, doors and windows. Basic electrical system and plumbing system. Special lighting fixtures. Bath detail. Permits and fees. Contractors' mark-up.

Funeral Homes – Wood Frame

Estimating Procedure

1. Use these figures to estimate mortuaries and buildings designed for care of the dead and funeral services.
2. Establish the building quality class by applying the information on page 138.
3. Compute the first floor area under the main roof and inside the exterior walls. Include lobby, chapel and living sections, but exclude garages, whether internal, external or attached.
4. Add to the first floor area between 1/3 and 1/4 of the actual garage area. Use the higher value for better quality garages.
5. Add to the first floor area all finished basement and second floor area if it is similar in quality to the first floor area. Inferior quality second floor area should be reclassified as to quality and calculated separately. Unfinished funeral home basements should be estimated from the figures on page 202.
6. Finished space within the area formed by the roof but with an exterior wall height less than 7 feet (half story area) should be included at 1/3 to 1/2 of the actual floor area. Exclude half story area when selecting a square foot cost from the table.
7. Multiply the sum from steps 3 to 6 above by the appropriate square foot cost below.
8. Add 8% if the exterior walls are brick or block and 12% if the exterior walls are concrete. Add 2% if the building has more than 16 corners or if the length is more than twice with width.
9. Multiply the building cost by the location factor listed on page 7.
10. Add the cost of heating and cooling systems, elevators, fireplaces, porches, pews, drapes, signs and yard improvement. See the section beginning on page 201.

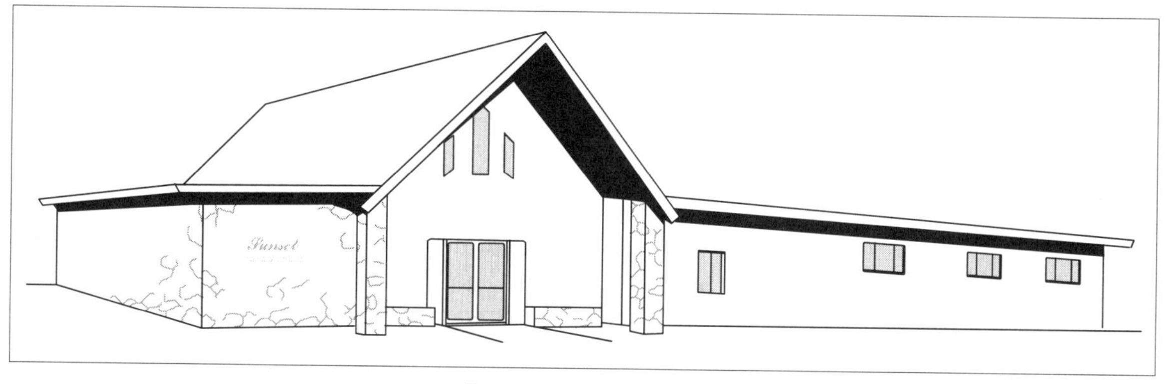

Funeral Home, Class 5

Square Foot Area

Quality	2,000	2,500	3,000	4,000	5,000	7,500	10,000	12,500	15,000	20,000	25,000	30,000
1, Best	—	—	—	—	106.57	102.68	100.63	99.31	98.40	97.20	96.43	95.90
1 & 2	—	—	—	—	99.22	95.63	93.71	92.48	91.63	90.49	89.78	89.26
2, Good	—	—	—	—	94.05	90.64	88.82	87.65	86.88	85.80	85.13	84.62
2 & 3	—	—	98.30	92.08	88.72	85.21	84.16	83.02	82.49	81.97	81.22	80.51
3, Average	—	—	92.90	87.03	83.85	79.20	78.86	78.23	77.96	77.80	77.89	78.05
3 & 4	95.67	90.57	87.10	82.79	80.19	76.65	74.85	73.74	73.01	72.07	71.52	71.12
4, Low Avg.	90.45	85.61	82.38	78.27	75.81	72.47	70.79	69.74	69.04	68.16	67.64	67.24
4 & 5	85.30	80.73	77.66	73.83	71.48	68.31	66.75	65.75	65.08	64.27	63.76	63.40
5, Low	79.14	74.92	72.04	68.47	66.34	63.39	61.93	61.00	60.38	59.63	59.15	58.82

Self Service Restaurants

Quality Classification

	Class 1 Best Quality	Class 2 Good Quality	Class 3 Average Quality	Class 4 Low Quality
Foundation (15% of total cost)	Reinforced concrete.	Reinforced concrete.	Reinforced concrete.	Reinforced concrete.
Floor Structure (7% of total cost)	Reinforced concrete or standard wood frame.	Reinforced concrete or standard wood frame.	Reinforced concrete or standard wood frame.	Reinforced concrete or standard wood frame.
Wall Structure (12% of total cost)	Standard wood frame.	Standard wood frame.	Standard wood frame.	Standard wood frame.
Roof & Cover (9% of total cost)	Complex angular structure, with concrete tile, mission tile, or heavy shake.	Standard wood frame structure, low or medium pitch, with composition tar and gravel or medium shake.	Standard wood frame structure, low pitch shed or gable type with composition tar and gravel.	Standard wood frame structure, low pitch shed or gable type with composition or composition shingle.
Floor Finish (7% of total cost)	Terrazzo in dining areas. Quarry tile in kitchen.	Vinyl tile in dining area. Terrazzo or quarry tile in kitchen.	Resilient tile in dining area. Grease proof tile in kitchen.	Composition tile in dining area. Grease proof tile in kitchen.
Walls, *Exterior Finish* (12% of total cost)	Very good wood siding; ornamental brick or natural stone veneer. Minimum 60% of front and side walls have float glass in heavy metal frames or rabbeted timbers.	Good wood siding or plywood; ornamental brick or concrete block; good stucco with color finish and with 2' to 4' brick, or natural stone veneer bulkheads. 50% to 75% of front and side walls have float glass set in good wood or metal frames.	Stucco or average wood siding with 2' to 4' brick or stone veneer bulkheads. Float glass front from waist high to roof line running along service front. Glassed area from 40% to 70% of front and side walls.	Stucco, low cost wood siding or plywood. 4' to 6' fixed glass waist high with sliding aluminum window opening running along service front. Glassed area from 30% to 40% of front and side walls.
Interior Finish (11% of total cost)	Plaster with putty coat finish; plastic coated paneling.	Plaster with putty coat finish; plastic coated paneling.	Plaster putty coat finish; textured gypsum wallboard with portions of wall finished with natural wood or plastic coated veneers.	Painted plaster or gypsum wallboard.
Ceilings (10% of total cost)	Complex angular structure with acoustical plaster.	Multi-level structure with good plaster or acoustical tile.	Flat or low slope with textured gypsum wallboard or acoustical tile.	Flat or low slope with textured gypsum wallboard or acoustical tile.
Plumbing (9% of total cost)	Good fixtures with metal or marble toilet partitions. Full wall height ceramic tile wainscot. Numerous floor drains.	Commercial fixtures with metal toilet partitions. Ceramic tile wainscot. Numerous floor drains in kitchen and dishwasher area.	4 to 6 standard commercial fixtures with metal toilet partitions. Floor drains in kitchen and dishwashing area.	4 to 6 standard commercial fixtures with wood toilet partitions. Minimum floor drainage in kitchen area.
Lighting (8% of total cost)	Triple encased fluorescent fixtures or decorative and ornate incandescent fixtures.	Triple encased fluorescent fixtures or decorative incandescent fixtures.	Triple open or double encased louvered fluorescent fixtures with some decorative incandescent fixtures.	Double open strip fluorescent fixtures or low cost incandescent fixtures.

Note: Use the percent of total cost to help identify the correct quality classification.

Square foot costs include the following components: Foundations as required for normal soil conditions. Floor, wall and roof structures. Interior floor, wall and ceiling finishes. Exterior wall finish and roof cover. All glass and glazing. Interior partitions. Basic electrical systems and lighting fixtures. Rough and finish plumbing. Permits and fees. Contractors' mark-up.

The in-place cost of these extra components should be added to the basic building cost to arrive at the total structure cost. See the section beginning on page 201. Heating and air conditioning systems. Booths and counters. Kitchen equipment. Fire sprinklers. Exterior signs. Paving and curbing. Yard improvements. Canopies and overhang.

Self Service Restaurants – Masonry

Estimating Procedure

1. Use these figures to estimate the cost of self-service or drive-in restaurants. These restaurants specialize in rapid service, may or may not have an interior eating area, have a large amount of glass on the front and side walls and usually are well lighted.
2. Establish the structure quality class by applying the information on page 140.
3. Compute the building floor area. This should include everything within the exterior walls and all insets outside the main walls but under the main roof.
4. Multiply the floor area by the appropriate cost below.
5. Multiply the total cost by the factor listed on page 7.
6. Add the cost of heating and air conditioning systems, booths and counters, kitchen equipment, fire sprinklers, exterior signs, paving and curbing, yard improvements and canopies. See the section beginning on page 201.

Self Service Restaurant, Class 2

Self Service Restaurant, Class 3

Square Foot Area

Quality Class	400	500	600	700	800	900	1,000	1,100	1,200	1,300	1,400
1, Best	296.64	273.81	256.85	243.52	232.73	223.72	216.08	209.47	203.71	198.62	194.03
1 & 2	256.62	236.93	222.20	210.67	201.32	193.55	186.95	181.21	176.24	171.81	167.87
2, Good	227.47	210.02	196.97	186.76	178.44	171.56	165.70	160.67	156.20	152.30	148.81
2 & 3	196.76	181.67	170.38	161.56	154.37	148.42	143.33	138.95	135.15	131.71	128.73
3, Average	173.55	160.23	150.30	142.50	136.16	130.89	126.45	122.56	119.20	116.20	113.52
3 & 4	150.12	138.59	129.98	123.25	117.78	113.22	109.36	106.03	103.08	100.50	98.20
4, Low	130.49	120.45	112.97	107.12	102.36	98.40	95.05	92.15	89.61	87.34	85.35

Square Foot Area

Quality Class	1,500	1,600	1,700	1,800	2,000	2,200	2,400	2,600	2,800	3,000	3,400
1, Best	190.40	188.04	185.78	183.66	179.04	176.18	172.98	170.04	167.36	164.91	160.50
1 & 2	164.80	162.72	160.78	158.94	154.94	152.48	149.69	147.17	144.85	142.73	138.91
2, Good	145.99	144.17	142.45	140.80	137.27	135.12	132.64	130.40	128.35	126.45	123.09
2 & 3	126.30	124.71	123.23	121.81	118.76	116.86	114.73	112.80	111.01	109.41	106.48
3, Average	111.37	109.99	108.69	107.43	104.76	103.07	101.17	99.47	97.91	96.46	93.91
3 & 4	96.38	95.16	94.00	92.93	90.60	89.16	87.58	86.07	84.73	83.46	81.23
4, Low	83.74	82.73	81.69	80.78	78.76	77.48	76.11	74.81	73.62	72.52	70.60

Self Service Restaurants – Wood Frame

Estimating Procedure

1. Use these figures to estimate the cost of self-service or drive-in restaurants. These restaurants specialize in rapid service, may or may not have an interior eating area, have a large amount of glass on the front and side walls and usually are well lighted.
2. Establish the structure quality class by applying the information on page 140.
3. Compute the building floor area. This should include everything within the exterior walls and all insets outside the main walls but under the main roof.
4. Multiply the floor area by the appropriate cost below.
5. Multiply the total cost by the factor listed on page 7.
6. Add the cost of heating and air conditioning systems, booths and counters, kitchen equipment, fire sprinklers, exterior signs, paving and curbing, yard improvements and canopies. See the section beginning on page 201.

Self Service Restaurant, Class 1 & 2

Self Service Restaurant, Class 1 & 2

Square Foot Area

Quality Class	400	500	600	700	800	900	1,000	1,100	1,200	1,300	1,400
1, Best	300.46	277.40	260.15	246.67	235.72	226.61	218.88	212.19	206.34	201.21	196.56
1 & 2	259.93	239.96	225.05	213.36	203.90	196.05	189.33	183.58	178.50	174.03	170.03
2, Good	229.18	211.58	198.46	188.13	179.81	172.88	166.96	161.88	157.39	153.47	149.96
2 & 3	198.16	182.91	171.56	162.64	155.44	149.43	144.33	139.92	136.08	132.67	129.64
3, Average	173.72	160.36	150.42	142.62	136.29	131.02	126.53	122.70	119.31	116.30	113.65
3 & 4	150.45	138.87	130.27	123.52	118.00	113.46	109.58	106.26	103.33	100.74	98.41
4, Low	130.68	120.63	113.14	107.29	102.52	98.56	95.23	92.30	89.76	87.51	85.49

Square Foot Area

Quality Class	1,500	1,600	1,700	1,800	2,000	2,200	2,400	2,600	2,800	3,000	3,400
1, Best	193.69	190.73	188.09	185.65	181.37	177.74	174.60	171.85	169.41	167.24	163.50
1 & 2	167.38	164.83	162.53	160.43	156.74	153.60	150.90	148.52	146.41	144.53	141.29
2, Good	147.75	145.47	143.46	141.59	138.33	135.57	133.18	131.08	129.22	127.55	124.72
2 & 3	127.77	125.82	124.05	122.45	119.66	117.25	115.17	113.34	111.74	110.30	107.84
3, Average	112.08	110.40	108.84	107.46	104.99	102.88	101.07	99.48	98.04	96.78	94.66
3 & 4	96.99	95.50	94.18	92.94	90.79	89.01	87.43	86.06	84.79	83.75	81.88
4, Low	84.25	82.96	81.83	80.75	78.89	77.32	75.95	74.76	73.70	72.76	71.13

Coffee Shop Restaurants

Quality Classification

	Class 1 Best Quality	Class 2 Good Quality	Class 3 Average Quality	Class 4 Low Quality
Foundation (15% of total cost)	Reinforced concrete.	Reinforced concrete.	Reinforced concrete.	Reinforced concrete.
Floor Structure (7% of total cost)	Reinforced concrete or standard wood frame.	Reinforced concrete or standard wood frame.	Reinforced concrete or standard wood frame.	Reinforced concrete or standard wood frame.
Wall Structure (12% of total cost)	Standard wood frame.	Standard wood frame.	Standard wood frame.	Standard wood frame.
Roof & Cover (9% of total cost)	Complex angular structure with concrete tile, mission tile, or heavy shake.	Standard wood frame structure, low or medium pitch, with composition tar and gravel roofing or medium shake.	Standard wood frame, low pitch shed or gable type with composition tar and gravel roofing	Standard wood frame, low pitch shed or gable type with composition tar and gravel or roofing.
Floor Finish (7% of total cost)	Terrazzo or good carpeting in dining area. Quarry tile in kitchen. Terrazzo or natural flagstone in entry vestibule.	Good tile or average carpeting in dining area. Terrazzo or quarry tile in kitchen. Terrazzo entry vestibule.	Good tile in dining area. Grease proof tile in kitchen area. Terrazzo in entry vestibule.	Composition tile in dining area. Grease proof tile in kitchen area.
Exterior Wall Finish (12% of total cost)	Very good wood siding; ornamental brick or natural stone veneer. Minimum 60% of front and side walls in float glass in heavy metal frames or rabbeted timbers. Entry vestibule with decorative screens of ornamental concrete block or natural stone panels.	Good wood siding; ornamental brick or concrete blocks. Good stucco with color finish and with 2' to 4' brick, or natural stone veneer bulkheads. 50% to 75% of front and side walls have float glass set in good wood or metal frames. Double entry doors, recessed entries with decorative screen of concrete block.	Stucco or average wood siding with 2' to 4' brick or stone veneer bulkheads. 50% to 60% of front and side walls in float glass set in average quality wood or aluminum frames. Double entry doors of metal or wood.	Stucco, low cost wood siding or plywood. 40% to 60% of front and side walls in crystal or float glass set in low-cost wood or metal frames. Double entry doors.
Interior Finish (11% of total cost)	Select and matched wood paneling, decorative wallpaper or synthetic fabrics.	Plaster with putty coat finish. Wood or plastic coated veneer paneling. Decorative wallpapers or fabrics.	Plaster putty coat finish textured gypsum wallboard with portions of wall finished with natural wood or plastic coated veneers.	Painted plaster or gypsum wallboard.
Ceilings (10% of total cost)	Complex angular structure with acoustical plaster.	Multi-level structure with good plaster or acoustical tile.	Flat or low slope with textured gypsum wallboard or acoustical tile.	Flat or low slope with textured gypsum wallboard or acoustical tile.
Plumbing (9% of total cost)	Good fixtures with metal or marble toilet partitions. Full wall height ceramic tile wainscot. Numerous floor drains.	Commercial fixtures with metal toilet partitions; ceramic tile wainscot. Numerous floor drains in kitchen and dishwasher area.	4 to 6 standard commercial fixtures with metal toilet partitions. Floor drains in kitchen and dishwashing area.	4 to 6 standard commercial fixtures with wood toilet partitions. Minimum floor drainage in kitchen.
Lighting (8% of total cost)	Triple encased fluorescent fixtures or decorative and ornate incandescent fixtures.	Triple encased fluorescent fixtures or decorative incandescent fixtures.	Triple open or double encased louvered fluorescent fixtures with some decorative incandescent fixtures.	Double open strip fluorescent fixtures or low cost incandescent fixtures.

Note: Use the percent of total cost to help identify the correct quality classification.

Square foot costs include the following components: Foundations as required for normal soil conditions. Floor, wall and roof structures. Interior floor, wall and ceiling finishes. Exterior wall finish and roof cover. All glass and glazing. Interior partitions. Basic electrical systems and lighting fixtures. Rough and finish plumbing. Permits and fees. Contractors' mark-up.

The in-place cost of these extra components should be added to the basic building cost to arrive at the total structure cost. See the section beginning on page 201: Heating and air conditioning systems. Booths and counters. Kitchen equipment. Fire sprinklers. Exterior signs. Paving and curbing. Yard improvements. Canopies and overhang.

Coffee Shop Restaurants – Masonry

Estimating Procedure

1. Use these figures to estimate the cost of restaurants with counters, or with counter, booth and table seating. A lounge or bar may be part of the building. Higher quality coffee shops usually have cut-up or complex sloping ceiling and roof structures. Coffee shops have extensive exterior glass and good lighting.
2. Establish the structure quality class by applying the information on page 143.
3. Compute the building floor area. This should include everything within the exterior walls and all insets outside the main walls but under the main roof.
4. Multiply the floor area by the appropriate cost below.
5. Multiply the total cost by the factor listed on page 7.
6. Add the cost of heating and air conditioning systems, booths and counters, kitchen equipment, fire sprinklers, exterior signs, paving and curbing, yard improvements and canopies. See the section beginning on page 201.

Coffee Shop Restaurant, Class 1

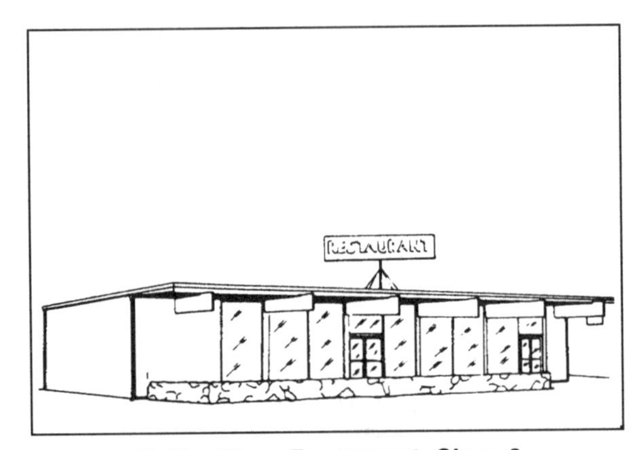

Coffee Shop Restaurant, Class 3

Square Foot Area

Quality Class	1,000	1,200	1,400	1,600	1,800	2,000	2,200	2,400	2,600	2,800	3,000
1, Best	205.36	190.10	179.11	170.72	164.17	158.81	154.43	150.72	147.55	144.84	142.42
1 & 2	177.96	164.77	155.22	147.95	142.24	137.64	133.79	130.59	127.88	125.49	123.43
2, Good	156.71	145.12	136.68	130.29	125.29	121.21	117.85	115.02	112.61	110.53	108.70
2 & 3	136.42	126.31	119.00	113.44	109.05	105.51	102.59	100.11	98.03	96.22	94.63
3, Average	120.17	111.26	104.82	99.92	96.05	92.93	90.36	88.18	86.33	84.74	83.34
3 & 4	103.95	96.26	90.68	86.44	83.12	80.40	78.19	76.31	74.69	73.32	72.10
4, Low	90.33	83.65	78.80	75.09	72.21	69.86	67.92	66.30	64.90	63.71	62.65

Square Foot Area

Quality Class	3,200	3,400	3,600	3,800	4,000	4,400	4,800	5,200	5,600	6,000	8,000
1, Best	138.68	137.12	135.76	134.46	133.28	131.18	129.34	127.73	126.27	125.00	120.09
1 & 2	119.96	118.65	117.43	116.33	115.29	113.48	111.89	110.48	109.24	108.12	103.89
2, Good	105.91	104.76	103.68	102.71	101.81	100.20	98.79	97.55	96.44	95.48	91.74
2 & 3	92.10	91.06	90.14	89.29	88.49	87.10	85.88	84.81	83.86	83.00	79.76
3, Average	81.23	80.32	79.50	78.76	78.06	76.81	75.75	74.81	73.95	73.21	70.33
3 & 4	70.25	69.48	68.77	68.15	67.54	66.47	65.55	64.72	63.97	63.33	60.86
4, Low	61.07	60.38	59.77	59.21	58.69	57.76	56.96	56.24	55.59	55.04	52.89

Coffee Shop Restaurants – Wood Frame

Estimating Procedure

1. Use these figures to estimate the cost of restaurants with counters, or with counter, booth and table seating. A lounge or bar may be part of the building. Higher quality coffee shops usually have cut-up or complex sloping ceiling and roof structures. Coffee shops have extensive exterior glass and good lighting.
2. Establish the structure quality class by applying the information on page 143.
3. Compute the building floor area. This should include everything within the exterior walls and all insets outside the main walls but under the main roof.
4. Multiply the floor area by the appropriate cost below.
5. Multiply the total cost by the factor listed on page 7.
6. Add the cost of heating and air conditioning systems, booths and counters, kitchen equipment, fire sprinklers, exterior signs, paving and curbing, yard improvements and canopies. See the section beginning on page 201.

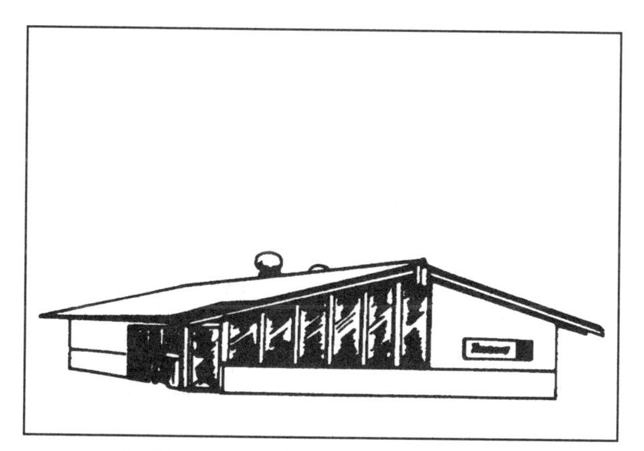

Coffee Shop Restaurant, Class 3

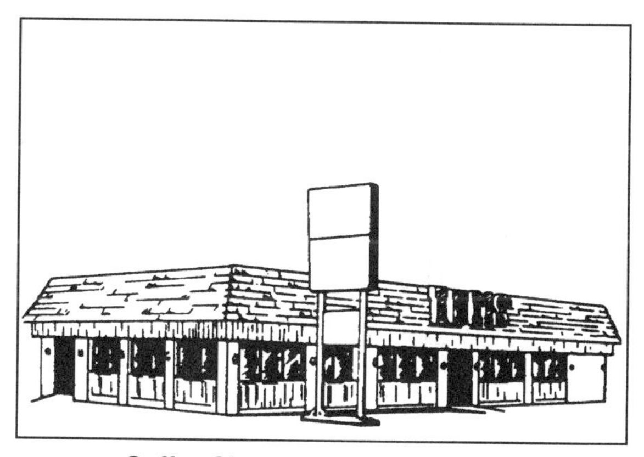

Coffee Shop Restaurant, Class 2

Square Foot Area

Quality Class	1,000	1,200	1,400	1,600	1,800	2,000	2,200	2,400	2,600	2,800	3,000
1, Best	211.17	195.38	184.00	175.36	168.59	163.13	158.60	154.83	151.58	148.79	146.35
1 & 2	182.82	169.15	159.28	151.85	145.97	141.25	137.33	134.06	131.25	128.82	126.76
2, Good	161.23	149.20	140.47	133.86	128.69	124.53	121.08	118.20	115.75	113.61	111.75
2 & 3	139.38	128.99	121.44	115.76	111.27	107.66	104.68	102.19	100.05	98.23	96.60
3, Average	122.31	113.17	106.59	101.58	97.65	94.47	91.88	89.67	87.80	86.19	84.78
3 & 4	105.79	97.88	92.16	87.83	84.45	81.69	79.47	77.54	75.95	74.52	73.31
4, Low	91.89	85.02	80.04	76.33	73.35	70.96	69.00	67.35	65.96	64.75	63.67

Square Foot Area

Quality Class	3,200	3,400	3,600	3,800	4,000	4,400	4,800	5,200	5,600	6,000	8,000
1, Best	142.69	141.11	139.71	138.39	137.17	135.02	133.14	131.47	129.99	128.65	123.59
1 & 2	123.43	122.08	120.85	119.72	118.69	116.80	115.17	113.73	112.46	111.29	106.93
2, Good	108.81	107.63	106.57	105.55	104.63	102.99	101.55	100.28	99.16	98.14	94.28
2 & 3	94.12	93.10	92.16	91.29	90.47	89.07	87.84	86.70	85.74	84.86	81.55
3, Average	82.58	81.66	80.86	80.09	79.40	78.13	77.05	76.07	75.24	74.46	71.54
3 & 4	71.41	70.65	69.91	69.28	68.68	67.56	66.64	65.79	65.07	64.40	61.86
4, Low	62.00	61.33	60.72	60.13	59.64	58.71	57.87	57.15	56.50	55.93	53.74

Conventional Restaurants

Quality Classification

	Class 1 Best Quality	Class 2 Good Quality	Class 3 Average Quality	Class 4 Low Quality
Foundation (10% of total cost)	Reinforced concrete.	Reinforced concrete.	Reinforced concrete.	Reinforced concrete.
Floor Structure (7% of total cost)	Reinforced concrete or standard wood frame.	Reinforced concrete or standard wood frame.	Reinforced concrete or standard wood frame.	Reinforced concrete or standard wood frame.
Wall Structure (12% of total cost)	Standard wood frame.	Standard wood frame.	Standard wood frame.	Standard wood frame.
Roof & Cover (9% of total cost)	Complex angular structure, with concrete tile, mission tile, or heavy shake.	Standard wood frame structure, low or medium pitch, with composition tar and gravel roofing or medium shake.	Standard wood frame, low pitch shed or gable type. composition tar and gravel roofing	Standard wood frame, low pitch shed or gable type, with composition tar and gravel roofing.
Floor Finish (7% of total cost)	Terrazzo or good carpeting in dining area. Quarry tile in kitchen. Terrazzo or natural flagstone in entry vestibule.	Vinyl tile or average carpeting in dining area. Terrazzo or quarry tile in kitchen. Terrazzo in entry vestibule.	Good tile in dining area. Grease proof tile in kitchen area. Terrazzo in entry vestibule.	Composition tile in dining area. Grease proof tile in kitchen area.
Exterior Wall Finish (12% of total cost)	Very good wood siding, ornamental brick or natural stone veneer. 10% of front and side walls in float glass in heavy metal frames or rabbeted timbers. Entry vestibule with decorative screens of ornamental concrete block or natural stone panels.	Good wood siding or plywood; ornamental brick or concrete block. Good stucco with color finish and with 2' to 4' brick, or natural stone veneer bulkheads. 10% to 20% of front and side walls in float glass set in good wood or metal frames. Double entry doors, recessed entries with decorative screen of concrete block.	Stucco or average wood siding with 2' to 4' brick or stone veneer bulkheads. 10% to 20% of front and side walls in float glass set in wood or aluminum extrusions. Double entry doors of metal or wood.	Stucco, low cost wood siding or plywood. 10% to 20% of front & side walls in crystal float glass set in low-cost wood or metal frames. Double entry doors.
Interior Finish (11% of total cost)	Select and matched wood paneling, decorative wallpaper or synthetic fabrics.	Plaster with putty coat finish; wood or plastic coated veneer paneling; decorative wallpapers or fabrics.	Plaster putty coat finish; textured gypsum wallboard with portions of wall finished with natural wood or plastic coated veneer.	Painted plaster or gypsum wallboard.
Ceilings (10% of total cost)	Multi-level structure with acoustical plaster.	Multi-level structure with good plaster or acoustical tile.	Flat or low slope with textured gypsum wallboard or acoustical tile.	Flat or low slope with textured gypsum wallboard or acoustical tile
Plumbing (9% of total cost)	Good fixtures with metal or marble toilet partitions. Full wall height ceramic tile wainscot. Numerous floor drains.	Commercial fixtures with metal toilet partitions. Ceramic tile wainscot. Numerous floor drains in kitchen and dishwasher area.	4 to 6 standard commercial fixtures with metal toilet partitions. Floor drains in kitchen and dishwashing area.	4 to 6 standard commercial fixtures with wood toilet partitions. Minimum floor drainage in kitchen area.
Lighting (8% of total cost)	Decorative and ornate incandescent fixtures.	Decorative incandescent fixtures.	Double encased louvered fluorescent fixtures with decorative incandescent fixtures.	Double open strip fluorescent fixtures or low cost incandescent fixtures.

Note: Use the percent of total cost to help identify the correct quality classification.

Square foot costs include the following components: Foundations as required for normal soil conditions. Floor, wall and roof structures. Interior floor, wall and ceiling finishes. Exterior wall finish and roof cover. All glass and glazing. Interior partitions. Basic electrical systems and lighting fixtures. Rough and finish plumbing. Permits and fees. Contractors' mark-up.

The in-place cost of these extra components should be added to the basic building cost to arrive at the total structure cost. See the section beginning on page 201: Heating and air conditioning systems. Booths and counters. Kitchen equipment. Fire sprinklers. Exterior signs. Paving and curbing. Yard improvements. Canopies and overhang.

Conventional Restaurants - Masonry or Wood Frame

Estimating Procedure

1. Use these figures to estimate the cost of restaurants with limited use of glass as exterior walls and subdued lighting. Roof and ceiling structures are simple and conventional. Seating may be any combination of counter, stools, booths or tables. A lounge or bar is often a part of a building of this type.
2. Establish the structure quality class by applying the information on page 146.
3. Compute the building floor area. This should include everything within the exterior walls and all insets outside the main walls but under the main roof.
4. Multiply the floor area by the appropriate cost below.
5. Multiply the total cost by the factor listed on page 7.
6. Add the cost of heating and air conditioning systems, booths and counters, kitchen equipment, fire sprinklers, exterior signs, paving and curbing, yard improvements and canopies. See the section beginning on page 201.

Conventional Restaurant, Class 2

Masonry - Square Foot Area

Quality Class	1,000	1,500	2,000	2,500	3,000	3,500	4,000	5,000	6,000	8,000	10,000
1, Best	200.11	178.95	167.17	159.43	153.93	149.74	146.43	141.51	137.97	133.09	129.87
1 & 2	173.09	154.80	144.58	137.90	133.11	129.51	126.66	122.36	119.31	115.11	112.34
2, Good	152.66	136.53	127.52	121.64	117.41	114.24	111.71	107.97	105.24	101.57	99.08
2 & 3	132.01	118.06	110.29	105.18	101.57	98.79	96.64	93.37	91.02	87.83	85.68
3, Average	115.84	103.61	96.75	92.31	89.11	86.66	84.77	81.92	79.88	77.05	75.18
3 & 4	100.17	89.56	83.67	79.83	77.05	74.95	73.30	70.82	69.05	66.63	65.02
4, Low	87.05	77.85	72.70	69.34	66.94	65.12	63.70	61.52	60.00	57.88	56.51

Wood Frame - Square Foot Area

Quality Class	1,000	1,500	2,000	2,500	3,000	3,500	4,000	5,000	6,000	8,000	10,000
1, Best	205.69	183.95	171.82	163.88	158.20	153.91	150.50	145.44	141.81	136.81	133.50
1 & 2	177.98	159.11	148.59	141.76	136.87	133.12	130.20	125.81	122.65	118.35	115.46
2, Good	156.90	140.32	131.06	125.03	120.66	117.39	114.79	110.96	108.16	104.36	101.83
2 & 3	135.73	121.36	113.34	108.11	104.38	101.58	99.31	95.96	93.57	90.29	88.07
3, Average	119.11	106.49	99.47	94.88	91.59	89.12	87.16	84.19	82.10	79.22	77.31
3 & 4	102.98	92.10	86.03	82.04	79.22	77.06	75.38	72.84	71.03	68.50	66.83
4, Low	89.47	80.01	74.74	71.29	68.83	66.94	65.46	63.28	61.71	59.50	58.08

"A-Frame" Restaurants

Quality Classification

	Class 1 Best Quality	Class 2 Good Quality	Class 3 Average Quality	Class 4 Low Quality
Foundation (19% of total cost)	Reinforced concrete.	Reinforced concrete.	Reinforced concrete.	Reinforced concrete.
Floor Structure (8% of total cost)	Reinforced concrete or standard wood frame.	Reinforced concrete or standard wood frame.	Reinforced concrete or standard wood frame.	Reinforced concrete or standard wood frame.
Roof & Cover (12% of total cost)	Wood frame or post and beam, high pitch with concrete tile or heavy wood shake.	Wood frame or post and beam, high pitch with aluminum shake or medium wood shake.	Standard wood frame or post and beam, high pitch with wood or composition shingles.	Standard wood frame or post and beam, high pitch with composition shingles.
Floor Finish (10% of total cost)	Terrazzo or good carpeting in dining area. Quarry tile in kitchen. Terrazzo or natural flagstone in entry vestibule.	Vinyl tile or average carpeting in dining area Terrazzo or quarry tile in kitchen, terrazzo in entry vestibule.	Good tile in dining area. Grease proof tile in kitchen area. Terrazzo in entry vestibule.	Composition tile in dining area. Grease proof tile in kitchen area.
Interior Finish (12% of total cost)	Select and matched wood paneling, decorative wallpapers or synthetic fabrics.	Plaster with putty coat finish, wood or plastic coated veneer paneling, decorative wallpapers or fabrics.	Plaster, putty coat finish; textured gypsum wallboard with portions of wall finished with natural wood or plastic coated veneer.	Painted plaster or gypsum wallboard.
End Walls (10% of total cost)	Very good wood siding; ornamental brick or natural stone veneer. 60% on front in float glass in heavy metal extrusions or rabbeted timbers. Entry vestibule with decorative screens of ornamental concrete block or natural stone panels.	Good wood siding; ornamental brick or concrete blocks. Good stucco with color finish and with 2' to 4' brick or natural stone veneer bulkheads. 50% of front in float glass set in good wood or metal frames. Double entry doors, recessed entries with decorative screen of concrete block.	Stucco or average wood siding with 2' to 4' brick or stone veneer bulkheads. 50% to 60% of front in float glass set in average quality wood or aluminum frames. Double entry doors of metal or wood.	Stucco, low cost wood siding or plywood. 40% to 60% of front in crystal glass set in low cost wood or metal frames.
Ceilings (9% of total cost)	Decorative wood.	Natural wood, good plaster or acoustical tile.	Painted wood, textured gypsum wallboard or tile.	Painted wood, textured gypsum wallboard or acoustical tile.
Plumbing (10% of total cost)	Good fixtures with metal or marble toilet partititions. Full wall height ceramic tile wainscot. Numerous floor drains.	Commercial fixtures with metal toilet partitions. Ceramic tile wainscot. Numerous floor drains in kitchen and dishwashing area.	4 to 6 standard commercial fixtures with metal toilet partitions. Floor drains in kitchen and dishwashing area.	4 to 6 standard commercial fixtures with wood toilet partitions. Minimum floor drainage in kitchen area.
Lighting (10% of total cost)	Triple encased fluorescent fixtures or decorative and ornate incandescent fixtures.	Triple encased fluorescent or decorative incandescent fixtures.	Triple open, double encased louvered fluorescent fixtures or average incandescent fixtures.	Double open strip fluorescent fixtures or low cost incandescent fixtures.

Note: Use the percent of total cost to help identify the correct quality classification.

Square foot costs include the following components: Foundations as required for normal soil conditions. Floor, wall and roof structures. Interior floor, wall and ceiling finishes. Exterior wall finish and roof cover. All glass and glazing. Interior partitions. Basic electrical systems and lighting fixtures. Rough and finish plumbing. Permits and fees. Contractors' mark-up.

The in-place cost of these extra components should be added to the basic building cost to arrive at the total structure cost. See the section beginning on page 201: Heating and air conditioning systems. Booths and counters. Kitchen equipment. Fire sprinklers. Exterior signs. Paving and curbing. Yard improvements. Canopies and overhang.

"A-Frame" Restaurants – Wood Frame

Estimating Procedure

1. Use these figures to estimate the cost of restaurants with a sloping roof that forms two or more exterior walls. Either conventional or self service may be used and a lounge or bar may be part of the building.
2. Establish the structure quality class by applying the information on page 148.
3. Compute the building floor area. This should include everything within the exterior walls and all insets outside the main walls but under the main roof.
4. Multiply the floor area by the appropriate cost below.
5. Multiply the total cost by the factor listed on page 7.
6. Add the cost of heating and air conditioning systems, booths and counters, kitchen equipment, fire sprinklers, exterior signs, paving and curbing, yard improvements and canopies. See the section beginning on page 201.

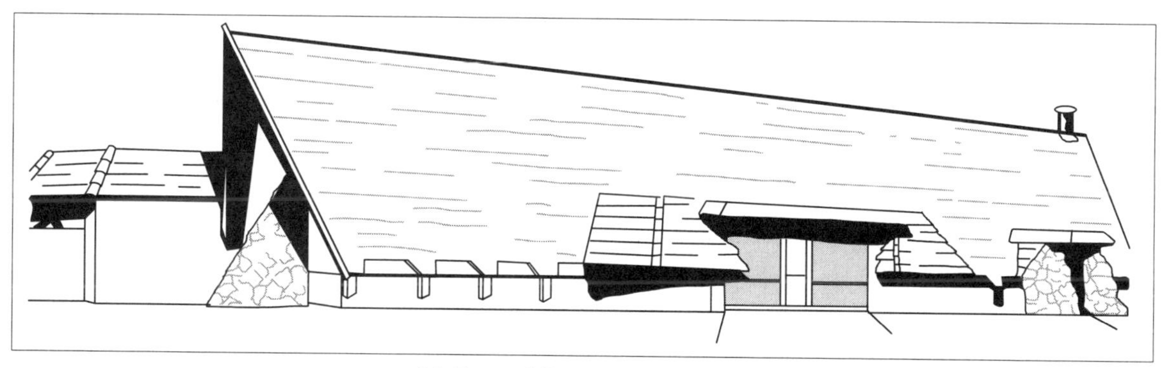

"A-Frame" Restaurant, Class 1

Square Foot Area

Quality Class	400	500	600	800	1,000	1,200	1,400	1,600	1,800	2,000	2,400
1, Best	269.08	239.22	219.08	193.48	177.83	167.21	159.51	153.64	148.99	145.25	139.60
1 & 2	232.70	206.93	189.51	167.37	153.82	144.62	137.94	132.88	128.87	125.64	120.71
2, Good	205.26	182.50	167.15	147.61	135.66	127.56	121.67	117.20	113.67	110.82	106.49
2 & 3	177.56	157.88	144.61	127.69	117.37	110.34	105.27	101.40	98.33	95.87	92.10
3, Average	155.86	138.58	126.93	112.09	103.00	96.86	92.40	89.01	86.32	84.15	80.85
3 & 4	134.64	119.75	109.64	96.84	89.04	83.67	79.83	76.88	74.58	72.70	69.85
4, Low	117.06	104.07	95.32	84.18	77.38	72.74	69.39	66.83	64.81	63.18	60.72

Square Foot Area

Quality Class	2,800	3,200	3,600	4,000	4,500	5,000	6,000	7,000	8,000	9,000	10,000
1, Best	135.47	132.40	129.90	127.85	125.74	123.37	121.22	119.16	117.53	116.20	115.13
1 & 2	117.18	114.50	112.36	110.57	108.73	107.22	104.84	103.06	101.66	100.50	99.58
2, Good	103.36	101.01	99.11	97.53	95.92	94.58	92.49	90.89	89.66	88.65	87.82
2 & 3	89.33	87.44	85.81	84.46	83.06	81.89	80.07	78.70	77.64	76.74	76.04
3, Average	78.49	76.68	75.26	74.06	72.83	71.84	70.23	69.00	68.07	67.33	66.68
3 & 4	67.79	66.27	64.98	63.96	62.90	62.03	60.67	59.64	58.81	58.17	57.61
4, Low	58.99	57.65	56.56	55.67	54.74	53.98	52.79	51.90	51.17	50.61	50.11

Theaters - Masonry or Concrete

Quality Classification

	Class 1 Best w/ Balcony	Class 2 Very Good w/ Balcony	Class 3 Good Quality	Class 4 Average Quality	Class 5 Low Quality
Foundation (12% of total cost)	Reinforced concrete.	Reinforced concrete.	Reinforced concrete.	Reinforced concrete.	Reinforced concrete.
Floor Structure (7% of total cost)	6" reinforced concrete on 6" rock fill or 2" x 12" joists, 16" o.c.	4" to 6" reinforced concrete on 6" rock fill or 2" x 10" joists, 16" o.c.	4" reinforced concrete on 6" rock fill or 2" x 10" joists, 16" o.c.	4" reinforced concrete on 6" rock fill or 2" x 8" joists, 16" o.c.	4" reinforced concrete on 6" rock fill or 2" x 6" joists, 16" o.c.
Wall Structure (15% of total cost)	8" reinforced concrete or 12" common brick.	8" reinforced concrete or 12" common brick.	8" reinforced concrete or 12" common brick.	8" reinforced concrete or 12" common brick.	8" reinforced concrete block or 6" reinforced concrete.
Roof Framing (7% of total cost)	2" x 10" joists 16" o.c. Trusses on heavy pilasters 20' o.c. 2" x 12" rafters or purlins 16" o.c.	2" x 10" joists 16" o.c. Trusses on heavy pilasters 20' o.c. 2" x 12" rafters or purlins 16" o.c.	2" x 10" joists 16" o.c. Trusses on heavy pilasters 20' o.c. 2" x 12" rafters or purlins 16" o.c.	2" x 10" joists 16" o.c. Trusses on heavy pilasters 20' o.c. 2" x 10" rafters or purlins 16" o.c.	2" x 10" joists 16" o.c. Trusses on pilasters 20' o.c. 2" x 10" rafters or purlins 16" o.c.
Roof Covering (5% of total cost)	5 ply composition roof on 1" sheathing with insulation.	5 ply composition roof on 1" sheathing with insulation.	5 ply composition roof on 1" sheathing with insulation.	5 ply composition roof on 1" sheathing with insulation.	4 ply composition roof on 1" sheathing.
Front (3% of total cost)	Highly ornamental stucco or plaster finishes, custom select stone veneers or highly ornamental terra cotta.	Ornamental stucco, custom select brick, natural stone veneers or marble.	Highly ornamental stucco or custom brick or natural stone or terra cotta veneers.	Ornamental stucco or select brick veneers or partial terra cotta veneers.	Plain or colored stucco or common brick or carrara glass veneers.
Floors, Entry (3% of total cost)	Custom terrazzo with intricate and ornamental designs, custom select natural stone, marble or very good carpet.	Custom terrazzo with highly ornamental designs, natural stone veneers, marble or very good carpet.	Custom designed terrazzo with ornamental designs with portions natural stone veneers, marble or good carpet.	Colored terrazzo with designs, or average carpet.	Colored concrete with some plain colored terrazzo.
Floors, Interior (5% of total cost)	Concrete with carpet throughout.	Concrete with carpet throughout.	Concrete with carpet throughout.	Concrete with carpet throughout.	Concrete with carpet runners at aisles.
Restrooms (8% of total cost)	Per code requirement capacity. Custom select ornamental ceramic tile, terrazzo or natural stone on floors and walls.	Per code requirement capacity. Custom ceramic tile or terrazzo on floor and walls.	Per code requirement capacity. Ceramic tile on floors and walls or terrazzo floors and walls.	Per code requirement capacity. Ceramic tile on walls and floors or terrazzo on floors.	Per code requirement capacity. Ceramic or vinyl tile on floors.
Walls, Interior (11% of total cost)	Painted or custom select canvas backed or custom molded cloth tapestry wallpapers or ornamental plaster finished with detailed sirocco type moldings and trim, walls with selected matched wood veneers.	Painted or custom canvas backed, custom molded cloth tapestry wallpapers or ornamental plaster finished with custom moldings and trimmings. Portions of wall select matched wood veneers.	Painted or finished with custom canvas backed wallpapers or molded tapestry finished wallpapers or select wood veneer matched full height at lobby.	Painted or finished with durable canvas or select quality wood veneers on gypsum board or plaster.	Painted and papered with durable canvas materials with portion wood veneers on gypsum board or plaster.
Ceiling (6% of total cost)	Suspended acoustical with highly ornate moldings and trim with acoustical baffles.	Suspended acoustical with highly ornate moldings and trim with acoustical baffles.	Suspended acoustical with ornate plaster cove moldings and trim with acoustical baffles.	Suspended acoustical with plaster moldings and sound baffles.	Suspended acoustical tile.
Lighting (10% of total cost)	Incandescent or recessed lights throughout. Interior with ornate chandelier type fixtures throughout. All dimmer controlled.	Incandescent or recessed lights throughout. Interior with ornate chandelier type fixtures throughout. All dimmer controlled.	Incandescent fixtures with chandelier and recessed lights at theater area, dimmer controlled.	Incandescent fixtures in lobby with fluorescent or chandelier type fixtures in theater area, recessed lighting, dimmer controlled.	Incandescent fixtures with recessed fixtures, dimmer controlled.
Seating (8% of total cost)	Main floor and balcony seating.	Main floor and balcony seating.	May or may not have balcony.	May or may not have balcony.	May or may not have balcony.

Note: Use the percent of total cost to help identify the correct quality classification.

Square foot costs include the following components: Foundations as required for normal soil conditions. Floor, wall, and roof structures. Interior floor, wall and ceiling finishes. Exterior wall finish and roof cover. Display fronts. Interior partitions. All doors. Ticket booth. Basic lighting and electrical systems. Rough and finish plumbing. A mezzanine floor projection booth. A frame-work for mounting a picture screen. A balcony in auditorium type theaters. Permits and fees. Contractors' mark-up.

Theaters with Balcony – Masonry or Concrete

Estimating Procedure

1. Establish the structure quality class by applying the information on page 150.
2. Compute the building floor area. This should include everything within the main walls and all insets outside the main walls but under the main roof.
3. Add to or subtract from the square foot cost below the appropriate amount from the Wall Height Adjustment Table on page 155 if the wall height is more or less than 28 feet.
4. Multiply the adjusted square foot cost by the building floor area.
5. Deduct, if appropriate, for common walls, using the figures on page 155.
6. Multiply the total cost by the location factor listed on page 7.
7. Add the cost of heating and air conditioning systems, fire extinguishers, exterior signs, paving and curbing. See the section beginning on page 201.

Length less that twice width - Square Foot Area

Quality Class	5,000	6,000	7,000	8,000	9,000	10,000	12,000	15,000	20,000	24,000	30,000
1 Best	157.29	151.54	147.15	143.68	140.89	138.55	134.86	130.84	126.43	123.98	121.36
1 & 2	151.38	145.82	141.61	138.31	135.59	133.33	129.78	125.91	121.67	119.33	116.79
2, Very Good	150.07	144.54	140.36	137.09	134.41	132.16	128.65	124.81	120.61	118.27	115.76
2 & 3	143.98	138.69	134.72	131.55	128.97	126.81	123.45	119.77	115.73	113.51	111.09
3, Good	140.62	135.46	131.53	128.45	125.92	123.86	120.55	116.97	113.01	110.84	108.49
3 & 4	133.98	129.07	125.36	122.43	120.01	118.02	114.87	111.43	107.69	105.63	103.36
4, Average	130.22	125.42	121.78	118.96	116.62	114.69	111.61	108.30	104.65	102.64	100.45
4 & 5	123.12	118.58	115.16	112.45	110.26	108.43	105.54	102.41	98.96	97.06	94.98
5, Low	116.82	112.52	109.32	106.72	104.63	102.90	100.14	97.16	93.91	92.11	90.14

Length between 2 and 4 times width - Square Foot Area

Quality Class	5,000	6,000	7,000	8,000	9,000	10,000	12,000	15,000	20,000	24,000	30,000
1, Best	167.49	161.30	156.64	152.98	150.00	147.50	143.54	139.30	134.57	132.00	129.19
1 & 2	161.18	155.27	150.77	147.22	144.35	141.96	138.16	134.06	129.54	127.04	124.34
2, Very Good	159.82	153.94	149.49	146.00	143.12	140.73	136.98	132.91	128.44	125.99	123.29
2 & 3	153.30	147.65	143.39	140.03	137.28	135.01	131.37	127.49	123.18	120.81	118.25
3, Good	149.73	144.22	140.05	136.77	134.07	131.86	128.32	124.54	120.31	118.01	115.50
3 & 4	141.52	136.31	132.39	129.28	126.76	124.64	121.33	117.72	113.74	111.58	109.20
4, Average	138.73	133.64	129.77	126.72	124.24	122.16	118.92	115.37	111.50	109.36	107.03
4 & 5	131.14	126.30	122.67	119.76	117.43	115.48	112.39	109.07	105.37	103.35	101.16
5, Low	124.39	119.84	116.39	113.64	111.41	109.55	106.64	103.47	99.97	98.06	95.98

Length more than 4 times width - Square Foot Area

Quality Class	5,000	6,000	7,000	8,000	9,000	10,000	12,000	15,000	20,000	24,000	30,000
1, Best	178.36	171.81	166.88	163.00	159.80	157.16	152.94	148.35	143.33	140.53	137.50
1 & 2	171.63	165.37	160.62	156.87	153.79	151.25	147.16	142.79	137.93	135.25	132.34
2, Very Good	170.13	163.94	159.21	155.51	152.44	149.90	145.92	141.53	136.76	134.07	131.18
2 & 3	163.36	156.60	152.85	149.29	146.39	143.93	140.08	135.89	131.28	128.72	125.93
3, Good	159.38	153.56	149.13	145.64	142.81	140.41	136.65	132.58	128.07	125.58	122.91
3 & 4	151.92	146.36	142.17	138.84	136.13	133.87	130.27	126.37	122.08	119.71	117.13
4, Average	148.18	142.76	138.66	135.43	132.76	130.55	127.06	123.28	119.07	116.77	114.26
4 & 5	139.56	134.46	130.57	127.55	125.06	122.97	119.67	116.12	112.16	109.99	107.59
5, Low	132.31	127.59	123.92	121.02	118.65	116.70	113.57	110.18	106.42	104.39	102.12

Theaters without Balcony – Masonry or Concrete

Length Less Than Twice Width

Estimating Procedure

1. Establish the structure quality class by applying the information on page 150.
2. Compute the building floor area. This should include everything within the main walls and all insets outside the main walls but under the main roof.
3. Add to or subtract from the square foot cost below the appropriate amount from the Wall Height Adjustment Table on page 155 if the wall height is more or less than 20 feet.
4. Multiply the adjusted square foot cost by the building floor area.
5. Deduct, if appropriate, for common walls, using the figures on page 155.
6. Multiply the total cost by the location factor listed on page 7.
7. Add the cost of heating and air conditioning systems, fire extinguishers, exterior signs, paving and curbing. See the section beginning on page 201.

Theater Without Balcony, Class 4

Square Foot Area

Quality Class	3,000	3,500	4,000	5,000	6,000	7,000	8,000	10,000	12,000	15,000	20,000
3, Good	91.36	88.46	86.21	82.90	80.59	78.84	77.49	75.50	74.09	72.57	70.96
3 & 4	87.69	84.89	82.75	79.58	77.34	75.69	74.37	72.49	71.10	69.66	68.07
4, Average	86.28	83.54	81.41	78.28	76.11	74.47	73.19	71.31	69.98	68.53	66.98
4 & 5	83.31	80.67	78.61	75.59	73.46	71.87	70.64	68.84	67.55	66.18	64.69
5, Low	80.59	78.04	76.06	73.13	71.09	69.56	68.38	66.60	65.36	64.02	62.61

Theaters without Balcony - Masonry or Concrete

Length Between 2 and 4 Times Width

Estimating Procedure

1. Establish the structure quality class by applying the information on page 150.
2. Compute the building floor area. This should include everything within the main walls and all insets outside the main walls but under the main roof.
3. Add to or subtract from the square foot cost below the appropriate amount from the Wall Height Adjustment Table on page 155 if the wall height is more or less than 20 feet.
4. Multiply the adjusted square foot cost by the building floor area.
5. Deduct, if appropriate, for common walls, using the figures on page 155.
6. Multiply the total cost by the location factor listed on page 7.
7. Add the cost of heating and air conditioning systems, fire extinguishers, exterior signs, paving and curbing. See the section beginning on page 201.

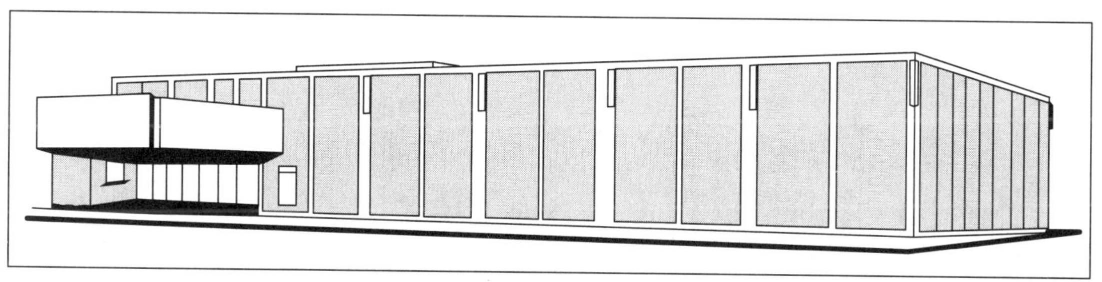

Theater Without Balcony, Class 3

Square Foot Area

Quality Class	3,000	3,500	4,000	5,000	6,000	7,000	8,000	10,000	12,000	15,000	20,000
3, Good	98.51	94.34	91.31	87.23	84.65	82.90	81.64	80.03	79.08	78.23	77.53
3 & 4	94.60	90.58	87.66	83.72	81.26	79.61	78.42	76.86	75.91	75.09	74.40
4, Average	93.18	89.23	86.34	82.49	80.05	78.39	77.23	75.71	74.80	73.98	73.32
4 & 5	89.94	86.13	83.36	79.63	77.30	75.68	74.57	73.07	72.18	71.41	70.76
5, Low	87.13	83.41	80.72	77.12	74.84	73.30	72.19	70.78	69.93	69.17	68.52

Theaters without Balcony – Masonry or Concrete

Length More Than 4 Times Width

Estimating Procedure

1. Establish the structure quality class by applying the information on page 150.
2. Compute the building floor area. This should include everything within the main walls and all insets outside the main walls but under the main roof.
3. Add to or subtract from the square foot cost below the appropriate amount from the Wall Height Adjustment Table on page 155 if the wall height is more or less than 20 feet.
4. Multiply the adjusted square foot cost by the building floor area.
5. Deduct, if appropriate, for common walls, using the figures on page 155.
6. Multiply the total cost by the location factor listed on page 7.
7. Add the cost of heating and air conditioning systems, fire extinguishers, exterior signs, paving and curbing. See the section beginning on page 201.

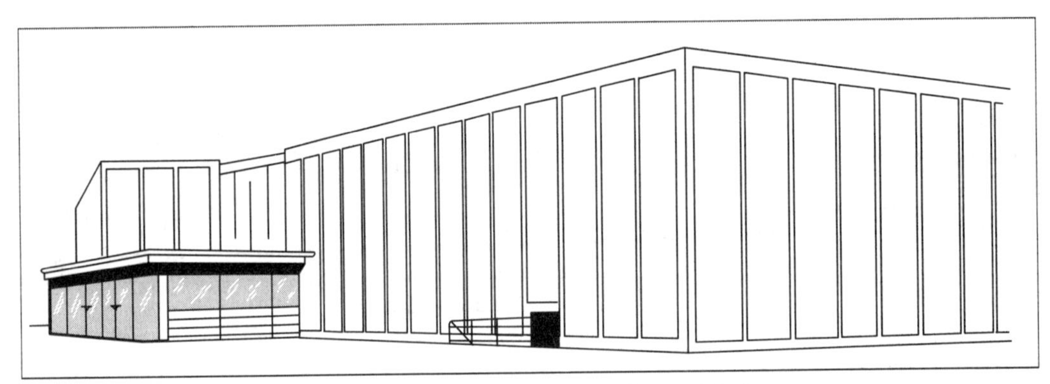

Theater Without Balcony, Class 3 & 4

Square Foot Area

Quality Class	3,000	3,500	4,000	5,000	6,000	7,000	8,000	10,000	12,000	15,000	20,000
3, Good	102.94	99.60	96.99	93.22	90.55	88.63	87.10	84.85	83.30	81.64	79.87
3 & 4	98.37	95.16	92.68	89.07	86.56	84.72	83.24	81.11	79.62	78.02	76.29
4, Average	97.23	94.07	91.59	88.05	85.55	83.69	82.27	80.16	78.67	77.13	75.45
4 & 5	93.86	90.78	88.43	84.97	82.59	80.77	79.44	77.37	75.96	74.45	72.82
5, Low	90.83	87.85	85.57	82.25	79.89	78.18	76.84	74.91	73.49	72.06	70.45

Theaters – Masonry or Concrete

Wall Height Adjustment

Add or subtract the appropriate amount listed in this table to the square foot of floor cost for each foot of wall height more or less than 28 feet, if adjusting for a theater with balcony, or 20 feet, if adjusting for a theater without a balcony.

Square Foot Area

Quality Class	3,000	3,500	4,000	5,000	6,000	7,000	8,000	10,000	12,000	15,000	20,000
1, Best	2.71	2.61	2.56	2.46	2.40	2.35	2.30	2.24	2.21	2.16	2.13
2, Very Good	2.56	2.50	2.45	2.35	2.26	2.21	2.19	2.14	2.11	2.04	1.99
3, Good	2.46	2.39	2.32	2.22	2.15	2.12	2.09	2.03	2.01	1.96	1.89
4, Average	2.31	2.20	2.16	2.10	2.04	2.01	1.93	1.89	1.86	1.82	1.79
5, Low	2.15	2.09	2.03	1.95	1.91	1.85	1.81	1.77	1.74	1.72	1.66

Perimeter Wall Adjustment

A common wall exists when two buildings share one wall. Adjust for common walls by deducting the linear foot costs below from the total structure cost. In some structures one or more walls are not owned at all. In this case, deduct the "No Ownership" cost per linear foot of wall not owned. For Common Wall, deduct $210.00 per linear foot. For no Wall Ownership, deduct $420.00 per linear foot.

Theaters - Wood Frame

Quality Classification

	Class 1 Best Quality	Class 2 Good Quality	Class 3 Average Quality	Class 4 Low Quality
Foundation (12% of total cost)	Reinforced concrete.	Reinforced concrete.	Reinforced concrete.	Reinforced concrete.
Floor Structure (7% of total cost)	4" reinforced concrete on 6" rock fill or 2" x 10" joists, 16" o.c.	4" reinforced concrete on 6" rock fill or 2" x 8" joists, 16" o.c.	4" reinforced concrete on 6" rock fill or 2" x 6" joists, 16" o.c.	4" reinforced concrete on 4" rock fill or 2" x 6" joists, 16" o.c.
Wall Structure (15% of total cost)	2" x 6", 16" o.c.	2" x 4" or 2" x 6", 16" o.c.	2" x 4", 16" o.c. up to 14' high, 2" x 6", 16" o.c. over 14' high.	2" x 4", 16" o.c. up to 14' high, 2" x 6", 16" o.c. over 14' high.
Roof Framing (7% of total cost)	2" x 10" joists, 16" o.c. Trusses on heavy pilasters, 20' o.c. 2" x 12" rafters or purlins, 16" o.c.	2" x 10" joists, 16" o.c. Trusses on heavy pilasters, 20' o.c. 2" x 10" rafters or purlins.	2" x 10" joists, 16" o.c. Steel trusses on pilasters, 20' o.c. 2" x 10" rafters or purlins, 16" o.c.	2" x 10" joists, 16" o.c. Wood trusses, 2" x 8" purlins, 16" o.c.
Roof Covering (5% of total cost)	5 ply composition roof on 1" x 6" sheathing with insulation.	5 ply composition roof on 1" x 6" sheathing with insulation.	4 ply composition roof on 1" x 6" sheathing.	4 ply composition roof on 1" x 6" sheathing.
Front (3% of total cost)	Highly ornamental stucco or custom brick or natural stone or terra cotta veneers.	Ornamental stucco or select brick veneers or partial terra cotta veneers.	Plain or colored stucco or common brick or ornamental wood.	Plain stucco.
Floors, Entry & Lobby (3% of total cost)	Custom designed terrazzo with ornamental designs with portions natural stone veneers, marble or good carpet.	Colored terrazzo with designs or average carpet.	Colored concrete and portions terrazzo, plain colored.	Plain or colored concrete.
Floors, Interior (5% of total cost)	Concrete with carpet throughout.	Concrete with carpet throughout.	Concrete with carpet runners at aisles.	Plain or colored concrete.
Restrooms (8% of total cost)	As per code requirement capacity. Ceramic tile on floors and walls or terrazzo floors and walls.	As per code requirement capacity. Ceramic tile on walls and floors or terrazzo on floors.	As per code requirement capacity. Ceramic or vinyl tile on floors.	As per code requirement capacity. Plain concrete floors and walls, painted.
Walls, Interior (11% of total cost)	Painted and finished with custom backed wallpapers or molded tapestry finished wallpapers or select wood veneer, matched full height at lobby.	Painted and finished with durable canvas or wood veneers, select quality on gypsum wallboard or plaster.	Painted and papered with durable canvas materials with portion wood veneers on gypsum wallboard or plaster.	Painted with or without stencil type painted molded designs on gypsum wallboard taped and textured.
Ceilings (6% of total cost)	Suspended acoustical, ornate cove moldings and trim with acoustical baffles.	Suspended acoustical with plaster moldings and sound baffles.	Suspended acoustical tile.	Gypsum wallboard taped, textured and painted.
Lighting (10% of total cost)	Incandescent fixtures with chandelier fixtures, recessed at theater area, dimmer controlled.	Incandescent fixtures in lobby with fluorescent or chandelier type fixtures in theater area, recessed lighting, dimmer controlled.	Incandescent recessed fixtures, dimmer controlled.	Plain incandescent fixtures, with dimmers.

Note: Use the percent of total cost to help identify the correct quality classification.

Square foot costs include the following components: Foundations as required for normal soil conditions. Floor, wall, and roof structures. Interior floor, wall and ceiling finishes. Exterior wall finish and roof cover. Display fronts. Interior partitions. All doors. Ticket booth. Basic lighting and electrical systems. Rough and finish plumbing. A mezzanine floor projection booth. A frame-work for mounting a picture screen. A balcony in auditorium type theaters. Permits and fees. Contractors' mark-up.

Theaters – Wood Frame

Length Less Than Twice Width

Estimating Procedure

1. Establish the structure quality class by applying the information on page 156.
2. Compute the building floor area. This should include everything within the main walls and all insets outside the main walls but under the main roof.
3. Add to or subtract from the square foot cost below the appropriate amount from the Wall Height Adjustment Table on page 160 if the wall height is more or less than 20 feet.
4. Multiply the adjusted square foot cost by the building floor area.
5. Deduct, if appropriate, for common walls, using the figures on page 160.
6. Multiply the total cost by the location factor listed on page 7.
7. Add the cost of heating and air conditioning systems, fire extinguishers, exterior signs, paving and curbing. See the section beginning on page 201.

Theater, Class 4 Front, Class 3 Rear

Square Foot Area

Quality Class	3,000	3,500	4,000	5,000	6,000	7,000	8,000	10,000	12,000	15,000	20,000
1, Best	77.64	75.18	73.25	70.44	68.51	67.02	65.88	64.18	63.00	61.70	60.30
1 & 2	75.23	72.81	70.95	68.25	66.35	64.90	63.78	62.16	60.99	59.76	58.40
2, Good	74.21	71.85	70.01	67.35	65.45	64.06	62.96	61.35	60.17	58.96	57.63
2 & 3	71.84	69.54	67.77	65.22	63.36	62.01	60.95	59.37	58.24	57.07	55.77
3, Average	70.37	68.16	66.41	63.88	62.07	60.76	59.72	58.18	57.09	55.94	54.66
3 & 4	68.26	66.11	64.42	61.98	60.23	58.94	57.91	56.45	55.37	54.25	53.02
4, Low	66.31	64.20	62.57	60.14	58.49	57.24	56.25	54.79	53.77	52.68	51.50

Theaters – Wood Frame

Length Between 2 and 4 Times Width

Estimating Procedure

1. Establish the structure quality class by applying the information on page 156.
2. Compute the building floor area. This should include everything within the main walls and all insets outside the main walls but under the main roof.
3. Add to or subtract from the square foot cost below the appropriate amount from the Wall Height Adjustment Table on page 160 if the wall height is more or less than 20 feet.
4. Multiply the adjusted square foot cost by the building floor area.
5. Deduct, if appropriate, for common walls, using the figures on page 160.
6. Multiply the total cost by the location factor listed on page 7.
7. Add the cost of heating and air conditioning systems, fire extinguishers, exterior signs, paving and curbing. See the section beginning on page 201.

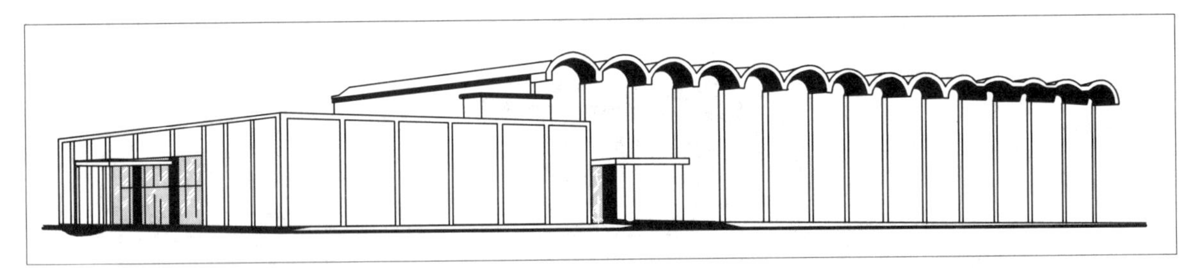

Theater, Class 3

Square Foot Area

Quality Class	3,000	3,500	4,000	5,000	6,000	7,000	8,000	10,000	12,000	15,000	20,000
1, Best	82.73	80.14	78.07	75.07	73.00	71.42	70.18	68.38	67.08	65.72	64.22
1 & 2	80.09	77.55	75.59	72.70	70.65	69.14	67.93	66.17	64.93	63.61	62.17
2, Good	79.00	76.49	74.51	71.69	69.69	68.17	66.99	65.26	64.03	62.73	61.32
2 & 3	76.44	74.01	72.11	69.36	67.41	65.98	64.84	63.16	61.95	60.66	59.30
3, Average	74.99	72.61	70.77	68.05	66.14	64.72	63.62	61.98	60.80	59.54	58.20
3 & 4	72.71	70.40	67.96	65.87	64.13	62.75	61.65	60.07	58.96	57.71	56.45
4, Low	70.48	68.25	66.50	63.95	62.16	60.82	59.77	58.22	57.11	55.96	54.69

Theaters – Wood Frame

Length More Than 4 Times Width

Estimating Procedure

1. Establish the structure quality class by applying the information on page 156.
2. Compute the building floor area. This should include everything within the main walls and all insets outside the main walls but under the main roof.
3. Add to or subtract from the square foot cost below the appropriate amount from the Wall Height Adjustment Table on page 160 if the wall height is more or less than 20 feet.
4. Multiply the adjusted square foot cost by the building floor area.
5. Deduct, if appropriate, for common walls, using the figures on page 160.
6. Multiply the total cost by the location factor listed on page 7.
7. Add the cost of heating and air conditioning systems, fire extinguishers, exterior signs, paving and curbing. See the section beginning on page 201.

Theater, Class 3

Square Foot Area

Quality Class	3,000	3,500	4,000	5,000	6,000	7,000	8,000	10,000	12,000	15,000	20,000
1, Best	88.00	85.18	83.03	79.86	77.62	75.94	74.62	72.74	71.35	69.90	68.29
1 & 2	85.21	82.53	80.41	77.33	75.19	73.58	72.30	70.44	69.13	67.71	66.17
2, Good	84.10	81.43	79.36	76.33	74.18	72.60	71.35	69.50	68.19	66.83	65.32
2 & 3	81.41	78.82	76.84	73.87	71.82	70.28	69.07	67.27	66.05	64.69	63.23
3, Average	79.90	77.33	75.38	72.49	70.49	68.95	67.76	66.05	64.79	63.46	62.04
3 & 4	77.42	74.95	73.05	70.25	68.29	66.83	65.67	63.95	62.77	61.49	60.11
4, Low	75.61	73.21	71.35	68.65	66.73	65.27	64.16	62.53	61.35	60.08	58.72

Theaters - Wood Frame

Wall Height Adjustment

Add or subtract the amount listed in this table to the square foot of floor cost for each foot of wall height more or less than 20 feet.

Square Foot Area

Quality Class	3,000	3,500	4,000	5,000	6,000	7,000	8,000	10,000	12,000	15,000	20,000
1, Best	2.07	2.01	1.97	1.87	1.84	1.81	1.77	1.73	1.68	1.64	1.61
2, Good	1.99	1.94	1.90	1.79	1.73	1.71	1.68	1.64	1.62	1.58	1.55
3, Average	1.89	1.83	1.80	1.71	1.68	1.64	1.62	1.58	1.56	1.53	1.48
4, Low	1.74	1.68	1.64	1.60	1.57	1.55	1.53	1.47	1.44	1.42	1.35

Perimeter Wall Adjustment

A common wall exists when two buildings share one wall. Adjust for common walls by deducting the linear foot costs below from the total structure cost. In some structures one or more walls are not owned at all. In this case, deduct the "No Ownership" cost per linear foot of wall not owned.

For common wall, deduct $116.00 per linear foot. For no wall ownership, deduct $232.00 per linear foot.

Mobile Home Parks

Quality Classification

	Class 1 Best Quality	Class 2 Good Quality	Class 3 Average Quality	Class 4 Low Quality
Engineering, Plans, Permits, Surveying (10% of total cost)	Good planning, necessary permits, good engineering; designed by architect.	Good planning, necessary permits, good engineering; designed by architect.	Average planning, necessary permits, engineered and designed.	Fair planning, necessary permits, minimum surveying.
Grading (10% of total cost)	Fully graded.	Fully graded.	Fully graded.	Minimum site leveling; grades not engineered; road grading.
Street Paving (10% of total cost)	2" thick asphalt surface on good base, concrete curbs, 30' width.	2" thick asphalt surface on good base, concrete curbs, 25' width.	20' roads, 2" asphalt on rock base; concrete or wood edging.	Narrow streets, 2" asphalt on ground; no curbs or edging.
Patios & Walks (5% of total cost)	Patios, 300 to 500 S.F. of good concrete. Walks to utility rooms, pools and recreation areas.	Patios, 200 to 300 S.F. of good concrete. Walks to utility rooms, pools and recreation areas.	Patios, approximately 150 S.F. average concrete or average grade asphalt. Walks to utility buildings.	Some patios, concrete or asphalt paving. No walks.
Trailer Pad & Parking (3% of total cost)	Concrete or good asphalt pad and driveway.	Asphalt under trailer and extended to one side for driveway.	Gravel under trailer and small asphalt driveway.	Gravel under trailer.
Sewer (10% of total cost)	8" lines, 10" mains. Meets all code requirements. Storm drain system.	8" lines, 10" mains. Meets all code requirements.	4" to 6" and 8" lines. Meets code requirements in most areas.	3" to 6" lines, inadequate. Below good code requirements.
Water (10% of total cost)	Engineered system for equalized pressure throughout park. Sprinkler system in common areas.	Adequate line size, designed and properly sized for equalized pressure.	Adequate line size; has required valves at each space.	Small lines; has required valves at each space.
Gas (10% of total cost)	Supplied to each space, sized to code requirements.	Supplied to each space, sized to code requirements.	None except in utility buildings and recreation buildings	None except in utility buildings.

Mobile Home Parks

Quality Classification continued

	Class 1 Best Quality	Class 2 Good Quality	Class 3 Average Quality	Class 4 Low Quality
Electric (10% of total cost)	Underground service, designed for larger modern trailers with adequate size to enlarge to take care of future needs. Approximately 100 amp or more. Speaker system, underground television system to each space.	Underground service, designed for larger modern trailers with facilities to enlarge capacity to 100 amp. Approximately 70 amp service or more. Speaker system, underground television system to each space.	Underground services, not designed for more capacity. Approximately 30 amp service or more. Speaker system.	Overhead system wired for 15 amp service at each space.
Outdoor Lighting (4% of total cost)	Lamp post each five spaces, ornate type.	Lamp post each five spaces, inexpensive type.	Overhead street lights at each corner.	Few overhead street lights.
Telephone (8% of total cost)	Underground to each space.	Underground to each space.	None.	None.
Sign (1% of total cost)	Large expensive sign.	Good sign.	Average sign.	Inexpensive sign.
Garbage (1% of total cost)	Built-in ground.	Built-in ground.	None.	None.
Mail Boxes (1% of total cost)	Good mail box and post each space.	Inexpensive mail box and post each space.	None.	None.
Fences and Gates (3% of total cost)	Good wood or cyclone. Ornamental fence or wall in front.	Good wood or cyclone. Block wall in front.	Inexpensive wood or wire.	None or inexpensive wire.
Pools (4% of total cost)	Good quality.	Good quality, adequate size for park.	Small with few extras.	None or inexpensive.
Utility Building (See page 162)	Wood frame and good stucco. Board batt redwood siding or concrete block exterior. Best composition shingle or tar and rock roofing. Good interior plaster or gypsum wallboard. Well finished concrete floors with vinyl tile. Good lighting. Good heating. Showers ceramic tile or fiberglass walls with ceramic tile floor. Glass shower doors. Good quality plumbing fixtures. Good workmanship throughout.	Wood frame and good stucco or concrete block exterior. Thick butt composition shingles or tar and gravel roof. Good exterior or plaster or gypsum wallboard. Well finished concrete floors. Good lighting. Good heating. Showers ceramic tile walls with ceramic tile base. Good quality plumbing fixtures. Good workmanship throughout.	Wood frame, average stucco exterior. Composition shingle or roll roofing. Gypsum wallboard taped and textured or plaster interior. Average concrete floors. Average lighting. Fair heating. Metal stall with showers or showers with enameled cement plaster walls and tile floor with tile base. Average plumbing fixtures. Average workmanship throughout.	Wood frame, fair stucco or fair siding exterior. Plastic interior. Composition roll roofing. Fair concrete finish. Fair lighting. Inexpensive heating. Showers enameled cement plaster walls and tile floors. Fair plumbing fixtures and fair workmanship throughout.
Recreation Building (See page 162)	Wood frame and stucco. Board and batt redwood siding or concrete block exterior. Best composition shingles or tar and rock roofing. Good interior plaster or gypsum wallboard taped, textured and painted. Well finished concrete floors with vinyl tile. Good heating. Good lighting. Rest room for each sex containing at least one each of the following fixtures: Shower, water closet & lavatory. Showers ceramic tile floors and walls or fiberglass walls and tilefloor with glass shower doors. Good quality plumbing fixtures. Kitchen sink, range, refrigerator, cabinets and drainboard of formica or equal material. Large glass area in community room.	Wood frame and good stucco or concrete block exterior. Thick butt composition shingles or tar and gravel roof. Good exterior or plaster or gypsum wallboard. Well finished concrete floors. Good heating. Good lighting. Ceramic tile stall showers with ceramic tile base. Good quality plumbing fixtures. Kitchen with tile drain board and some hardwood cabinets. Small office area. Large glass windows in community room.	Wood frame, average stucco or siding exterior. Composition shingle or roll roofing. Gypsum wallboard taped and textured or plaster interior. Average concrete floors. Average lighting. Average heating. Showers with enameled cement plaster walls and ceramic tile base. Water closets and lavatories. One rest room for each sex with at least 1 each shower, water closet and lavatory. Average grade of plumbing fixtures. Ceiling of gypsum wallboard. Average workmanship throughout.	None.

Note: Use the percent of total cost to help identify the correct quality classification.

Mobile Home Parks

Estimating Procedure

1. Establish the park quality class by applying the information on pages 160 and 161.
2. Compute the square foot area per home space. This should include the mobile home space, streets, recreation and other community use areas but exclude excess land not improved or not in use as a part of the park operation. Divide this total area by the number of home spaces. The result is the average area per home space.
3. Multiply the appropriate cost below by the number of home spaces.
4. Determine the quality class and area of recreation and utility buildings. Compute the total cost of these buildings and add this amount to or subtract it from the total from step 3 to adjust for more or fewer buildings than included in the quality specification.
5. Multiply the total building costs by the location factor listed on page 7.
6. Add the cost of septic tank systems, wells, and covered areas built at the individual spaces.

Space costs with community facilities include the cost of the following components: Grading associated with a level site under normal soil conditions. Street paving and curbs. Patios and walks. Pads and parking paving. Sewer, electrical, gas and water systems including normal hook-up costs. Outdoor lighting. Signs. Mail boxes. Fences and gates. Contractors' mark-up.

Space costs with community facilities and buildings include the cost of all the above components plus these components in amounts proportionate to the size of the park: Recreation, administrative and utility buildings adequate for the size of the park. Recreation facilities such as pools, shuffle board courts, playground equipment, fire pits, etc. Telephones. Restrooms.

The cost of the following components are not included in the basic building cost: Septic tank systems. Wells. Structures or covered areas on individual spaces. The cost of grading beyond that associated with a level site.

Parks Without Community Facilities or Buildings - Square Foot Area

Quality Class	1,500	2,000	2,500	3,000	3,500	4,000	4,500	5,000	5,500	6,000	6,500
1, Best	—	—	—	—	—	8,962	9,129	9,480	10,454	10,039	10,039
2, Good	—	—	—	8,113	8,527	8,702	8,863	9,076	9,239	9,239	9,239
3, Average	—	—	6,138	6,674	6,509	7,280	7,427	7,428	7,427	7,427	—
4, Low	3,741	4,115	4,503	4,791	5,055	5,276	5,276	5,276	5,276	—	—

Parks With Community Facilities and Buildings - Square Foot Area

Quality Class	1,500	2,000	2,500	3,000	3,500	4,000	4,500	5,000	5,500	6,000	6,500
1, Best	—	—	—	—	—	13,115	13,744	14,161	14,563	14,967	15,207
2, Good	—	—	—	12,394	12,695	12,987	13,236	13,463	13,975	13,975	13,975
3, Average	—	—	9,989	9,989	10,482	10,649	10,774	10,774	10,774	10,774	—
4, Low	6,786	6,881	7,077	7,385	7,645	7,801	7,785	7,785	7,785	—	—

Square Foot Costs for Building Alone

	Recreational Buildings	Utility Buildings
1, Best	$46.30 to $67.00	$44.50 to $53.60
2, Good	41.00 to 53.80	33.00 to 44.80
3, Average	33.20 to 47.70	31.20 to 33.60
4, Low	—	28.50 to 32.60

Service Stations - Wood, Masonry or Painted Steel

Quality Classification

	Wood Frame	Masonry or Concrete	Painted Steel, Good	Painted Steel, Average	Painted Steel, Low
Foundation & Floor (25% of total cost)	Reinforced concrete.	Reinforced concrete.	Reinforced concrete.	Reinforced concrete.	Reinforced concrete.
Walls (15% of total cost)	Wood frame 2 x 4, 16" o.c.	8" concrete block.	Steel frame.	Steel frame.	Steel frame.
Roof Structure (6% of total cost)	Light wood frame, flat or shed type.	Light wood frame, flat or shed type.	Steel frame, flat or shed type.	Steel frame, flat or shed type.	Steel frame, flat or shed type.
Exterior Finish (10% of total cost)	Painted wood siding or stucco.	Painted concrete block.	Painted steel.	Painted steel.	Painted steel.
Roof Cover (5% of total cost)	Composition.	Composition.	Steel deck.	Steel deck.	Steel deck.
Glass Area (5% of total cost)	Small area, painted wood frames.	Small area, painted steel frames.	Large area, painted steel frames.	Average area, painted steel frames.	Small area, painted steel frames.
Lube Room Doors (5% of total cost)	Folding steel gate.	Folding steel gate.	Painted steel sectional roll up.	Painted steel sectional roll up.	Folding steel gate.
Floor Finish (5% of total cost)	Concrete.	Concrete.	Concrete, colored concrete in office.	Concrete.	Concrete.
Interior Wall Finish (5% of total cost)	Exposed studs, painted.	Concrete block, painted.	Exposed structure painted. Painted steel panels in office.	Exposed structure painted. Painted steel panels in office.	Exposed structure painted. Painted steel panels in office.
Ceiling Finish (3% of total cost)	Exposed structure painted.	Exposed structure painted.	Exposed structure painted. Painted steel panels in office.	Exposed structure painted. Painted steel panels in office.	Exposed structure painted.
Rest Room Finish (5% of total cost)	Wallboard and paint walls and ceilings.	Concrete block and paint walls, wallboard and paint ceilings.	Ceramic tile floors, 4' ceramic tile wainscot, painted steel ceilings.	Ceramic tile floors, 4' ceramic tile wainscot, painted steel ceilings.	Concrete floors, painted steel walls, painted steel ceilings.
Rest Room Fixtures (8% of total cost)	4 low cost fixtures.	4 low cost fixtures.	5 average cost fixtures.	5 average cost fixtures.	4 low cost fixtures.
Exterior Appointments (3% of total cost)	None.	None.	2' overhang on 3 sides, 3' raised walk on 3 sides, fluorescent soffit lights on 3 sides.	1' overhang on 2 sides, 3' raised walk on 2 sides.	None.

Note: Use the percent of total cost to help identify the correct quality classification.

Square foot costs include the cost of the following components: Foundations as required for normal soil conditions. Floor, wall and roof structure. Interior floor, wall and ceiling finishes as described above. Interior partitions. Exterior finish and roof cover. A built-in work bench, tire rack and shelving. Electrical services and fixtures contained within the building. Air and water lines within the building. That portion of rough plumbing serving the building and plumbing fixtures within the building. Roof overhangs and raised walks as described above. Lube room doors. Permits and fees. Contractor's mark-up.

The in-place cost of these extra components should be added to the basic building cost to arrive at the total structure cost. See the section "Additional Costs for Service Stations" on Page 169. Canopies. Pumps, dispensers and turbines. Air and water services outside the building. Island lighters. Gasoline storage tanks. Hoists. Compressors. Yard lights. Signs. Paving. Curbs and fences. Miscellaneous equipment and accessories. Island office and storage buildings. Site improvements. Heating and cooling systems

Land improvement costs: Most service stations sites require an expenditure of $10,000 or more for items such as leveling, excavation, curbs, driveways, relocation of power poles, replacement of sidewalks with reinforced walks and street paving.

Service Stations – Wood, Masonry or Painted Steel

Estimating Procedure

1. Establish the structure quality class by applying the information on page 163.
2. Compute the building floor area.
3. Multiply the square foot cost by the building floor area.
4. Multiply the total cost by the location factor listed on page 7.
5. Add the cost of appropriate equipment and fixtures from the section "Additional Costs for Service Stations" beginning on page 169.

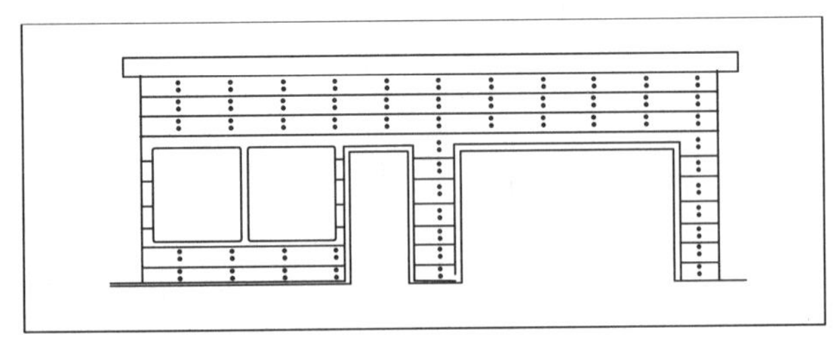

Wood Frame

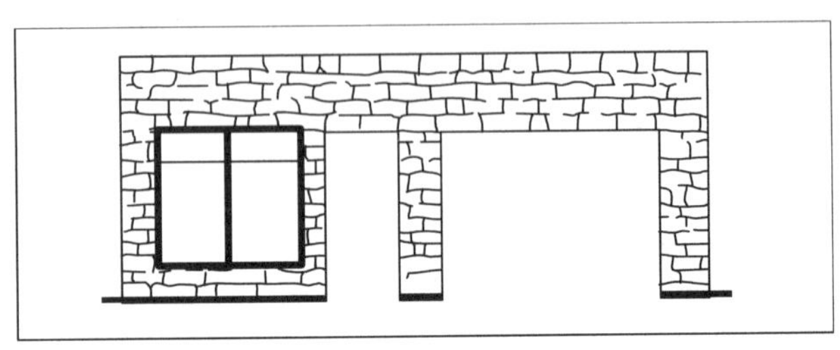

Masonry

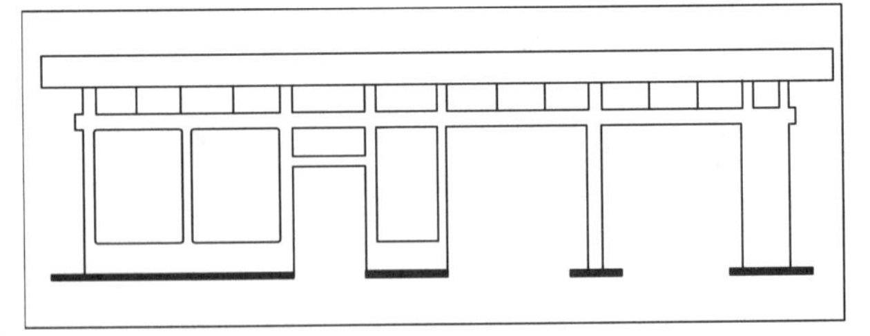

Painted Steel, Good

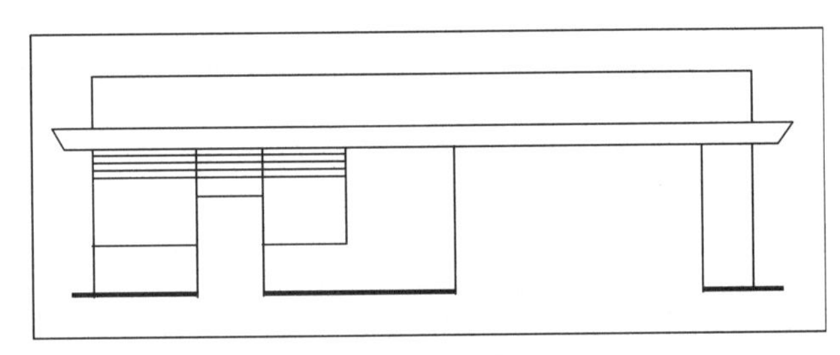

Painted Steel, Average

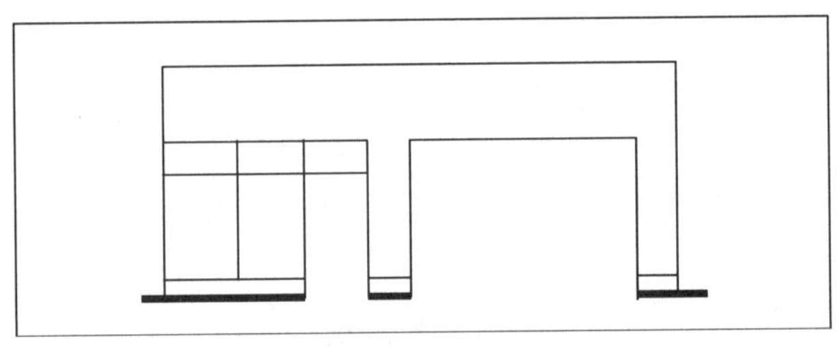

Painted Steel, Low

Square Foot Area

Quality Class	500	600	700	800	900	1,000	1,100	1,200	1,300	1,400	1,800
Wood Frame	75.45	69.30	64.81	61.57	58.98	56.93	55.24	53.86	52.66	51.62	48.68
Masonry or Concrete	86.21	78.98	73.94	70.16	67.23	64.90	63.00	61.38	60.03	58.87	55.51
Painted Steel, Good	130.44	119.73	112.11	106.37	101.93	98.40	95.46	93.05	91.00	89.26	84.16
Painted Steel, Avg.	116.86	108.04	101.66	96.00	91.97	88.25	85.17	81.77	79.71	77.72	75.96
Painted Steel, Low	102.36	93.97	88.00	83.51	80.01	77.22	74.94	73.05	71.42	69.66	66.08

Service Stations – Porcelain Finished Steel

Quality Classification

	Good Quality	Average Quality	Low Quality
Foundation & Floor (20% of total cost)	Reinforced concrete.	Reinforced concrete.	Reinforced concrete.
Walls (15% of total cost)	Steel frame.	Steel frame.	Steel frame.
Roof Structure (8% of total cost)	Steel frame, flat or shed type.	Steel frame, flat or shed type.	Steel frame, flat or shed type.
Exterior Finish (10% of total cost)	Porcelain and steel.	Porcelain and steel.	Porcelain and steel.
Roof Cover (6% of total cost)	Steel deck.	Steel deck.	Steel deck.
Glass Area (7% of total cost)	Large area, aluminum frames.	Large area, aluminum frames.	Average area, painted steel frames.
Lube Room Doors (3% of total cost)	Aluminum and glass sectional roll up.	Aluminum and glass sectional roll up.	Painted steel and glass sectional roll up.
Floor Finish (5% of total cost)	Concrete floors, ceramic tile in office.	Concrete floors, colored concrete in office.	Concrete floors, colored concrete in office.
Interior Wall Finish (5% of total cost)	Porcelain steel panels. Painted steel panels in office.	Exposed structure painted. Painted steel panels in office.	Exposed structure painted.
Ceiling Finish (3% of total cost)	Exposed structure painted. Porcelain steel panels in office.	Exposed structure painted. Painted steel panels in office.	Exposed structure painted. Painted steel panels in office.
Rest Room Finish (5% of total cost)	Ceramic tile floors, 8' ceramic tile or porcelain panel. Painted steel ceiling.	Ceramic tile floors, 5' ceramic tile wainscot. Painted steel ceiling.	Ceramic tile floors, 5' ceramic tile wainscot. Painted steel ceiling.
Rest Room Fixtures (10% of total cost)	5 good fixtures.	5 good fixtures.	5 good fixtures.
Exterior Appointments (3% of total cost)	3' to 4' overhang on 3 sides, 6' x 8' sign pylon, 3' raised walk on 3 sides, fluorescent soffit lights on 3 sides.	3' to 4' overhang on 3 sides, 6' x 8' sign pylon, 3' raised walk on 3 sides, fluorescent soffit lights on 3 sides.	3' raised walk on 3 sides, fluorescent soffit lights on 3 sides.

Note: Use the percent of total cost to help identify the correct quality classification.

Square foot costs include the cost of the following components*:* Foundations as required for normal soil conditions. Floor, wall and roof structure. Interior floor, wall and ceiling finishes as described above. Interior partitions. Exterior finish and roof cover. A built-in work bench, tire rack and shelving. Electrical services and fixtures contained within the building. Air and water lines within the building. That portion of rough plumbing serving the building and plumbing fixtures within the building. Roof overhangs and raised walks as described above. Lube room doors. Permits and fees. Contractor's mark-up.

The in-place cost of these extra components should be added to the basic building cost to arrive at the total structure cost. See the section "Additional Costs for Service Stations" on page 169. Canopies. Pumps, dispensers and turbines. Air and water services outside the building. Island lighters. Gasoline storage tanks. Hoists. Compressors. Yard lights. Signs. Paving. Curbs and fences. Miscellaneous equipment and accessories. Island office and storage buildings. Site improvements. Heating and cooling systems

Land improvement costs: Most service station sites require an expenditure of $10,000 or more for items such as leveling, excavation, curbs, driveways, relocation of power poles, replacement of sidewalks with reinforced walks and street paving.

Service Stations – Porcelain Finished Steel

Estimating Procedure

1. Establish the structure quality class by applying the information on page 165.
2. Compute the building floor area.
3. Multiply the square foot cost by the building floor area.
4. Multiply the total cost by the location factor listed on page 7.
5. Add the cost of appropriate equipment and fixtures from the section "Additional Costs for Service Stations" beginning on page 169.

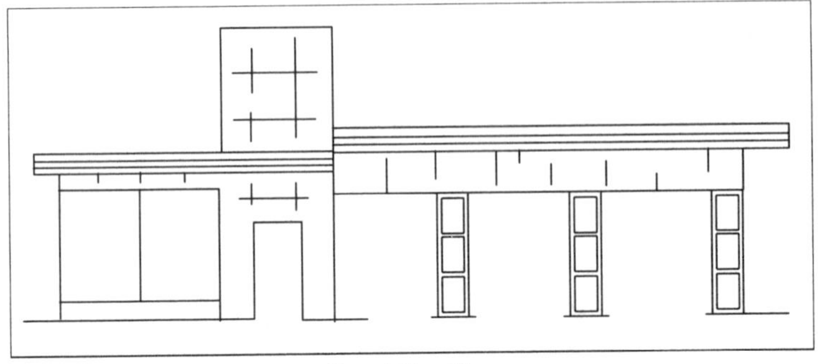

Good Quality

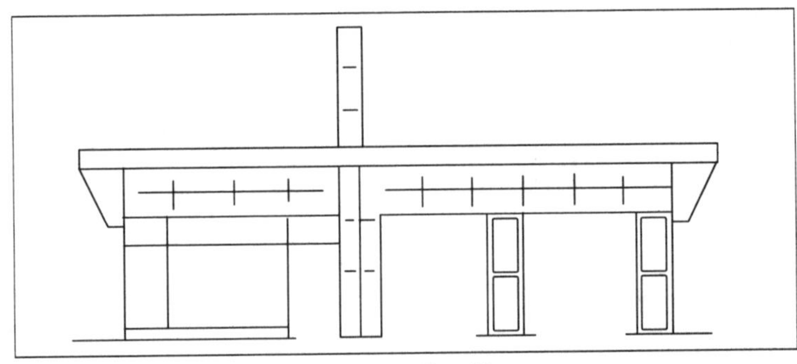

Average Quality

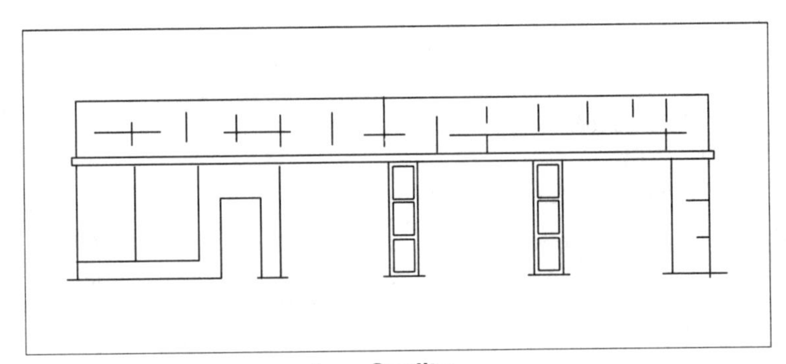

Low Quality

Square Foot Area

Quality Class	1,000	1,100	1,200	1,300	1,400	1,500	1,600	1,700	1,800	2,000	2,400
Good	121.11	116.91	113.58	110.88	108.69	106.94	105.41	104.16	103.12	101.49	99.40
Average	115.87	111.84	108.65	106.09	104.01	102.27	100.84	99.64	98.64	97.07	95.11
Low	105.27	101.60	98.68	96.36	94.49	92.90	91.61	90.55	89.63	88.20	86.41

Service Stations - Ranch or Rustic Type

Quality Classification

	Best Quality	Good Quality	Average Quality	Low Quality
Foundation & Floor (20% of total cost)	Reinforced concrete.	Reinforced concrete.	Reinforced concrete.	Reinforced concrete.
Walls (12% of total cost)	Steel frame.	Steel frame.	Steel frame, wood frame or masonry.	Steel frame, wood frame or masonry.
Roof Structure (8% of total cost)	Steel frame, hip or gable type.	Steel frame, hip or gable type.	Steel or wood frame, hip or gable type.	Steel or wood frame, hip or gable type.
Exterior Finish (10% of total cost)	Natural stone veneer.	Used brick veneer.	Painted steel and masonry veneer.	Painted steel or wood siding.
Roof Cover (6% of total cost)	Shingle tile or mission tile.	Heavy wood shakes or shingle tile.	Wood shakes or tar and rock.	Composition shingle or tar and gravel.
Glass Area (7% of total cost)	Large area float glass in heavy aluminum frame.	Large area float glass in heavy aluminum frame.	Large area, painted steel frame.	Average area, painted steel frame.
Lube Room Doors (5% of total cost)	Painted steel or aluminum and glass sectional roll up.	Painted steel or aluminum and glass sectional roll up.	Painted steel sectional roll up.	Painted steel sectional roll up.
Floor Finish (5% of total cost)	Concrete floors, ceramic tile in office.	Concrete floors, ceramic tile in office.	Concrete floors.	Concrete floors.
Interior Wall Finish (5% of total cost)	Painted steel panels or gypsum wallboard and paint.	Painted steel panels or gypsum wallboard and paint.	Painted steel panels or gypsum wallboard and paint.	Painted steel panels or gypsum wallboard and paint.
Ceiling Finish (3% of total cost)	Painted steel panels.	Painted steel panels.	Painted steel panels, gypsum wallboard, or "V" rustic and paint.	Painted steel panels, gypsum wallboard or "V" rustic and paint.
Restroom Finish (5% of total cost)	Ceramic tile floors, ceramic tile walls, painted steel ceiling.	Ceramic tile floors, ceramic tile walls, painted steel ceiling.	Ceramic tile floors, 5' ceramic tile wainscot, painted steel ceiling.	Ceramic tile floors, 5' ceramic tile wainscot, painted steel ceiling.
Restroom Fixtures (10% of total cost)	5 good fixtures.	5 good fixtures.	5 good fixtures.	5 good fixtures.
Exterior Appointments (4% of total cost)	3' to 6' overhang on all sides, 3' raised walk on 3 sides, fluorescent soffit lights on all sides.	3' to 6' overhang on all sides, 3' raised walk on 3 sides, fluorescent soffit lights on all sides.	3' to 6' overhang on 3 sides, 6' x 8' sign pylon, 3' raised walk on 3 sides, fluorescent soffit lights on 3 sides.	2' to 3' overhang on 3 sides, 3' raised walk on 3 sides, fluorescent soffit lights on 3 sides.

Note: Use the percent of total cost to help identify the correct quality classification.

Square foot costs include the cost of the following components: Foundations as required for normal soil conditions. Floor, wall and roof structure. Interior floor, wall and ceiling finishes as described above. Interior partitions. Exterior finish and roof cover. A built-in work bench, tire rack and shelving. Electrical services and fixtures contained within the building. Air and water lines within the building. That portion of rough plumbing serving the building and plumbing fixtures within the building. Roof overhangs and raised walks as described above. Lube room doors. Permits and fees. Contractor's mark-up.

The in-place cost of these extra components should be added to the basic building cost to arrive at the total structure cost. See the section "Additional Costs for Service Stations" on page 169. Canopies. Pumps, dispensers and turbines. Air and water services outside the building. Island lighters. Gasoline storage tanks. Hoists. Compressors. Yard lights. Signs. Paving. Curbs and fences. Miscellaneous equipment and accessories. Island office and storage buildings. Site improvements. Heating and cooling systems

Land improvement costs: Most service stations sites require an expenditure of $10,000 or more for items such as leveling, excavation, curbs, driveways, relocation of power poles, replacement of sidewalks with reinforced walks and street paving.

Service Stations - Ranch or Rustic Type

Estimating Procedure

1. Establish the structure quality class by applying the information on page 167.
2. Compute the building floor area.
3. Multiply the square foot cost by the building floor area.
4. Multiply the total cost by the location factor listed on page 7.
5. Add the cost of appropriate equipment and fixtures from the section "Additional Costs for Service Stations" beginning on page 169.

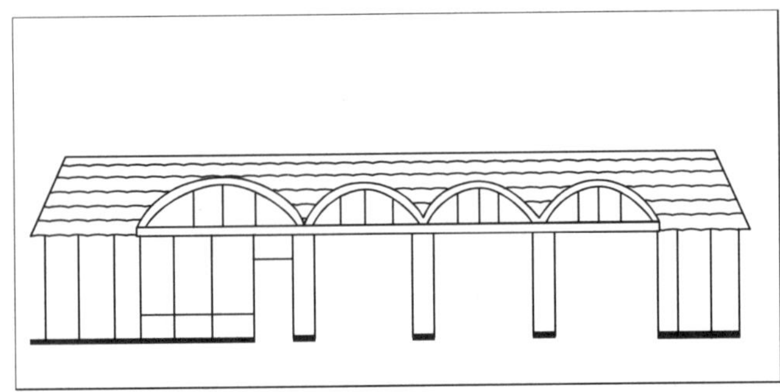

Best Quality

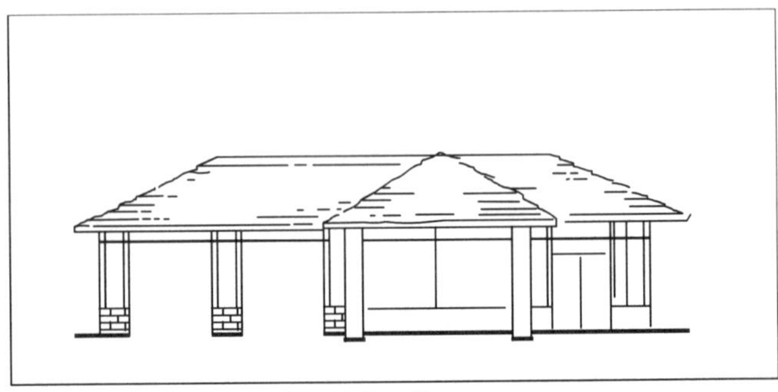

Good Quality

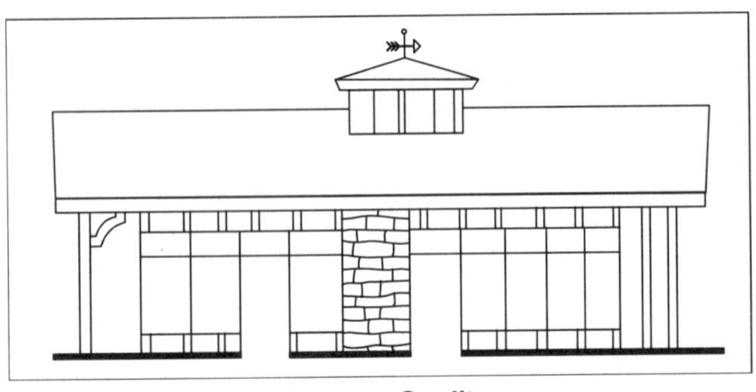

Average Quality

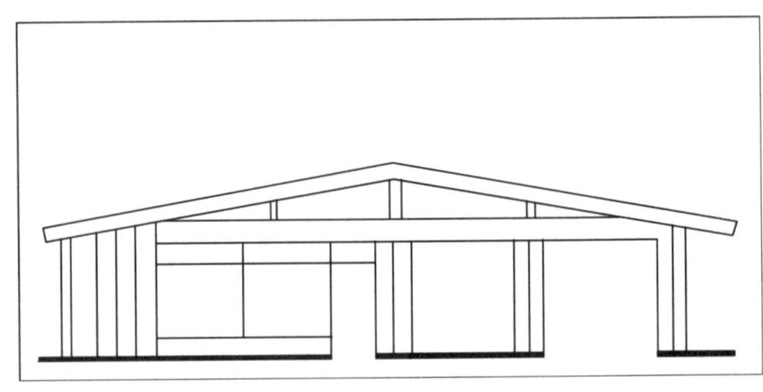

Low Quality

Square Foot Area

Quality Class	1,000	1,100	1,200	1,300	1,400	1,500	1,600	1,700	1,800	2,000	2,400
Best	130.87	126.24	122.56	119.64	117.28	115.31	113.72	112.36	111.25	109.50	107.36
Good	125.55	121.10	117.59	114.78	112.50	110.63	109.07	107.82	106.70	105.08	103.01
Average	120.49	116.20	112.85	110.14	107.94	106.16	104.70	103.43	102.43	100.80	98.85
Low	113.79	109.77	106.60	104.06	101.97	100.30	98.86	97.73	96.73	95.22	93.36

Additional Costs for Service Stations

A portion of the typical plumbing or electrical cost has been added to each item of equipment requiring these services. It will not be necessary except in rare instances to add extra cost for these items.

Canopies, cost per square foot

Type	Less than 500 S.F.	500 to 1,000 S.F.	Over 1,000 S.F.
Painted steel	$23.90 to $26.90	$21.80 to $26.90	$18.80 to $20.90
Porcelain and steel	27.50 to 28.60	23.90 to 26.70	21.70 to 23.70
Ranch style or gable roof type	31.10 to 33.90	24.50 to 29.00	22.10 to 24.90
Deluxe steel with illuminated plastic signs on sides or in gables. Also includes illuminated plastic island lighters.			
Complete,	$47.30 to $56.40		
Round type, good steel	45.70 to 53.60		

Costs include cost of foundation, steel support column or columns, complete canopy, painting or porcelaining, light fixtures, and electrical service. Ranch or gable roof types include the cost of a rock or shake roof cover. Concrete pads under canopies or masonry trim on support columns are not included in these costs.

Island Office and Storage Buildings, cost per square foot

	Area						
Type	Under 30	31 - 40	41 - 50	51 - 60	61 - 80	81 - 100	101 - 120
Steel and glass or concrete block	300.00	287.00	256.00	228.00	204.00	191.00	153.00
Wood frame with stucco and glass	259.00	244.00	201.00	188.00	144.00	134.00	128.00

These buildings are usually found at self-service stations. Add $1,500 per unit for any plumbing fixtures in these buildings. Steel island offices cost about $1,400

Pumps, Dispensers and Turbines, cost each

Type	Installed Cost	Type	Installed Cost
Single pump	$4,566	Blendomatic pump	$7,503
Twin pump	6,198	Blendomatic dispenser	6,155
Single dispenser	3,480	Turbine pump, 1/3 HP	1,295
Twin dispenser	5,818	Turbine pump, 3/4 HP	1,795

Installed cost includes the cost of the pump or dispenser, installation cost, electrical hookup cost, a portion of the piping cost and a portion of the island block cost. Concrete islands 4" to 6" thick cost from $9.25 to $10.50 per square foot.

All of the above pump and dispenser costs are for the computing type. Add for electronic remote control totalizer, per hose, $1,520. Add for vapor control system, per hose/dispenser, $1,580.

Dispenser cost does not include the cost of the pump. Turbine pump costs must be added. 1/3 HP turbines will serve a single product up to four dispensers. 3/4 HP turbines will serve a single product up to eight dispensers.

Additional Costs for Service Stations

Air and Water Services

Type	Air Only Equipment Cost	Air Only Installed Cost	Air and Water Equipment Cost	Air and Water Installed Cost
Underground disappearing hose type	$399	$588	$425	$890
Post type with auto inflator	$627	$925	$795	$1,260
Post type with auto inflator and disappearing hoses	$1,050	$1,430	$1,600	$1,960

Costs include cost of installation and a portion of the cost of air and water lines.

Island Lighters

Width	Length	4 Tubes	6 Tubes
42"	9'-5"	$1,690 ea.	$2,040 ea.
42"	11'-5"	2,100 ea.	2,665 ea.
42"	15'-6"	2,740 ea.	3,010 ea.
42"	19'-6"	3,330 ea.	3,590 ea.
36"	30'-0"	—	4,210 ea.

Cost includes foundation, davit poles or steel support columns and electrical service.

Cash Boxes complete, with pedestal $281 each

Gasoline Storage Tanks (Fiberglass)

Capacity in Gallons	Tank Cost	Installed Cost	Capacity in Gallons	Tank Cost	Installed Cost
110	$627	$1,100	**5,300**	$6,370	$8,275
150	773	1,320	**6,300**	6,680	8,540
280	930	1,510	**7,400**	8,265	9,230
550	1,230	1,925	**8,400**	8,525	9,310
1,000	2,320	3,980	**10,500**	8,560	11,090
2,000	4,340	5,300	**12,600**	10,870	13,810
4,000	5,800	8,730	—	—	—

Installed cost includes cost of tank, excavation (4' bury and soil disposal), placing backfill, fill box (concrete slab over tank), tank piping and vent piping.

Miscellaneous Lube Room Equipment

Item	Cost	Item	Cost
Air hose reel	$925	Pneumatic tube changer	$9,950
Water hose reel	952	Automatic lube equipment	7,390
Grease pit for trucks	$364 to $420 per L.F.	5 hose reel assembly	9,240

Yard Lights

High pressure sodium luminaires. Costs include electrical connection and mounting on a building soffit. For pole mounted yard lights, add pole mounting costs from page 172. Cost per light fixture.

70 Watt	100 Watt	200 Watt	300 Watt	400 Watt
$382	$398	$414	$448	$532

Additional Costs for Service Stations

Vehicle Hoist

Type	Equipment Cost	Installed Cost
One post 8,000 lb. semi hydraulic hoist	$3,850.00	$ 8,370.00
One post 8,000 lb. fully hydraulic hoist	4,110.00	8,690.00
Two post 11,000 lb. semi hydraulic hoist	5,830.00	13,440.00
Two post 11,000 fully hydraulic hoist	5,880.00	13,500.00
Two post 11,000 lb. pneumatic hoist	7,500.00	12,320.00
Two post 24,000 lb. pneumatic hoist	10,610.00	18,150.00

Air Compressors

Horsepower	Equipment Cost	Installed Cost
1/2	$1,935.00	$2,185.00
3/4	1,960.00	2,230.00
1	1,990.00	2,270.00
1-1/2	1,860.00	2,320.00
2	$2,088.00	$2,340.00
3	2,200.00	2,830.00
5	2,340.00	2,880.00
7-1/2	4,000.00	4,240.00

Costs include compressor and tank only.

Paving, cost per S.F.

Asphalt, 2" with 4" base	$2.08 to $2.69
Concrete 4", with base	2.57 to 3.59
Concrete 6", with base	3.25 to 4.20
Oil macadam	1.95
Pea gravel	1.01

Site Improvement

Vertical curb and gutter	$5.88 to $18.90 LF
Concrete apron	7.29 to 16.20 SF
6" reinforced concrete sidewalks	4.48 to 5.88 SF
Standard 4" sidewalk	3.64 to 4.20 SF

The above costs are normally included in land value.

Fencing and Curbing, cost per L.F.

Heavy 2 rail fence, 2" x 6"	$9.31 to $11.28
Rails on 4" x 4" posts 6' to 8' o.c.	9.74 to 11.79
Chain link 3' to 4' high	8.58 to 12.71
Solid board 3' to 4' high	9.80 to 11.69
Log barrier	8.58 to 15.32
Metal guard rail on wood posts	31.78 to 57.71
6" x 6" doweled wood bumper strip	9.31 to 11.99
6" x 6" concrete bumper strip	7.07 to 10.05
Cable railing on wood posts	8.92 to 10.56
6" x 12" concrete curb and gutter	15.32 to 17.63
6" concrete block walls, per S.F.	6.24 to 8.92

plus $7.80/LF for foundation

Service Station Signs, cost per square foot of sign area on one side

Painted sheet metal with floodlights	$52.60 to $69.80
Porcelain enamel with floodlights	57.40 to 77.20
Plastic with interior lights	65.50 to 99.70
Simple rectangular neon with painted sheet metal faces and a moderate amount of plain letters	69.20 to 123.00
Round or irregular neon with porcelain enamel faces and more elaborate lettering	99.80 to 151.70

All of the above sign costs are for single faced signs. Add 50% to these costs for double faced signs. Sign costs include costs of installation and normal electrical hookup. They do not include the post cost. See page 172. These costs are intended for use on **service station signs only** and are based on volume production. Costs of custom-built signs will be higher.

Rotators, cost per sign for rotating mount

Small signs	Less than 50 S.F.	$2,080.00 to $2,240.00
Medium signs	50 to 100 S.F.	2,190.00 to 4,450.00
Large signs	100 to 200 S.F.	4,390.00 to 8,070.00
Extra large signs	Over 200 S.F.	$45.50 per S.F. of sign area

Additional Costs for Service Stations

Post Mounting Costs

Post Height	Pole Diameter at Base 4"	6"	8"	10"	12"	14"
15	$1,000	$1,186	$1,751	$2,360	$3,675	$3,675
20	1,186	1,414	2,022	2,448	4,372	4,916
25	1,282	1,598	2,099	2,914	4,871	5,372
30	1,479	1,903	2,175	3,187	5,144	5,948
35	—	2,077	2,448	3,470	5,829	6,350
40	—	2,197	2,991	3,753	6,047	7,123
45	—	—	3,502	4,273	6,601	7,372
50	—	—	3,753	4,871	7,025	8,080
55	—	—	—	5,144	7,372	8,798
60	—	—	—	5,438	7,906	9,199
65	—	—	—	—	8,472	9,777

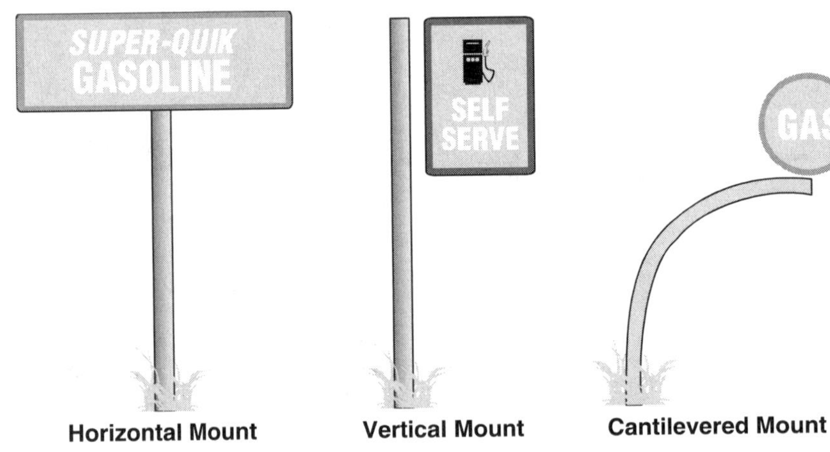

Horizontal Mount   Vertical Mount   Cantilevered Mount

If signs are mounted on separate posts, post mounting costs must be added. Post mounting costs include the installed cost of a galvanized steel post and foundation. On horizontally mounted signs, post height is the distance from the ground to the bottom of the sign. On vertically mounted signs, post height is the distance to the top of the post.

For cantilevered posts, use one and one-half to two times the conventional post cost.

All of the above post costs are for single posts. Use 90% of the single post costs for each additional post.

If signs are mounted on buildings or canopies and if, because of the extra weight of the sign, extra heavy support posts or foundations are required, 125% of the post mounting cost should be used.

For example, the cost of a 4' x 25' plastic sign mounted on a 15' by 6" post shared by an adjacent canopy might be estimated as follows:

Sign Cost (100 x $100.00)	$10,000.00
Post Cost ($1,186.00 x 1/2)	593.00
Total Cost	**$10,593.00**

If this sign were mounted on an 8" post 20' above the canopy with extra supports not needed, the cost might be estimated as follows:

Sign Cost (100 x $100.00)	$10,000.00
Post Cost ($2,022 x 1)	2,022.00
Total Cost	**$12,022.00**

Service Garage – Masonry or Concrete

Quality Classification

	Class 1 Best Quality	Class 2 Good Quality	Class 3 Average Quality	Class 4 Low Quality
Foundation (25% of total cost)	Reinforced concrete or masonry.	Reinforced concrete or masonry.	Reinforced concrete or masonry	Unreinforced concrete or masonry.
Floor Structure (15% of total cost)	6" rock fill, 4" concrete with reinforcing mesh.	6" rock fill, 4" concrete with reinforcing mesh.	4" rock fill, 4" concrete with reinforcing mesh.	Unreinforced 4" concrete.
Walls (15% of total cost)	8" reinforced concrete block, 12" common brick.	8" reinforced concrete block, 6" reinforced concrete.	8" reinforced concrete block, 6" reinforced concrete or 8" common brick.	8" unreinforced concrete block or 8" clay tile.
Roof Structure (12% of total cost)	Glu-lams or steel trusses on heavy pilasters 20' o.c. 2" x 10" purlins 16" o.c.	Glu-lams or steel trusses on pilasters 20' o.c., 2" x 10" purlins 16" o.c.	Glu-lams or wood trusses with 2" x 8" purlins 16" o.c.	Glu-lams or light wood trusses, 2" x 8" rafters 24" o.c.
Roof Cover (8% of total cost)	5 ply built-up roof on wood sheathing, with small rock.	4 ply built-up roof on wood sheathing, with small rock.	4 ply built-up roof on wood sheathing.	4 ply built-up roof on wood sheathing.
Restrooms (10% of total cost)	Two rest rooms with three average fixtures each.	Two rest rooms with two average fixtures each.	One rest room with two low cost fixtures.	One rest room with two low cost fixtures.
Lighting (10% of total cost)	One incandescent fixture per 300 square feet of floor area.	One incandescent fixture per 300 square feet of floor area.	One incandescent fixture per 300 square feet of floor area.	One incandescent fixture per 300 square feet of floor area.
Windows (5% of total cost)	3% to 5% of wall area.	3% to 5% of wall area.	3% to 5% of wall area.	3% to 5% of wall area.

Note: Use the percent of total cost to help identify the correct quality classification.

Square foot costs include the cost of the following components: Foundations as required for normal soil conditions. Floor, wall and roof structures. Exterior wall finish and roof cover. Entry doors. Basic lighting and electrical systems. Rough and finish plumbing. Permits and fees. Contractors' mark-up.

The in-place cost of these extra components should be added to the basic building cost to arrive at the total structure cost. See page 201. Heating and air conditioning systems. Fire sprinklers. Interior finish costs. Interior partitions. Drive-through doors. Canopies and walks. Exterior signs. Paving and curbing. Miscellaneous yard improvements. Hoists, gas pump and compressor costs are listed in the section "Additional Costs for Service Stations" on page 169.

Service Garage – Masonry or Concrete

Length Less Than Twice Width

Estimating Procedure

1. Use these figures to estimate buildings designed primarily for motor vehicle repair. Sales area should be figured separately. Use the costs for urban stores beginning on page 42.
2. Establish the building quality class by applying the information on page 173.
3. Compute the floor area.
4. If the wall height is more or less than 18 feet, add to or subtract from the square foot costs below the appropriate amount from the Wall Height Adjustment Table on page 177.
5. Multiply the adjusted square foot cost by the floor area.
6. Deduct for common walls or no wall ownership. Use the figures on page 177.
7. Multiply the total cost by the location factor on page 7.
8. Add the cost of heating and air conditioning systems, fire sprinklers, interior finish and partitions, drive-thru doors, canopies and walks, exterior signs, paving, curbing, and yard improvements. See page 201. Add the cost of hoists, pumps and compressors from pages 169 and 171.

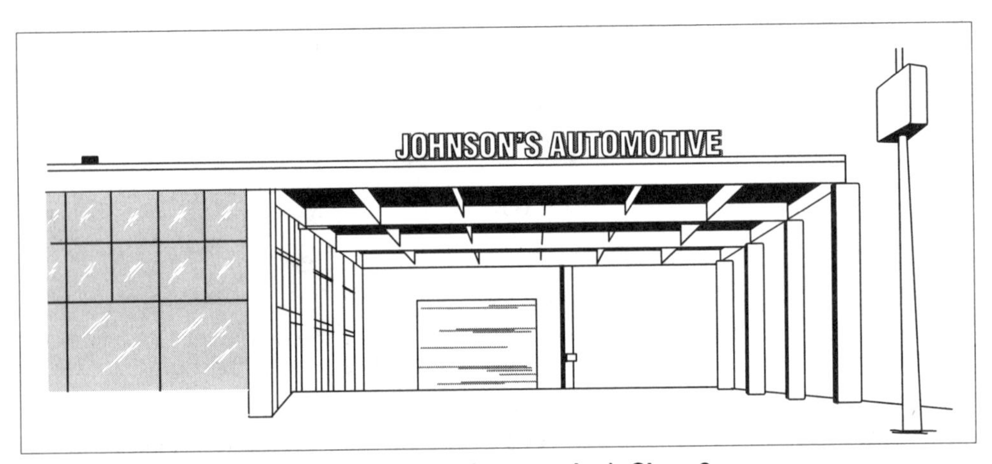

Service Garage (rear portion), Class 2

Square Foot Area

Quality Class	2,000	2,500	3,000	4,000	5,000	6,000	7,500	10,000	15,000	20,000	30,000
1, Best	43.09	39.14	36.34	32.57	30.09	28.35	26.45	24.37	22.03	20.67	19.14
1 & 2	41.37	37.58	34.85	31.25	28.88	27.19	25.37	23.42	21.14	19.86	18.34
2, Good	40.52	36.79	34.16	30.61	28.27	26.62	24.87	22.92	20.69	19.44	17.98
2 & 3	38.55	35.03	32.50	29.14	26.93	25.36	23.67	21.81	19.71	18.51	17.10
3, Average	37.43	34.00	31.56	28.26	26.13	24.60	22.97	21.16	19.13	17.97	16.61
3 & 4	35.41	32.15	29.83	26.75	24.72	23.28	21.74	20.02	18.08	16.96	15.69
4, Low	33.55	30.48	28.28	25.37	23.45	22.07	20.59	19.00	17.15	16.09	14.89

Service Garage – Masonry or Concrete

Length Between 2 and 4 Times Width

Estimating Procedure

1. Use these figures to estimate buildings designed primarily for motor vehicle repair. Sales area should be figured separately. Use the costs for urban stores beginning on page 42.
2. Establish the building quality class by applying the information on page 173.
3. Compute the floor area.
4. If the wall height is more or less than 18 feet, add to or subtract from the square foot costs below the appropriate amount from the Wall Height Adjustment Table on page 177.
5. Multiply the adjusted square foot cost by the floor area.
6. Deduct for common walls or no wall ownership. Use the figures on page 177.
7. Multiply the total cost by the location factor on page 7.
8. Add the cost of heating and air conditioning systems, fire sprinklers, interior finish and partitions, drive-thru doors, canopies and walks, exterior signs, paving, curbing, and yard improvements. See page 201. Add the cost of hoists, pumps and compressors from pages 169 and 171.

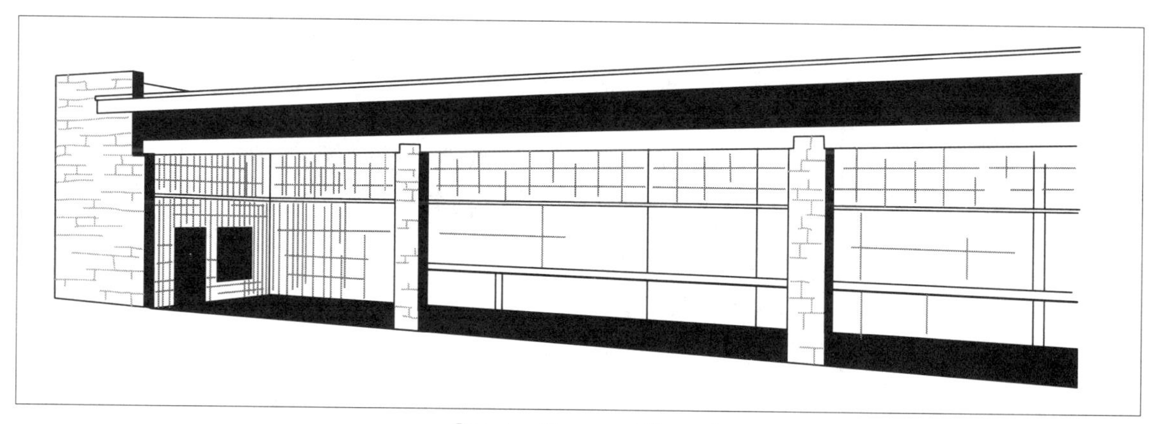

Service Garage, Class 3

Square Foot Area

Quality Class	2,000	2,500	3,000	4,000	5,000	6,000	7,500	10,000	15,000	20,000	30,000
1, Best	45.91	41.68	38.71	34.69	32.04	30.18	28.17	25.97	23.47	22.01	20.36
1 & 2	43.96	39.93	37.05	33.20	30.69	28.89	26.98	24.88	22.45	21.07	19.48
2, Good	43.04	39.08	36.31	32.52	30.03	28.29	26.40	24.33	21.98	20.64	19.07
2 & 3	40.94	37.19	34.51	30.92	28.60	26.90	25.09	23.13	20.89	19.61	18.14
3, Average	38.34	35.91	33.34	29.88	27.63	26.04	24.27	22.38	20.20	18.96	17.53
3 & 4	37.56	34.07	31.63	28.35	26.20	24.68	23.03	21.23	19.17	17.56	16.65
4, Low	35.57	32.31	29.97	26.89	24.88	23.39	21.82	20.13	18.17	17.04	15.75

Service Garage – Masonry or Concrete

Length More Than 4 Times Width

Estimating Procedure

1. Use these figures to estimate buildings designed primarily for motor vehicle repair. Sales area should be figured separately. Use the costs for urban stores beginning on page 42.
2. Establish the building quality class by applying the information on page 173.
3. Compute the floor area.
4. If the wall height is more or less than 18 feet, add to or subtract from the square foot costs below the appropriate amount from the Wall Height Adjustment Table on page 177.
5. Multiply the adjusted square foot cost by the floor area.
6. Deduct for common walls or no wall ownership. Use the figures on page 177.
7. Multiply the total cost by the location factor on page 7.
8. Add the cost of heating and air conditioning systems, fire sprinklers, interior finish and partitions, drive-thru doors, canopies and walks, exterior signs, paving curbing, and yard improvements. See page 201. Add the cost of hoists, pumps and compressors from pages 169 and 171.

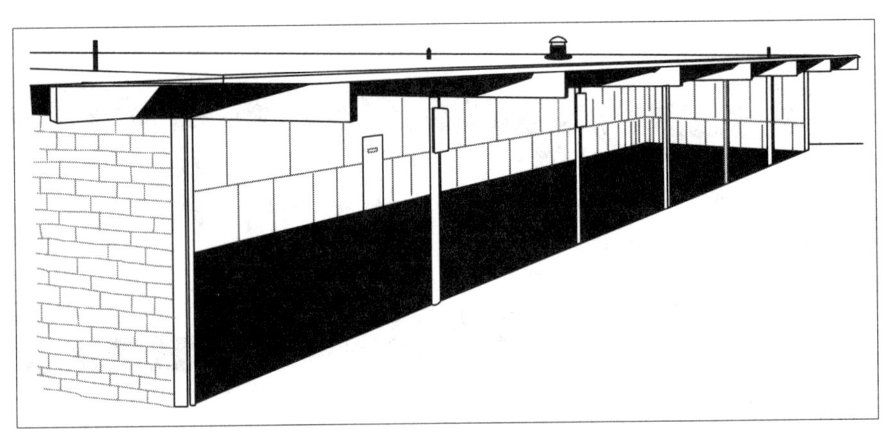

Service Garage, Class 2 & 3

Square Foot Area

Quality Class	2,000	2,500	3,000	4,000	5,000	6,000	7,500	10,000	15,000	20,000	30,000
1, Best	48.91	44.38	41.22	36.93	34.12	32.14	29.98	27.64	24.97	23.47	21.69
1 & 2	47.07	42.73	39.64	35.52	32.85	30.91	28.84	26.58	24.01	22.55	20.86
2, Good	45.88	41.64	38.63	34.60	32.00	30.09	28.13	25.91	23.41	21.98	20.33
2 & 3	43.71	39.66	36.83	32.97	30.49	28.71	26.78	24.68	22.33	20.96	19.40
3, Average	42.35	38.44	35.68	31.98	29.51	27.81	25.98	23.93	21.61	20.31	18.77
3 & 4	40.01	36.30	33.70	30.18	27.88	26.26	24.52	22.61	20.41	19.17	17.74
4, Low	39.92	34.36	31.89	28.59	26.40	24.89	23.20	21.37	19.31	18.14	16.78

Service Garage – Masonry or Concrete

Wall Height Adjustment

Add or subtract the amount listed in this table to the square foot of floor cost for each foot of wall height more or less than 18 feet.

Area	**2,000**	**2,500**	**3,000**	**4,000**	**5,000**	**6,000**	**7,500**	**10,000**	**15,000**	**20,000**	**30,000**
Cost	.46	.42	.39	.34	.31	.29	.27	.15	.12	.09	.06

Perimeter Wall Adjustment

A common wall exists when two buildings share one wall. Adjust for common walls by deducting the linear foot costs below from the total structure cost. In some structures, one or more walls are not owned at all. In this case, deduct the "No Ownership" cost per linear foot of wall not owned.

For common wall, deduct $130 per linear foot.

For no wall ownership, deduct $260 per linear foot.

Service Garage - Wood Frame

Quality Classification

	Class 1 Best Quality	Class 2 Good Quality	Class 3 Average Quality	Class 4 Low Quality
Foundation (25% of total cost)	Concrete, heavily reinforced.	Reinforced concrete.	Masonry or reinforced concrete.	Masonry or concrete.
Floor Structure (12% of total cost)	4" reinforced concrete on 6" rock fill.	4" reinforced concrete on 6" rock fill.	4" concrete on 6" rock fill.	4" concrete on 4" rock fill.
Walls (12% of total cost)	2" x 4" studs 16" o.c. in walls 14' high; 2" x 6" studs 16" o.c. in walls over 14' high; 3" sill, double plate, adequate blocking and bracing.	2" x 4" studs 16" o.c. in walls 14' high; 2" x 6" studs 16" o.c. in walls over 14' high; 2" sill, double plate, adequate blocking and bracing.	2" x 4" studs 16" o.c. in walls to 14' high; 2" x 6" studs 16" o.c. in walls over 14' high; 2" sill, double plate, minimum blocking and bracing.	2" x 4" studs 24" o.c.; 2" x 4" sill, double 2" x 4" plate, minimum diagonal bracing.
Exterior (9% of total cost)	Good corrugated iron or board and batt.	Good corrugated iron or board and batt.	Average corrugated iron or board and batt.	Light corrugated iron or board and batt.
Roof Structures (12% of total cost)	Glu-lams, trusses or tapered steel girders on steel intermediate columns; 2" x 10" rafters 16" o.c.	Glu-lams, average wood trusses, tapered steel girders on steel intermediate columns; 2" x 8" purlins or rafters 16" o.c.	Glu-lams or light wood trusses, on wood posts 18' o.c.; 2" x 8" rafters on purlins 24" o.c.	Light trussed rafters, clear span in small buildings, post and beam support in large buildings.
Roof Cover (5% of total cost)	Good quality 4 ply composition roofing on wood sheathing.	Average quality 4 ply composition roofing on wood sheathing.	Average quality 3 ply composition roofing on wood sheathing, or good corrugated aluminum.	Light weight 3 ply composition roofing on wood sheathing, or heavy corrugated iron.
Rest Rooms (10% of total cost)	Two restrooms with three average fixtures each.	Two restrooms with three average fixtures each.	One restroom with two low cost fixtures.	One restroom with two low cost fixtures.
Lighting (10% of total cost)	One incandescent fixture per 300 square feet of floor area.	One incandescent fixture per 300 square feet of floor area.	One incandescent fixture per 300 square feet of floor area.	One incandescent fixture per 300 square feet of floor area.
Windows (5% of total cost)	3% to 5% of wall area.	3% to 5% of wall area.	3% to 5% of wall area.	3% to 5% of wall area.

Note: Use the percent of total cost to help identify the correct quality classification.

Square foot costs include the cost of the following components: Foundations as required for normal soil conditions. Floor, wall and roof structure. Exterior wall finish and roof cover. Entry doors. Basic lighting and electrical systems. Rough and finish plumbing. Permits and fees. Contractors' mark-up.

The in-place cost of these extra components should be added to the basic building cost to arrive at the total structure cost. See page 201. Heating and air conditioning systems. Fire sprinklers. Interior finish costs. Interior partitions. Drive-through doors. Canopies and walks. Exterior signs. Paving and curbing. Miscellaneous yard improvements. Hoists, gas pump and compressor costs are listed in the section "Additional Costs for Service Stations" on page 169.

Service Garage - Wood Frame

Length Less Than Twice Width

Estimating Procedure

1. Use these figures to estimate buildings designed primarily for motor vehicle repair. Sales area should be figured separately. Use the costs for urban stores beginning on page 42.
2. Establish the building quality class by applying the information on page 178.
3. Compute the floor area.
4. If the wall height is more or less than 16 feet, add to or subtract from the square foot costs below the appropriate amount from the Wall Height Adjustment Table on page 182.
5. Multiply the adjusted square foot cost by the floor area.
6. Deduct for common walls or no wall ownership. Use the figures on page 182.
7. Multiply the total cost by the location factor on page 7.
8. Add the cost of heating and air conditioning systems, fire sprinklers, interior finish and partitions, drive-thru doors, canopies and walks, exterior signs, paving, curbing, and yard improvements. See page 201. Add the cost of hoists, pumps and compressors from pages 169 and 171.

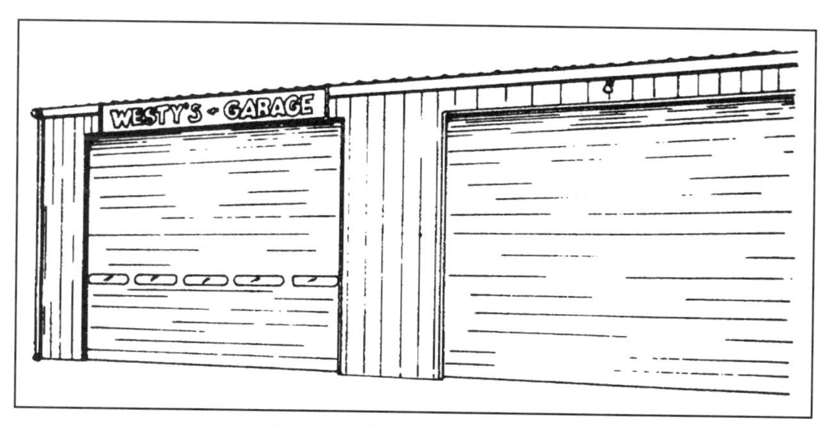

Service Garage, Class 3

Square Foot Area

Quality Class	2,000	2,500	3,000	4,000	5,000	6,000	7,500	10,000	15,000	20,000	30,000
1, Best	28.71	26.06	24.20	21.69	20.04	18.86	17.61	16.25	14.70	13.80	12.74
1 & 2	27.24	24.72	22.96	20.57	19.00	17.90	16.73	15.40	13.93	13.08	12.09
2, Good	26.30	23.87	22.16	19.87	18.36	17.30	16.13	14.88	13.44	12.65	11.70
2 & 3	24.74	22.44	20.83	18.67	17.28	16.24	15.17	13.98	12.66	11.87	10.99
3, Average	23.50	21.33	19.82	17.75	16.40	15.45	14.41	13.31	12.02	11.30	10.46
3 & 4	22.06	20.03	18.62	16.67	15.43	14.48	13.54	12.48	11.29	10.60	9.83
4, Low	20.76	18.83	17.51	15.67	14.48	13.65	12.74	11.72	10.58	9.98	9.24

Service Garage – Wood Frame

Length Between 2 and 4 Times Width

Estimating Procedure

1. Use these figures to estimate buildings designed primarily for motor vehicle repair. Sales area should be figured separately. Use the costs for urban stores beginning on page 42.
2. Establish the building quality class by applying the information on page 178.
3. Compute the floor area.
4. If the wall height is more or less than 16 feet, add to or subtract from the square foot costs below the appropriate amount from the Wall Height Adjustment Table on page 182.
5. Multiply the adjusted square foot cost by the floor area.
6. Deduct for common walls or no wall ownership. Use the figures on page 182.
7. Multiply the total cost by the location factor on page 7.
8. Add the cost of heating and air conditioning systems, fire sprinklers, interior finish and partitions, drive-thru doors, canopies and walks, exterior signs, paving, curbing, and yard improvements. See page 201. Add the cost of hoists, pumps and compressors from pages 169 and 171.

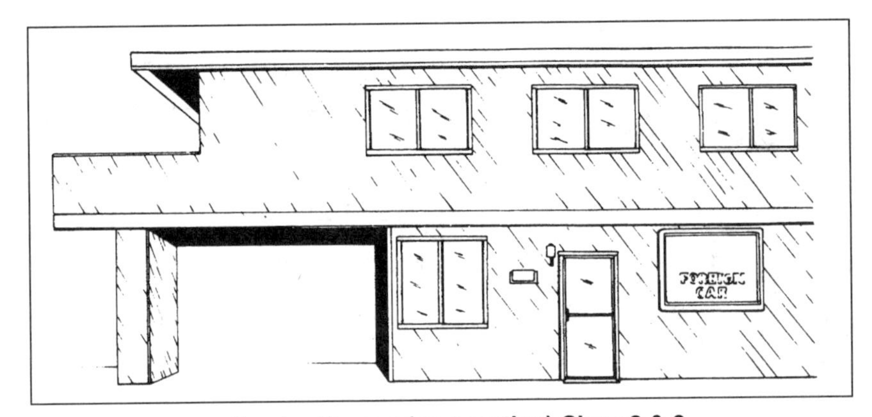

Service Garage (rear portion) Class 2 & 3

Square Foot Area

Quality Class	2,000	2,500	3,000	4,000	5,000	6,000	7,500	10,000	15,000	20,000	30,000
1, Best	30.53	27.73	25.74	23.08	21.33	20.09	18.75	17.30	15.61	14.67	13.57
1 & 2	29.00	26.35	24.44	21.92	20.26	19.08	17.82	16.41	14.85	13.93	12.88
2, Good	27.97	25.43	23.60	21.13	19.53	18.41	17.19	15.82	14.33	13.44	12.43
2 & 3	26.33	23.91	22.21	19.90	18.40	17.33	16.17	14.92	13.48	12.67	11.72
3, Average	25.13	22.84	21.19	19.00	17.57	16.54	15.44	14.23	12.85	12.08	11.16
3 & 4	23.43	21.26	19.76	17.72	16.37	15.43	14.38	13.26	11.97	11.25	10.40
4, Low	21.74	19.75	18.32	16.42	15.18	14.33	13.34	12.30	11.13	10.43	9.67

Service Garage – Wood Frame

Length More Than 4 Times Width

Estimating Procedure

1. Use these figures to estimate buildings designed primarily for motor vehicle repair. Sales area should be figured separately. Use the costs for urban stores beginning on page 42.
2. Establish the building quality class by applying the information on page 178.
3. Compute the floor area.
4. If the wall height is more or less than 16 feet, add to or subtract from the square foot costs below the appropriate amount from the Wall Height Adjustment Table on page 182.
5. Multiply the adjusted square foot cost by the floor area.
6. Deduct for common walls or no wall ownership. Use the figures on page 182.
7. Multiply the total cost by the location factor on page 7.
8. Add the cost of heating and air conditioning systems, fire sprinklers, interior finish and partitions, drive-thru doors, canopies and walks, exterior signs, paving, curbing, and yard improvements. See page 201. Add the cost of hoists, pumps and compressors from pages 169 and 171.

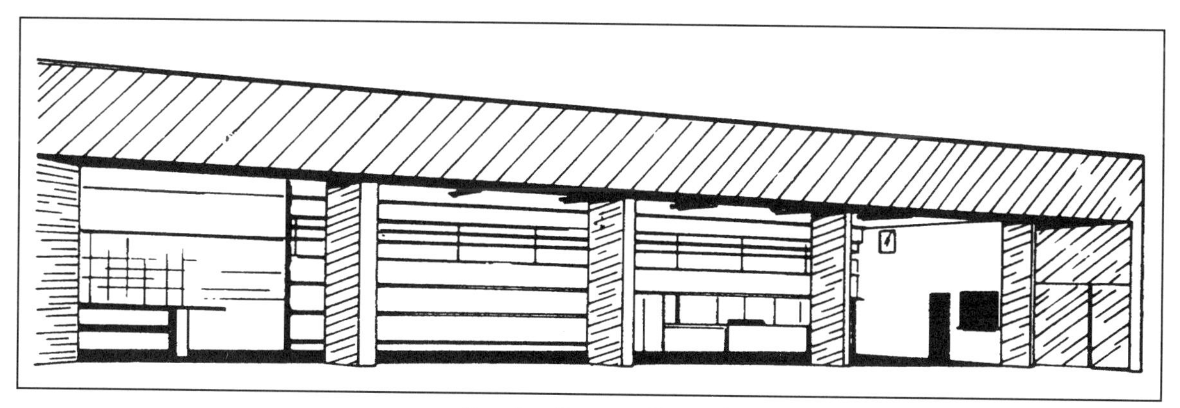

Service Garage, Class 2

Square Foot Area

Quality Class	2,000	2,500	3,000	4,000	5,000	6,000	7,500	10,000	15,000	20,000	30,000
1, Best	32.58	29.58	27.47	24.61	22.75	21.43	19.97	18.43	16.66	15.61	14.44
1 & 2	30.94	28.10	26.07	23.39	21.62	20.35	18.99	17.51	15.82	14.87	13.72
2, Good	29.92	27.14	25.20	22.58	20.88	19.66	18.36	16.92	15.28	14.35	13.26
2 & 3	28.15	25.56	23.75	21.29	19.67	18.50	17.29	15.94	14.39	13.52	12.48
3, Average	26.76	24.35	22.57	20.26	18.72	17.62	16.42	15.16	13.70	12.85	11.88
3 & 4	24.99	22.68	21.07	18.86	17.47	16.42	15.34	14.11	12.78	12.00	11.09
4, Low	23.32	21.18	19.65	17.63	16.28	15.33	14.30	13.21	11.93	11.19	10.34

Service Garage - Wood Frame

Wall Height Adjustment

Add or subtract the amount listed in this table to the square foot cost for each foot of wall height more or less than 16 feet.

Area	**2,000**	**2,500**	**3,000**	**4,000**	**5,000**	**6,000**	**7,500**	**10,000**	**15,000**	**20,000**	**30,000**
Cost	.38	.33	.30	.24	.21	.14	.12	.09	.07	.06	.05

Perimeter Wall Adjustment

A common wall exists when two buildings share one wall. Adjust for common walls by deducting the linear foot costs below from the total structure cost. In some structures one or more walls are not owned at all. In this case, deduct the "No Ownership" cost per linear foot of wall not owned.

For common wall, deduct $45.50 per linear foot.

For no wall ownership, deduct $91.00 per linear foot.

Auto Service Centers – Masonry or Concrete

Quality Classification

	Class 1 Best Quality	Class 2 Good Quality	Class 3 Average Quality	Class 4 Low Quality
Foundation (20% of total cost)	Reinforced concrete.	Reinforced concrete.	Reinforced concrete.	Reinforced concrete.
Floor Structure (10% of total cost)	4" reinforced concrete on 6" rock fill.	4" reinforced concrete on 6" rock fill.	4" reinforced concrete on 6" rock fill.	4" reinforced concrete on 6" rock fill.
Walls (15% of total cost)	8" reinforced decorative colored concrete block.	8" reinforced detailed concrete block.	8" reinforced concrete block.	8" reinforced concrete block.
Roof Structures (5% of total cost)	Steel open web joists, steel deck.	Glu-lams, 3" x 12" purlins 3' o.c., or wood open web joists (truss joists) 4' o.c.	Glu-lams, 3" x 12" purlins 3' o.c., or wood open web joists (truss joists) 4' o.c., 1/2" plywood sheathing.	Glu-lams, 3" x 12" purlins 3' o.c., or wood open web joists (truss joists) 4' o.c., 1/2" plywood sheathing.
Floor Finish (5% of total cost)	Concrete in work area, resilient tile in sales area.	Concrete in work area, resilient tile in sales area.	Concrete in work area, minimum grade tile in sales area.	Concrete.
Interior Wall Finish (5% of total cost)	Painted concrete block in work area; gypsum wallboard, texture and paint in sales area.	Painted concrete block in work area and sales area.	Unfinished in work area, painted concrete block in sales area	Unfinished.
Ceiling Finish (3% of total cost)	Open in work area, acoustical tile suspended in exposed grid in sales area.	Open in work area, acoustical tile suspended in exposed grid in sales area.	Open in work area. Celotex tile in sales area.	Open.
Exterior Finish (5% of total cost)	Decorative colored concrete block.	Colored or painted detailed block.	Painted concrete block.	Unpainted concrete block.
Display Front *(Covers about 25% of the exterior wall)* (7% of total cost)	1/4" float glass in good aluminum frame. Good aluminum and glass doors.	1/4" float glass in average aluminum frame. Aluminum and glass doors.	1/4" float glass in light aluminum frame. Wood and glass door.	Crystal glass in wood frame, wood door.
Roof and Cover (7% of total cost)	5 ply built-up roofing with insulation.	4 ply built-up roofing.	4 ply built-up roofing.	4 ply built-up roofing.
Plumbing (10% of total cost)	Two rest rooms with three good fixtures. Metal toilet partitions.	Two rest rooms with two average fixtures. Wood toilet partitions.	One rest room with two fixtures.	One rest room with two fixtures.
Electrical & Wiring (8% of total cost)	Conduit wiring with triple tube fluorescent strips, 8' o.c.	Conduit wiring with triple tube fluorescent strips, 8' o.c.	Conduit wiring, double tube fluorescent strips, 8' o.c.	Conduit wiring, incandescent fixtures, 10' o.c. or single tube fluorescent strips, 8' o.c.

Note: Use the percentage of total cost to help identify the correct quality classification.

Square foot costs include the cost of the following components: Foundations as required for normal soil conditions. Floor, wall and roof structures. Interior floor, wall and ceiling finishes (in sales area). Exterior wall finish and roof cover. Display windows. Interior partitions. Entry doors. Basic lighting and electrical systems. Rough and finish plumbing. Permits and fees. Contractors' mark-up.

Wall Height Adjustment: Add or subtract the amount listed in this table to the square foot of floor cost for each foot of wall height more or less than 16 feet.

Area	1,500	2,000	2,500	3,000	3,500	4,000	5,000	6,000	7,500	10,000	15,000
Cost	1.31	1.13	.96	.87	.82	.77	.71	.66	.54	.49	.44

Perimeter Wall Adjustment: For common wall, deduct $101.00 per linear foot. For no wall ownership, deduct $202.00 per linear foot.

Auto Service Centers – Masonry or Concrete

Length Less Than Twice Width

Estimating Procedure

1. Use these figures to estimate buildings designed for selling and installing automobile accessories. The square foot costs below allow for a sales area occupying 25% of the building space. The sales area has finished floors, walls and ceiling as described in the quality classification. The remaining 75% of the building is service area and has no interior finish.
2. Establish the building quality class by applying the information on page 183.
3. Compute the floor area. This should include everything within the exterior walls.
4. If the wall height is more or less than 16 feet, add to or subtract from the square foot costs below the appropriate amount from the Wall Height Adjustment Table on page 183.
5. Deduct for common walls or no wall ownership. See page 183.
6. Multiply the total cost by the location factor on page 7.
7. Add the cost of heating and air conditioning systems, fire sprinklers, canopies, walks, exterior signs, paving, curbing, loading docks, ramps, and yard improvements. See page 201. Add the cost of service station equipment from page 169.

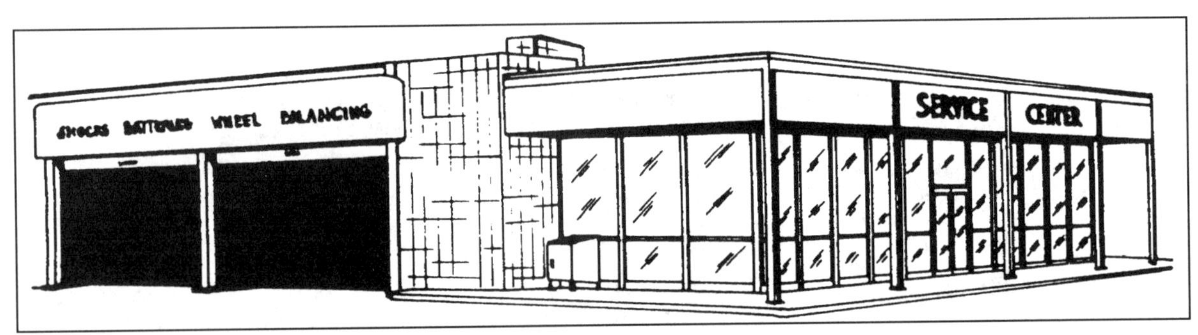

Auto Service Center, Class 2

Square Foot Area

Quality Class	1,500	2,000	2,500	3,000	3,500	4,000	5,000	6,000	7,500	10,000	15,000
Exceptional	84.35	77.17	72.45	69.07	66.46	64.45	61.37	59.14	56.73	54.06	50.94
1, Best	80.72	73.86	69.32	66.08	63.62	61.65	58.71	56.59	54.28	51.72	48.75
1 & 2	77.17	70.59	66.28	63.18	60.82	58.94	56.12	54.08	51.88	49.42	46.61
2, Good	74.88	68.50	64.31	61.31	59.02	57.21	54.46	52.50	50.36	47.98	45.20
2 & 3	71.40	65.30	61.32	58.45	56.26	54.53	51.94	50.04	48.00	45.75	43.13
3, Average	69.35	63.44	59.55	56.77	54.64	52.98	50.45	48.60	46.65	44.43	41.88
3 & 4	65.96	60.37	56.67	54.03	51.99	50.41	48.00	46.26	44.36	42.28	39.84
4, Low	63.22	57.84	54.29	51.78	49.83	48.30	46.01	44.34	42.53	40.52	38.17

Auto Service Centers - Masonry or Concrete

Length Between 2 and 4 Times Width

Estimating Procedure

1. Use these figures to estimate buildings designed for selling and installing automobile accessories. The square foot costs below allow for a sales area occupying 25% of the building space. The sales area has finished floors, walls and ceiling as described in the quality classification. The remaining 75% of the building is service area and has no interior finish.
2. Establish the building quality class by applying the information on page 183.
3. Compute the floor area. This should include everything within the exterior walls.
4. If the wall height is more or less than 16 feet, add to or subtract from the square foot costs below the appropriate amount from the Wall Height Adjustment Table on page 183.
5. Deduct for common walls or no wall ownership. See page 183.
6. Multiply the total cost by the location factor on page 7.
7. Add the cost of heating and air conditioning systems, fire sprinklers, canopies, walks, exterior signs, paving, curbing, loading docks, ramps, and yard improvements. See page 201. Add the cost of service station equipment from page 169.

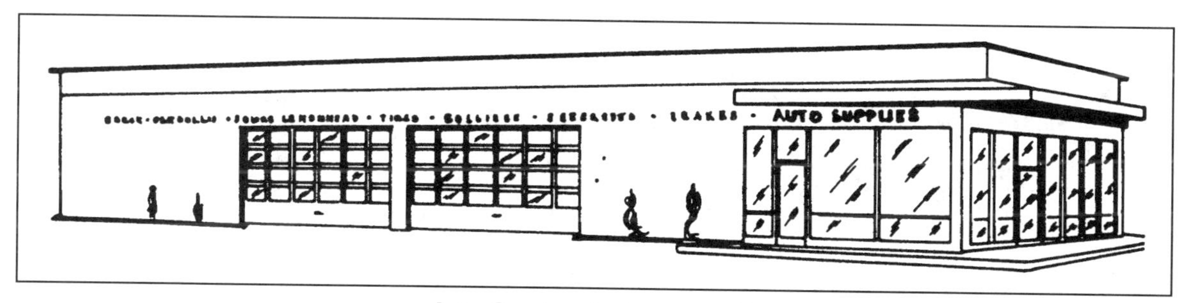

Auto Service Center, Class 2

Square Foot Area

Quality Class	1,500	2,000	2,500	3,000	3,500	4,000	5,000	6,000	7,500	10,000	15,000
Exceptional	89.64	81.60	76.31	72.54	69.65	67.36	63.96	61.52	58.84	55.88	52.48
1, Best	85.79	78.07	73.06	69.40	66.67	64.49	61.22	58.87	56.30	53.45	50.25
1 & 2	81.92	74.55	69.71	66.28	63.64	61.56	58.45	56.21	53.76	51.07	47.94
2, Good	79.39	72.24	67.56	64.22	61.67	59.67	56.65	54.48	52.12	49.51	46.48
2 & 3	75.82	68.99	64.55	61.35	58.91	57.01	54.11	52.03	49.77	47.27	44.39
3, Average	73.42	66.81	62.47	59.40	57.03	55.17	52.37	50.37	48.18	45.78	42.98
3 & 4	70.10	63.82	59.71	56.71	54.48	52.70	50.03	48.11	46.02	43.73	41.04
4, Low	66.85	60.86	56.92	54.11	51.96	50.26	47.73	45.87	43.91	41.68	39.14

Auto Service Centers – Masonry or Concrete

Length More Than 4 Times Width

Estimating Procedure

1. Use these figures to estimate buildings designed for selling and installing automobile accessories. The square foot costs below allow for a sales area occupying 25% of the building space. The sales area has finished floors, walls and ceiling as described in the quality classification. The remaining 75% of the building is service area and has no interior finish.
2. Establish the building quality class by applying the information on page 183.
3. Compute the floor area. This should include everything within the exterior walls.
4. If the wall height is more or less than 16 feet, add to or subtract from the square foot costs below the appropriate amount from the Wall Height Adjustment Table on page 183.
5. Deduct for common walls or no wall ownership. See page 183.
6. Multiply the total cost by the location factor on page 7.
7. Add the cost of heating and air conditioning systems, fire sprinklers, canopies, walks, exterior signs, paving, curbing, loading docks, ramps, and yard improvements. See page 201. Add the cost of service station equipment from page 169.

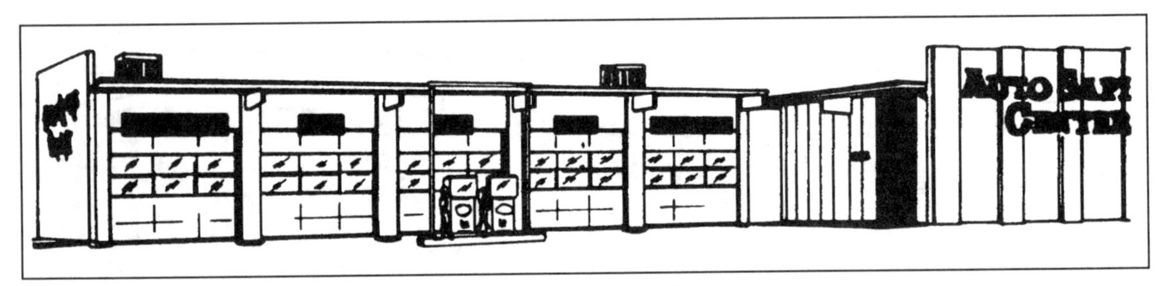

Auto Service Center, Class 3

Square Foot Area

Quality Class	1,500	2,000	2,500	3,000	3,500	4,000	5,000	6,000	7,500	10,000	15,000
Exceptional	96.68	87.89	82.05	77.83	74.58	72.00	68.06	65.22	62.07	58.58	54.47
1, Best	92.46	84.07	78.49	74.44	71.33	68.84	65.11	62.35	59.36	56.04	52.10
1 & 2	88.21	80.23	74.93	71.03	68.07	65.70	62.13	59.51	56.65	53.44	49.72
2, Good	85.61	77.88	72.71	68.95	66.07	63.77	60.27	57.75	55.00	51.88	48.25
2 & 3	81.65	74.25	69.33	65.73	63.01	60.81	57.48	55.05	52.43	49.49	46.02
3, Average	79.27	72.07	67.30	63.82	61.15	59.03	55.83	53.47	50.90	48.01	44.67
3 & 4	75.63	68.77	64.22	60.91	58.33	56.29	53.26	51.00	48.56	45.81	42.63
4, Low	72.11	65.58	61.21	58.07	55.65	53.71	50.77	48.63	46.31	43.70	40.65

Industrial Structures Section

Section Contents

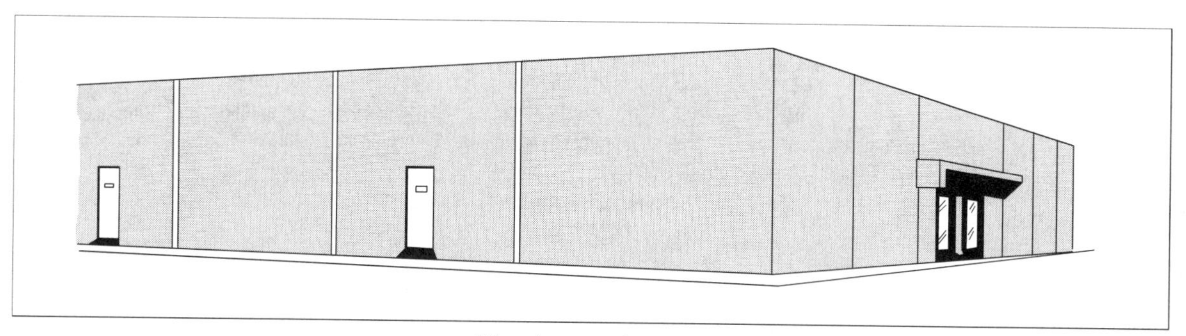

Warehouse, Class 3

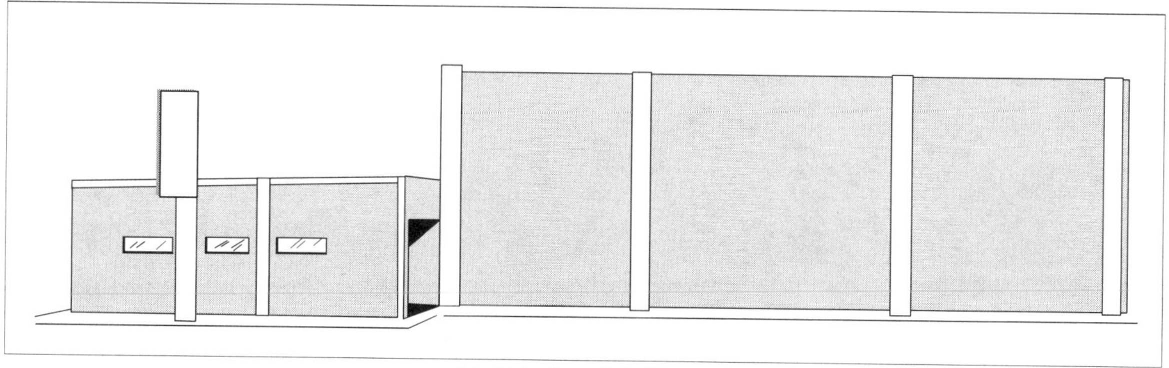

Light Industrial, Class 2

Industrial Buildings

Quality Classification

	Class 1 Best Quality	Class 2 Good Quality	Class 3 Average Quality	Class 4 Low Quality
Foundations (22% of total cost)	Continuous reinforced concrete.	Continuous reinforced concrete.	Continuous reinforced concrete.	Reinforced concrete pads under pilasters.
Floor Structure (15% of total cost)	6" rock base, 6" concrete with reinforcing mesh or bars.	6" rock base, 6" concrete with reinforcing mesh or bars.	6" rock base, 5" concrete with reinforcing mesh or bars.	6" rock base, 4" concrete with reinforcing mesh.
Wall Structure (25% of total cost)	8" reinforced concrete block or brick with pilasters 20' o.c., painted sides and rear exterior, front wall brick veneer.	8" reinforced concrete block or brick with pilasters 20' o.c., painted sides and rear exterior, stucco and some brick veneer on front.	8" reinforced concrete block or brick, unpainted.	8" reinforced concrete block or brick, unpainted.
Roof Structure (12% of total cost)	Glu-lams, wood or steel trusses on steel intermediate columns, span exceeds 70'.	Glu-lams, wood or steel trusses on steel intermediate columns, span exceeds 70'.	Glu-lams or steel beams on steel intermediate columns, short span.	Glu-lams on steel intermediate columns, short span.
Roof Cover (7% of total cost)	Panelized roof system, 1/2" plywood sheathing, 5 ply built-up roof.	Panelized roof system, 1/2" plywood sheathing, 4 ply built-up roof.	Panelized roof system, 1/2" plywood sheathing, 4 ply built-up roof.	Panelized roof system, 1/2" plywood sheathing, 4 ply built-up roof.
Skylights (1% of total cost)	48 S.F. of skylight per 2500 S.F. of floor area (1-6' x 8' skylight 40' to 50' o.c.).	32 S.F. of skylight per 2500 S.F. of floor area (1-4' x 8' skylight 40' to 50' o.c.).	24 S.F. of skylight per 2500 S.F. of floor area (1-4' x 6' skylight 40' to 50' o.c.).	10 S.F. of skylight per 2500 S.F. of floor area (1-2' x 4' skylight 40' to 50' o.c.).
Ventilators (2% of total cost)	1 large rotary vent per 2500 S.F. of floor area.	1 medium rotary vent per 2500 S.F. of floor area.	1 medium rotary vent per 2500 S.F. of floor area.	1 small rotary vent per 2500 S.F. of floor area.
Rest Rooms, Finish (3% of total cost)	Good vinyl asbestos tile floors enameled gypsum wallboard partitions.	Vinyl asbestos tile floors, enameled gypsum wallboard partitions.	Concrete floors, painted gypsum wallboard partitions.	Concrete floors, unfinished wallboard partitions.
Fixtures (5% of total cost)	2 rest rooms, 3 good fixtures in each.	2 rest rooms, 3 average fixtures in each.	2 rest rooms, 2 average fixtures in each.	1 rest room, 2 low cost fixtures in each.
Lighting (8% of total cost)	4' single tube fluorescent fixtures 10' x 12' spacing.	Low cost single tube fluorescent fixtures 12' x 20' spacing.	Low cost 4' single tube fluorescent fixtures 20' x 20' spacing.	Low cost incandescent fixtures 20' x 30' spacing.

Note: Use the percent of total cost to help identify the correct quality classification.

Square foot costs include the following components: Foundations as required for normal soil conditions. Floor, wall and roof structures. Exterior wall finish and roof cover. Basic lighting and electrical systems. Rough and finish plumbing. A usual or normal parapet wall. Walk-through doors. Contractors' mark-up.

Warehouses

Estimating Procedure

1. Establish the structure quality class by applying the information on page 188.
2. Compute the building floor area.
3. Add to or subtract from the square foot cost below the appropriate amount from the Wall Height Adjustment Table (at the bottom of page 191) if the wall height is more or less than 20 feet.
4. Multiply the adjusted square foot cost by the building floor area.
5. Deduct, if appropriate, for common walls, using the figures at the bottom of page 191.
6. Multiply the total cost by the location factor on page 7.
7. Add the cost of heating and air conditioning equipment, fire sprinklers, interior offices, drive-through or delivery doors, canopies, interior partitions, docks and ramps, paving and curbing, and miscellaneous yard improvements. See the section beginning on page 201.

Length less than twice width - Square Foot Area

Quality Class	3,000	4,000	5,000	7,500	10,000	15,000	20,000	30,000	50,000	100,000	200,000
1, Best	75.34	68.95	64.67	58.11	54.26	49.76	47.14	44.01	40.96	37.89	35.75
1 & 2	70.15	64.21	60.23	54.12	50.53	46.37	43.91	41.02	38.13	35.29	33.30
2, Good	66.28	60.64	56.88	51.12	47.74	43.77	41.45	38.72	36.01	33.31	31.44
2 & 3	61.99	56.70	53.19	47.82	44.64	40.97	38.78	36.22	33.70	31.16	29.40
3, Average	58.32	53.38	50.04	44.98	42.01	38.54	36.49	34.08	31.69	29.35	27.68
3 & 4	54.57	49.94	46.80	42.11	39.29	36.08	34.13	31.89	29.64	27.43	25.85
4, Low	50.83	46.52	43.63	39.21	36.62	33.58	31.81	29.70	27.65	25.56	24.14

Length between 2 and 4 times width - Square Foot Area

Quality Class	3,000	4,000	5,000	7,500	10,000	15,000	20,000	30,000	50,000	100,000	200,000
1, Best	80.37	73.17	68.39	61.09	56.85	51.85	48.94	45.53	42.18	38.84	36.52
1 & 2	75.05	68.35	63.86	57.03	53.05	48.43	45.70	42.52	39.38	36.26	34.09
2, Good	70.85	64.53	60.28	53.86	50.09	45.73	43.14	40.16	37.20	34.24	32.16
2 & 3	66.28	60.37	56.40	50.35	46.84	42.77	40.36	37.57	34.77	32.02	30.10
3, Average	62.36	56.79	53.05	47.40	44.08	40.23	37.99	35.35	32.73	30.16	28.30
3 & 4	58.34	53.13	49.65	44.35	41.27	37.65	35.54	33.06	30.62	28.20	26.52
4, Low	54.23	49.36	46.15	41.22	38.34	35.02	33.03	30.72	28.43	26.18	24.63

Length more than 4 times width - Square Foot Area

Quality Class	3,000	4,000	5,000	7,500	10,000	15,000	20,000	30,000	50,000	100,000	200,000
1, Best	87.07	79.12	73.76	65.53	60.68	54.99	51.62	47.70	43.76	39.86	37.08
1 & 2	83.52	73.89	68.86	61.19	56.65	51.36	48.21	44.55	40.86	37.22	34.63
2, Good	78.99	69.80	65.06	57.79	53.53	48.53	45.54	42.08	38.62	35.16	32.74
2 & 3	71.89	65.30	60.86	54.08	50.08	45.39	42.62	39.35	36.14	32.88	30.61
3, Average	67.64	61.47	57.30	50.89	47.12	42.72	40.12	37.03	34.02	30.92	28.80
3 & 4	63.22	57.42	53.53	47.54	44.06	39.91	37.46	34.60	31.76	28.91	26.95
4, Low	58.85	53.48	49.86	44.30	41.01	37.19	34.89	32.24	29.58	26.95	25.07

Light Industrial Buildings

Estimating Procedure

1. Establish the structure quality class by applying the information on page 188.
2. Compute the building floor area.
3. Add to or subtract from the square foot cost below the appropriate amount from the Wall Height Adjustment Table (at the bottom of page 191) if the wall height is more or less than 20 feet.
4. Multiply the adjusted square foot cost by the building floor area.
5. Deduct, if appropriate, for common walls, using the figures at the bottom of page 191.
6. Multiply the total cost by the location factor on page 7.
7. Add the cost of heating and air conditioning equipment, fire sprinklers, interior offices, drive-through or delivery doors, canopies, interior partitions, docks and ramps, paving and curbing, and miscellaneous yard improvements. See the section beginning on page 201.

Length less than twice width - Square Foot Area

Quality Class	3,000	4,000	5,000	7,500	10,000	15,000	20,000	30,000	50,000	100,000	200,000
1, Best	74.63	68.91	65.05	59.14	55.67	51.61	49.21	46.39	43.60	40.82	38.85
1 & 2	69.64	64.27	60.67	55.18	51.94	48.15	45.91	43.29	40.67	38.07	36.24
2, Good	65.14	60.15	56.77	51.62	48.61	45.06	42.97	40.50	38.07	35.62	33.92
2 & 3	61.56	56.85	53.65	48.80	45.94	42.58	40.60	38.28	35.97	33.67	32.04
3, Average	57.33	52.95	49.98	45.43	42.80	39.65	37.80	35.66	33.50	31.38	29.86
3 & 4	53.72	49.59	46.83	42.57	40.07	37.17	35.43	33.40	31.40	29.38	27.98
4, Low	50.00	46.17	43.57	39.63	37.30	34.58	32.98	31.10	29.21	27.34	26.04

Length between 2 and 4 times width - Square Foot Area

Quality Class	3,000	4,000	5,000	7,500	10,000	15,000	20,000	30,000	50,000	100,000	200,000
1, Best	79.38	72.88	68.54	61.95	58.08	53.56	50.91	47.81	44.75	41.71	39.57
1 & 2	74.23	68.17	64.12	57.93	54.33	50.10	47.61	44.73	41.87	39.01	37.00
2, Good	69.68	63.99	60.19	54.39	51.00	47.04	44.72	41.97	39.30	36.62	34.75
2 & 3	65.11	59.78	56.22	50.80	47.63	43.94	41.77	39.21	36.67	34.19	32.46
3, Average	61.01	56.04	52.72	47.63	44.65	41.21	39.16	36.79	34.41	32.05	30.41
3 & 4	57.16	52.51	49.36	44.61	41.82	38.58	36.65	34.44	32.25	30.02	28.50
4, Low	53.17	48.84	45.91	41.51	38.91	35.89	34.12	32.05	29.98	27.95	26.52

Length more than 4 times width - Square Foot Area

Quality Class	3,000	4,000	5,000	7,500	10,000	15,000	20,000	30,000	50,000	100,000	200,000
1, Best	85.50	78.20	73.31	65.84	61.47	56.30	53.26	49.71	46.18	42.65	40.18
1 & 2	80.05	73.26	68.65	61.65	57.55	52.75	49.90	46.56	43.26	39.95	37.64
2, Good	74.99	68.59	64.31	57.76	53.90	49.37	46.73	43.60	40.50	37.40	35.23
2 & 3	70.24	64.26	60.24	54.10	50.48	46.27	43.75	40.84	37.95	35.04	33.02
3, Average	65.88	60.30	56.51	50.75	47.37	43.39	41.05	38.32	35.58	32.85	30.98
3 & 4	61.53	56.29	52.77	47.40	44.22	40.53	38.35	35.79	33.23	30.69	28.91
4, Low	57.29	52.42	49.17	44.14	41.17	37.72	35.73	33.33	30.93	28.60	26.93

Factory Buildings

Estimating Procedure

1. Establish the structure quality class by applying the information on page 188.
2. Compute the building floor area.
3. Add to or subtract from the square foot cost below the appropriate amount from the Wall Height Adjustment Table (at the bottom of this page) if the wall height is more or less than 20 feet.
4. Multiply the adjusted square foot cost by the building floor area.
5. Deduct, if appropriate, for common walls, using the figures at the bottom of this page.
6. Multiply the total cost by the location factor on page 7.
7. Add the cost of heating and air conditioning equipment, fire sprinklers, interior offices, drive-through or delivery doors, canopies, interior partitions, docks and ramps, paving and curbing, and miscellaneous yard improvements. See the section beginning on page 201.

Length less than twice width - Square Foot Area

Quality Class	3,000	4,000	5,000	7,500	10,000	15,000	20,000	30,000	50,000	100,000	200,000
1, Best	73.89	68.63	65.11	59.72	56.57	52.88	50.70	48.13	45.60	43.07	41.32
1 & 2	69.37	64.46	61.14	56.07	53.12	49.64	47.57	45.18	42.82	40.46	38.81
2, Good	65.38	60.73	57.59	52.84	50.02	46.76	44.84	42.60	40.34	38.12	36.56
2 & 3	61.23	56.87	53.97	49.49	46.86	43.79	41.98	39.88	37.78	35.71	34.26
3, Average	57.56	53.44	50.70	46.49	44.05	41.16	39.48	37.46	35.51	33.55	32.21
3 & 4	53.75	49.93	47.38	43.43	41.13	38.45	36.87	35.03	33.17	31.34	30.04
4, Low	49.96	46.40	44.00	40.36	38.20	35.75	34.27	32.53	30.83	29.13	27.94

Length between 2 and 4 times width - Square Foot Area

Quality Class	3,000	4,000	5,000	7,500	10,000	15,000	20,000	30,000	50,000	100,000	200,000
1, Best	78.05	72.19	68.24	62.27	58.75	54.66	52.24	49.44	46.67	43.89	41.92
1 & 2	73.22	67.71	64.05	58.39	55.13	51.26	49.02	46.38	43.77	41.16	39.34
2, Good	69.16	63.93	60.45	55.17	52.05	48.43	46.31	43.78	41.32	38.87	37.17
2 & 3	64.74	59.86	56.61	51.63	48.72	45.34	43.34	41.01	38.69	36.40	34.79
3, Average	60.80	56.23	53.16	48.50	45.76	42.56	40.70	38.49	36.32	34.17	32.68
3 & 4	56.82	52.55	49.69	45.34	42.77	39.80	38.05	35.98	33.94	31.94	30.53
4, Low	52.76	48.81	46.13	42.08	39.72	36.95	35.34	33.43	31.54	29.63	28.35

Length more than 4 times width - Square Foot Area

Quality Class	3,000	4,000	5,000	7,500	10,000	15,000	20,000	30,000	50,000	100,000	200,000
1, Best	83.56	77.02	72.60	65.84	61.87	57.20	54.43	51.18	47.96	44.72	42.48
1 & 2	78.52	72.39	68.21	61.88	58.12	53.75	51.14	48.10	45.06	42.03	39.90
2, Good	74.03	68.25	64.32	58.33	54.81	50.67	48.22	45.34	42.50	39.64	37.62
2 & 3	69.37	63.94	60.26	54.65	51.36	47.47	45.18	42.50	39.82	37.17	35.25
3, Average	65.17	60.11	56.66	51.40	48.28	44.64	42.45	39.91	37.40	34.87	33.11
3 & 4	60.89	56.10	52.90	47.94	45.05	41.66	39.65	37.28	34.93	32.60	30.94
4, Low	56.50	52.09	49.08	44.53	41.82	38.67	36.82	34.60	32.41	30.25	28.72

Wall Height Adjustment: Industrial building costs are based on a 20' wall height, measured from the bottom of the floor slab to the top of the roof cover. Add or subtract the amount listed to the square foot cost for each foot more or less than 20 feet.

Area	3,000	4,000	5,000	7,500	10,000	15,000	20,000	30,000	50,000	100,000	200,000
Cost	.57	.50	.46	.40	.36	.31	.27	.23	.10	.06	.03

Perimeter Wall Adjustment: A common wall exists when two buildings share one wall. Adjust for common walls by deducting $75.70 per linear foot from the total structure cost. In some structures one or more walls are not owned at all. In this case, deduct $151.70 per liner foot of wall not owned.

Internal Offices

Internal offices are office areas built into the interior area of an industrial building. Add the square foot costs in this section to the basic building cost. The costs include floor finish, partition framing, wall finish, trim and doors, counter, ceiling structure, ceiling finish, and the difference between the cost of lighting and windows in an industrial building and the cost of lighting and windows in an office building.

Plumbing costs are not included in these internal office costs. if a building has fixtures in excess of those listed in the quality classification for the main building, add $1,260 to $1,560 for each extra fixture. For two-story internal offices, apply a square foot cost based on the total area of both floor levels.

	Best	Good	Average	Low Cost
Floor Finish (15% of total cost)	Carpet, some vinyl tile or terrazzo.	Resilient tile, some carpet.	Composition tile.	Colored concrete.
Wall Finish (25% of total cost)	Hardwood veneer, some textured cloth wall cover.	Gypsum board, texture and paint, some hardwood veneer.	Gypsum board, texture texture and paint.	Plywood and paint.
Ceiling Finish (15% of total cost)	Illuminated plastic ceilings.	Suspended "T" bar and acoustical tile.	Gypsum board, texture and paint.	Exposed or plywood and paint.
Doors - Interior (10% of total cost)	Good grade hardwood.	Good grade hardwood.	Low cost hardwood.	Standard paint grade.
Exterior (5% of total cost)	Pair aluminum entry doors with extensive sidelights.	Aluminum entry door with sidelights.	Wood store door.	Standard paint grade.
Windows (5% of total cost)	Average amount in good aluminum frame, fixed float glass in good frame in front.	Average amount in good aluminum frame, some fixed float glass.	Average amount of average cost aluminum sliding type.	Average amount of low cost aluminum sliding type.
Counters (15% of total cost)	Good grade hardwood counter, good shelving below.	Good grade plastic counter, average amount of shelving below.	Low cost plastic counter, average amount of shelving.	Paint grade counter, small amount of shelving.
Lighting (10% of total cost)	Illuminated ceilings.	Recessed fluorescent fixtures.	Average amount of fluorescent fixtures.	One incandescent per each 150 S.F.
S.F. Costs	$41.51 to $58.32	$32.70 to $42.33	$26.45 to $34.44	$12.92 to $18.96

Note: Use the percent of total cost to help identify the correct quality classification.

The range of costs is primarily due to the density of partitions in the office area. Use the lower costs for offices with larger rooms and fewer partitions.

Wall Height Adjustment*	$0.55 to $0.78	$0.50 to $0.72	$0.48 to $0.68	$0.44 to $0.66

***Wall Height Adjustment:** Add or subtract the amount listed in this column to the square foot cost for each foot when the office wall height is more or less than 8 feet.

External Offices

External offices outside the main building walls have one side that is common to the main building. Square foot costs for external office areas should be estimated with the figures for general office buildings with interior suite entrances. See pages 106 to 109 or 114 to 117.

In selecting a square foot cost for the external office area as well as the main building area, the area of each portion should be used separately for area classification. The area and the perimeter formed by these individual areas is used for shape classification of the main area and for calculating the area and perimeter relationship of the office area. A "no wall ownership" deduction based on the office wall cost should be made for the length of the common side. Wall height adjustments should be made for each area individually.

Steel Buildings

Engineered steel buildings are constructed to serve as warehouses, factories, airplane hangars, garages, and stores, among other uses. They are generally either high-profile (4 in 12 rise roof) or low-profile (1 in 12 rise roof).

The square foot costs for the basic building include: Foundations as required for normal soil conditions. A 4 inch concrete floor with reinforcing mesh and a 2 inch sand fill. A steel building made up of steel frames or bents set 20 or 24 feet on centers, steel roof purlins 4-1/2 to 5-1/2 feet on centers, steel wall girts 3-1/2 to 4-1/2 feet on centers, post and beam type end wall frames, 26-gauge galvanized steel on ends, sides, and roof, window area equal to 2% of the floor area. Basic wiring and minimum lighting fixtures. One small or medium gravity vent per 2,500 square feet of floor area.

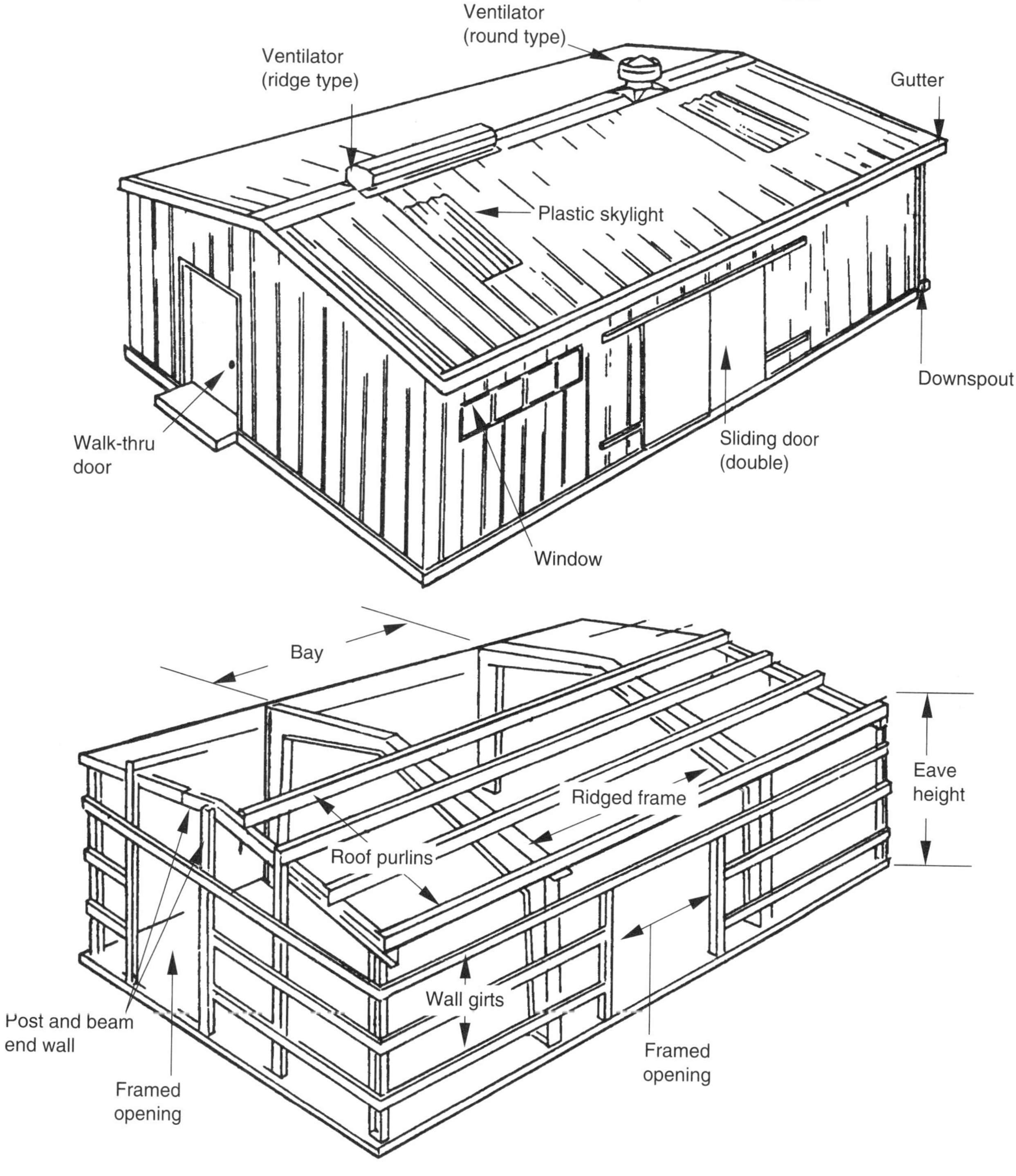

Steel Buildings

Estimating Procedure

1. Compute the building area.
2. Add to or subtract from the square foot cost below $.25 for each foot of wall height at the eave more or less than 14 feet. Subtract $.40 from the square foot cost below for low profile buildings.
3. Multiply the adjusted square foot cost by the building area.
4. Multiply the total by the appropriate live load adjustment. The basic building costs are for a 12 pound live load usually built where snow load is not a design factor. A 20 pound live load building is usually found in light snow areas, and a 30 pound live load building is usually found in heavy snow areas.
5. Add or subtract alternate costs from pages 195 to 197. Alternate costs reflect the difference between unit costs of components that are included in the basic building square foot costs and the unit costs of alternate components. Total alternate cost is found by multiplying the alternate cost by the area or number of alternate components.
6. Add the appropriate costs from pages 198 to 199, "Alternate Costs for Steel Buildings." These costs reflect the in-place cost of components that are not included in the basic square foot cost but are often found as part of steel buildings. The cost of items that alter or replace a portion of the building reflect the net added cost of the component in-place. The cost of the item that is replaced has been deducted from the total cost of the additive components. No further deduction is necessary.
7. Add the cost of heating and air conditioning systems, fire sprinklers, plumbing, additional electrical work, alternate interior partitions and offices, dock height additive costs, loading docks or ramps, paving and curbing, and miscellaneous yard improvements. See the section beginning on page 201.

High Profile Rigid Steel Buildings

Square Foot Cost

Width	Length 60'	80'	120'	Over 140'
20'	$23.17	$21.73	$21.73	$20.94
30'	21.53	20.50	19.17	17.68
40'	19.89	18.60	17.22	16.71
50' to 70'	17.58	15.79	14.66	14.56
80' to 140'	—	14.86	14.11	13.33

Deduct $.36 per square foot for low profile buildings.

Live Load Adjustment Factors (multiply base cost by factor shown)

Building Width	High Profile 20 lb. Live Load 20' Bays	High Profile 20 lb. Live Load 24' Bays	High Profile 30 lb. Live Load 20' Bays	High Profile 30 lb. Live Load 24' Bays	Low Profile 20 lb. Live Load 20' Bays	Low Profile 20 lb. Live Load 24' Bays	Low Profile 30 lb. Live Load 20' Bays	Low Profile 30 lb. Live Load 24' Bays
20'	1.00	1.01	1.05	1.07	1.00	1.00	1.03	1.04
24'	1.01	1.02	1.06	1.08	1.00	1.00	1.07	1.09
28'	1.01	1.02	1.06	1.08	—	—	—	—
30'	1.01	1.02	1.06	1.08	1.01	1.02	1.09	1.10
32'	1.01	1.02	1.07	1.09	1.01	1.02	1.10	1.11
36'	1.01	1.02	1.08	1.10	1.01	1.02	1.10	1.11
40'	1.02	1.03	1.08	1.11	1.01	1.02	1.10	1.11
50'	1.04	1.06	1.10	1.15	1.02	1.03	1.10	1.12
60'	1.05	1.07	1.12	1.18	1.03	1.04	1.15	1.18
70'	1.07	1.09	1.14	1.20	1.03	1.04	1.15	1.19
80'	1.08	1.10	1.15	1.21	1.04	1.05	1.16	1.20
90'	1.09	1.10	1.16	1.22	1.04	1.05	1.17	1.21
100'	1.09	1.10	1.17	1.23	1.04	1.05	1.17	1.22
110'	1.10	1.11	1.18	1.24	1.05	1.06	1.18	1.24
120'	1.11	1.12	1.19	1.25	1.06	1.07	1.19	1.27
130'	—	—	—	—	1.07	1.08	1.20	1.30
140'	1.13	1.14	1.20	1.27	1.08	1.09	1.22	1.33
150'	—	—	—	—	1.09	1.10	1.24	1.35

Alternate Costs for Steel Buildings

Floors, cost difference per square foot

4" concrete with reinforced mesh and 2" sand fill	0
5" concrete with reinforced mesh and 2" sand fill	+.34
6" concrete with reinforced mesh and 2" sand fill	+.66
6" concrete with 1/2" rebars, 12" x 12" o.c. and 2" sand fill	+1.31
3" plant mix asphalt on 4" untreated rock base	-.44
2" plant mix asphalt on 4" untreated rock base	-.56
Gravel (compacted)	-.98
Dirt (compacted)	-1.40

Roof Covering,
cost difference per square foot

26 gauge galvanized steel	0
24 gauge galvanized steel	+.16
22 gauge galvanized steel	+.56
26 gauge colored steel	+.43
24 gauge colored steel	+.67
.024 aluminum	+.39
.032 aluminum	+.79
.032 colored aluminum	+1.16
No roof cover (just purlins)	-1.37

Wall Covering,
cost difference per square foot

26 gauge colored steel	+.43
24 gauge colored steel	+.66
26 gauge galvanized steel	0
24 gauge galvanized steel	+.19
22 gauge galvanized steel	+.56
26 gauge aluminum and steel	+.36
24 gauge aluminum and steel	+.60
.024 aluminum	+.40
.032 aluminum	+.77
.032 colored aluminum	+1.20
26 gauge galvanized steel insulated panel	+.55
26 gauge colored steel insulated panel	+1.07
26 gauge aluminum steel insulated panel	+.84
26 gauge galvanized steel mono panel	+1.07
26 gauge colored steel mono panel	+1.64
No wall cover (just girts)	-1.54

These costs are to be added to or subtracted from the building square foot costs on page 194.

Open Endwall (includes ridged frame in lieu of post and beam), low profile building

	Cost Deduction Per Each Endwall						
	Eave Heights						
Width	10'	12'	14'	16'	18'	20'	24'
20'	$506	$612	$739	855	—	—	—
24'	739	771	982	1,014	—	—	—
30'	876	1,024	1,215	1,436	—	—	—
32'	940	1,098	1,320	1,520	1,689	1,964	2,787
36'	1,024	1,161	—	—	—	—	—
40'	1,098	1,288	1,552	1,763	1,922	2,270	2,787
50'	1,203	1,436	1,773	2,016	2,354	2,723	3,263
60'	1,362	1,584	1,922	2,249	2,798	3,263	4,107
70'	—	1,753	1,942	2,576	3,104	3,579	4,561
80'	—	1,827	2,249	2,798	3,352	3,833	5,321
90'	—	—	2,418	2,723	3,568	3,960	5,596
100'	—	—	2,038	2,723	3,357	3,980	5,775
110'	—	—	1,879	2,512	3,209	3,928	5,986
120'	—	—	1,679	2,185	3,030	3,928	5,986
130'	—	—	1,415	2,112	2,904	3,801	5,986
140'	—	—	1,056	1,753	2,809	3,548	5,669
150'	—	—	845	1,584	2,407	3,263	5,532

These costs are to be subtracted from the basic building cost.

Open Sidewalls, per linear foot of wall

Eave Height	10'	12'	14'	16'	18'	20'	24'
	$2.83	$3.43	$3.99	$4.55	$5.09	$5.66	$6.82

These costs are to be subtracted from the basic building cost.

Alternate Costs for Steel Buildings

Open Endwall (includes ridged frame in lieu of post and beam), high profile building

Width	Cost Deduction Per Each Endwall - Eave Heights						
	10'	12'	14'	16'	18'	20'	24'
20'	620	653	925	—	—	—	—
24'	718	946	1,163	—	—	—	—
28'	1,055	1,186	1,370	—	—	—	—
30'	1,153	1,262	1,501	—	—	—	—
32'	1,206	1,327	1,578	1,795	2,088	2,349	2,697
36'	1,370	1,479	1,784	—	—	—	—
40'	1,631	1,751	2,023	2,360	2,686	2,925	3,414
50'	2,066	2,349	2,653	2,925	3,187	3,414	4,416
60'	2,697	2,925	3,414	3,644	4,122	4,633	5,612
70'	3,154	3,882	4,002	4,686	5,122	5,612	6,927
80'	3,720	4,393	4,633	5,122	5,612	6,329	7,536
90'	—	5,383	6,090	6,503	6,471	7,699	9,385
100'	—	5,601	6,524	7,036	7,786	8,613	10,308
110'	—	—	8,472	9,385	10,331	10,993	12,886

Ridged Frame Endwall in lieu of post and beam, low profile building

Width	Cost Deduction Per Each Endwall - Eave Heights						
	10'	12'	14'	16'	18'	20'	24'
20'	587	630	675	—	—	—	—
24'	630	663	686	—	—	—	—
28'	675	739	739	—	—	—	—
30'	707	750	794	—	—	—	—
32'	750	816	837	946	1,055	1,120	1,295
36'	892	946	979	—	—	—	—
40'	979	1,055	1,120	1,186	1,305	1,327	1,642
50'	1,327	1,414	1,512	1,598	1,686	1,772	2,066
60'	1,675	1,784	1,871	2,023	2,121	2,142	2,458
70'	2,317	2,370	2,741	2,621	2,718	2,751	2,980
80'	2,621	2,697	2,980	2,980	3,208	3,480	3,664
90'	—	3,361	3,730	3,947	3,947	4,219	4,762
100'	—	4,197	4,513	4,762	5,057	5,219	5,721
110'	—	—	5,231	5,721	5,721	6,014	6,644

Ridged Frame Endwall in lieu of post and beam, high profile building

Width	Cost Deduction Per Each Endwall - Eave Heights						
	10'	12'	14'	16'	18'	20'	24'
20'	686	718	750	750	—	—	—
24'	729	739	750	816	—	—	—
30'	750	772	969	—	—	—	—
32'	946	946	1,000	1,022	1,077	1,088	1,206
36'	1,022	1,055	—	—	—	—	—
40'	1,045	1,120	1,174	1,229	1,295	1,327	1,512
50'	1,446	1,545	1,566	1,653	1,762	1,805	2,088
60'	1,838	1,914	2,023	1,653	2,121	2,197	2,468
70'	—	2,468	2,675	2,741	2,741	2,904	3,230
80'	—	2,980	3,251	3,361	3,480	3,589	3,730
90'	—	2,980	3,251	3,361	3,480	4,557	4,762
100'	—	—	5,002	5,122	5,231	5,318	5,644
110'	—	—	6,188	6,188	6,340	6,448	6,937
120'	—	—	7,220	7,394	7,677	7,862	8,133
130'	—	—	8,133	8,558	8,765	8,373	9,625
140'	—	—	9,820	10,059	10,472	10,505	11,005

Alternate Costs for Steel Buildings

Canopies (sidewall location), cost per linear foot

Type and Size	5'	6'	8'	10'	12'	15'
High profile 20' bays	$31.67	$36.08	$45.17	$52.40	$57.58	$65.23
High profile 24' bays	30.86	34.75	43.44	49.71	54.40	62.29
Low profile 20' bays	24.86	28.60	36.75	43.57	51.18	58.95
Low profile 24' bays	23.90	27.27	35.14	41.71	48.13	56.01
			Cost Per Each Canopy			
Basic end cost	$238.00	$289.00	$400.00	$533.00	$719.00	$919.00

The above costs are for 26 gauge galvanized metal canopy.

Canopies (endwall location), add the following to sidewall location costs above.

Eave Height	10'	12'	14'	16'	18'	20'	24'
Cost	$275.00	$359.00	$359.00	$401.00	$447.00	$471.00	$637.00

Sliding Sidewall Doors, cost per opening, including framed opening.

Single Doors	10'	12'	14'	16'	18'	20'	24'
8' x 9'	$1,066	$1,066	$1,046	$1,183	$1,193	$1,288	$1,288
10' x 9'	1,119	1,193	1,161	1,288	1,331	1,341	1,362
10' x 11'	—	1,161	1,172	1,331	1,362	1,362	1,383
10' x 13'	—	—	1,130	1,341	1,383	1,383	1,404
12' x 9'	1,520	1,573	1,520	1,785	1,731	1,753	1,721
12' x 11'	—	—	1,552	1,743	1,785	1,805	1,805
12' x 13'	—	—	1,520	1,785	1,805	1,816	1,837
12' x 15'	—	—	—	1,773	1,837	1,922	1,974
14' x 9'	1,753	1,711	1,668	1,879	1,922	1,942	1,942
14' x 11'	—	1,858	1,552	1,858	1,942	1,900	2,038
14' x 13'	—	—	1,657	1,974	2,090	1,922	2,070
14' x 15'	—	—	—	1,879	2,080	2,143	2,154
16' x 9'	1,879	2,038	2,016	2,143	2,175	2,175	2,154
16' x 11'	—	1,942	2,038	2,175	2,207	2,227	2,227
16' x 13'	—	—	1,911	2,239	—	—	—

The above costs are for 26 gauge colored sliding. Doors in eave heights listed at the top of each column. If door is in an endwall, use the lowest cost of size desired and add $253.00.

Hollow Metal Walk-Thru Doors, cost per opening including framed opening.

Door Size	1-3/8" Thick				1-3/4" Thick			
	Flush		Half Glass		Flush		Half Glass	
	Galvanized	Colored	Galvanized	Colored	Galvanized	Colored	Galvanized	Colored
2'0" x 6'8" single	$506	$549	$623	$665	—	—	—	—
2'6" x 6'8" single	518	549	623	665	—	—	—	—
3'0" x 6'8" single	506	591	644	739	$739	$823	$845	$898
3'0" x 7'0" single	506	591	644	739	739	823	845	940
3'4" x 7'0" single	—	—	—	—	761	866	887	972
4'0" x 7'0" single	—	—	—	—	845	972	1,004	1,098
6'0" x 6'8" pair	855	1,004	1,046	1,172	1,172	1,331	1,446	1,542
6'0" x 7'0" pair	908	1,077	1,088	1,245	1,257	1,404	1,436	1,615
8'0" x 7'0" pair	—		—	—	1,341	1,488	1,615	1,816

Half glass doors include glazing.

Additives: When walk-thru door is used as a pilot in a sliding door, **add** $81.80 to the costs above.

Heavy duty lockset	Add $114 each	Panic hardware, pair	Add $1,088pair
Door closer	Add $158 each	Weatherstripping, bronze and neoprene	Add $13.50/LF
Panic hardware, single	Add $544 each		

Alternate Costs for Steel Buildings

These costs are to be added to the basic building cost.

Framed Openings, cost per opening

Location	10'	12'	14'	16'	18'	20'	24'
			Eave Height				
Sidewalls	$359	$370	$393	$409	$537	$576	$873
Endwalls	516	544	555	583	620	711	807

These costs are for the standard widths of 8', 10', 12', 14', 16', 18', and 20'. If the width is other than standard, add $84.00 per opening. If there is only one opening per building, add $116.00.

Gutters and Downspouts, cost per linear foot

Eave gutter (4")	
24 gauge galvanized steel	$8.73
24 gauge colored steel	9.07
.032 aluminum	8.01
Valley gutter (6")	
12 gauge galvanized steel	17.36
24 gauge galvanized steel	14.86
Downspouts	
4" x 4" x 26 gauge galvanized steel	6.05
4" x 4" x 26 gauge colored steel	6.44
6" x 6" x 24 gauge galvanized steel	6.72
6" x 6" x 24 gauge colored steel	7.73

Door Hoods, cost each

Width of Door	Galvanized	Colored
12'	$174.00	$196.00
16'	224.00	247.00
20'	257.00	281.00

Louvers, cost each, including screens

Size	Fixed Galvanized	Fixed Color	Adjustable Galvanized	Adjustable Color
3' x 2'	$302	$314	$393	$408
3' x 3'	314	409	420	462
5' x 5'	381	381	588	647

Insulation, cost per square foot of surface area

Size and Type	Roof Thickness 1"	Roof Thickness 1-1/2"	Roof Thickness 2"	Wall Thickness 1"	Wall Thickness 1-1/2"	Wall Thickness 2"
.6 pound density						
White vinyl faced	$1.06	$1.11	$1.16	$1.11	$1.11	$1.33
Colored vinyl faced	1.21	1.28	1.38	1.33	1.38	1.55
Aluminum faced	1.33	1.38	1.55	1.45	1.55	1.67
.75 pound density						
White vinyl faced	1.11	1.16	1.28	1.16	1.28	1.38
Colored vinyl faced	1.33	1.33	1.45	1.38	1.45	1.67
Aluminum faced	1.45	1.45	1.67	1.55	1.67	1.77

Overhangs (sidewall location), cost per linear foot of overhang, 26 gauge galvanized

Size and Type	20' Bays 3'	20' Bays 4'	20' Bays 5'	24' Bays 3'	24' Bays 4'	24' Bays 5'
No soffit	19.97	24.20	27.91	18.35	21.91	24.49
Soffit	43.04	50.66	56.11	41.69	47.71	56.69
Color, soffit	48.16	50.66	57.70	43.33	50.66	56.11
Color, no soffit	21.32	26.06	30.32	19.97	23.92	28.06
Color, galvanized soffit	45.05	54.20	59.28	43.33	50.66	57.70
Color, color soffit	46.27	53.72	60.61	66.52	50.66	59.31

Overhangs (endwall location), cost per linear foot of overhang, 26 gauge galvanized

Size and Type	3'	4'	5'	6'
No soffit	12.81	21.24	20.77	24.36
Galvanized soffit	31.04	38.35	45.39	51.00
Color, soffit	38.73	37.69	47.24	52.71
Color, no soffit	14.04	18.05	22.89	27.40
Color, galvanized soffit	31.83	41.28	47.24	54.15
Color, color soffit	33.68	41.28	48.59	52.71

Alternate Costs for Steel Buildings

These costs are to be added to the basic building cost

Skylights, Polycarbonate, with curb

	2' x 2'	4' x 4'	4' x 8'
Single dome	$116.00	$405.00	$488.00
Double dome	138.00	449.00	555.00
Triple dome	172.00	506.00	772.00
Double, ventilating	337.00	682.00	865.00

Partitions, Interior, 26 gauge steel, cost per square foot of partition with two sides finished

Painted drywall finish	$1.75
Painted plywood, fire retardant	6.55

Ventilators, round type, includes screen (gravity type), cost each

Diameter	Stationary Galvanized	Stationary Colored	Stationary Aluminum	Rotary Galvanized	Rotary Colored	Rotary Aluminum
12"	$179.00	$190.00	$314.00	$286.00	$314.00	$393.00
16"	252.00	257.00	364.00	359.00	381.00	487.00
20"	302.00	314.00	414.00	437.00	448.00	555.00
24"	325.00	336.00	464.00	487.00	504.00	634.00

Ventilators, ridge type, includes screen and damper

Throat Size	4"	9"	12"	14"
Cost per linear foot, galvanized	$46.95	$69.70	$88.66	$101.78
Cost per linear foot, colored	48.59	76.77	95.84	109.68

Ventilator-Dampers, cost each

Diameter	12"	16"	20"	24"
Damper only	$66.40	$90.00	$94.40	$109.00
Dampers with cords and pulleys	133.00	166.00	205.00	236.00

Continuous Ridge Ventilator, includes screen and damper, cost per 10 foot unit

Size & Type	First 10' Galvanized	First 10' Color	Each Additional 10' Galvanized	Each Additional 10' Color
9" throat	$661	$757	$622	$700
10" throat	757	829	672	757
12" throat	487	1,088	897	952

Steel Sliding Windows, includes glass and screens, cost per window

3' x 2'6"	$353
6' x 2'6"	436
6' x 3'8"	503

Smoke and Heat Vents, automatic control, cost per 10 foot unit

Size & Type	First 10' Galvanized	First 10' Color	Each Additional 10' Galvanized	Each Additional 10' Color
9" ridge mounted	$2,576	$2,723	$2,133	$2,185
9" slope mounted	2,661	2,861	2,185	2,291

Add for operators:

One or two 10 foot sections	Add $83.00
Two to seven 10 foot sections	Add $174.00

Aluminum Industrial Windows, includes glass and screens, cost per window

Size and Type	Project Out	Fixed
3' x 2'6"	$314	$230
2' x 2'8"	336	252
6' x 2'6"	464	309
6' x 3'8"	549	393

Aluminum Sliding Windows, includes glass and screens, cost per window

Width	Height 2'	2'6"	3'	3'6"	4'
2'	$244	$267	$273	$288	$306
3'	267	283	288	306	317
4'	—	299	317	317	350
5'	—	317	327	405	417
6'	—	350	405	433	465

If window is fixed, deduct $4.38 per window. For mullions add $9.02 each.

Commercial and Industrial Structures

Typical Physical Lives in Years by Quality Class

Building Type	Masonry or Concrete 1	2	3	4	5	6	Wood or Wood and Steel Frame 1	2	3	4	5
Urban Stores	70	70	70	60	60	60	70	60	60	60	–
Suburban Stores	70	60	60	60	60	–	60	50	50	45	–
Supermarkets	70	60	60	60	–	–	60	50	50	45	–
Small Food Stores	60	60	60	60	–	–	50	50	45	45	–
Discount Houses	70	60	60	60	–	–	60	50	50	45	–
Banks and Savings Offices	70	70	70	60	60	–	60	60	60	50	50
Department Stores	70	60	60	60	–	–	60	50	50	45	–
General Office Buildings	60	60	60	60	–	–	60	50	50	45	–
Medical-Dental Buildings	60	60	60	50	–	–	60	50	50	45	–
Convalescent Hospitals	60	60	60	50	–	–	55	50	50	45	–
Funeral Homes	70	70	70	60	60	–	60	60	60	50	50
Restaurants	70	60	60	60	–	–	60	50	50	45	–
Theaters	50	50	50	50	–	–	50	45	45	40	–
Service Garages	60	60	50	50	–	–	45	45	40	40	–
Auto Service Centers	50	50	45	45	–	–	–	–	–	–	–
Warehouses	55	55	50	50	–	–	–	–	–	–	–
Light Industrial Buildings	55	55	50	50	–	–	–	–	–	–	–
Factory Buildings	40	40	35	35	–	–	–	–	–	–	–

Service Stations located on main highways or in high land value areas can be expected to become obsolete in 20 years. Other service stations can be expected to become obsolete in 25 years. Reinforced Concrete Department Stores have a typical physical life of 80 years for class one structures and 70 years for lower quality class structures. Steel Buildings have a typical physical life of 50 years

Normal Percent Good Table

Average Life in Years

Age	20 Years Rem. Life	20 Years % Good	25 Years Rem. Life	25 Years % Good	30 Years Rem. Life	30 Years % Good	35 Years Rem. Life	35 Years % Good
0	20	100	25	100	30	100	35	100
1	19	95	24	97	29	98	34	99
2	18	90	23	93	28	96	33	97
3	17	85	22	90	27	93	32	95
4	16	79	21	86	26	90	31	93
5	15	73	20	82	25	88	30	91
6	14	67	19	78	24	85	29	89
7	13	61	18	74	23	82	28	87
8	12	56	17	70	22	79	27	85
9	11	51	16	65	21	75	26	83
10	10	49	15	60	20	72	25	80
11	9	48	14	56	19	68	24	78
12	9	46	13	52	18	65	23	75
13	8	44	12	50	17	61	22	72
14	7	43	11	48	16	58	21	69
15	6	43	10	47	15	54	20	66
16	6	41	9	46	14	50	19	63
17	5	39	8	45	13	49	18	60
18	5	38	8	44	12	48	17	57
19	5	37	7	43	12	47	16	54
20	4	35	7	42	11	47	15	51
21	4	34	6	41	11	46	14	50
22	4	33	6	40	10	45	13	49
23	3	32	5	39	10	44	13	48
24	3	30	5	38	9	43	12	47
25	3	29	5	37	9	43	12	47
26	3	28	4	36	8	42	11	46
27	2	27	4	35	8	41	11	45
28	2	25	4	34	7	40	10	44
29	2	24	4	33	7	39	10	43
30	2	22	3	32	6	38	9	43
31	2	21	3	31	6	37	9	42
32	1	20	3	30	5	36	8	42
33	–	–	3	29	5	35	8	41
34	–	–	3	28	5	35	7	40
35	–	–	2	27	5	34	7	39
36	–	–	2	26	4	33	6	38
38	–	–	2	24	4	32	6	37
40	–	–	2	22	3	30	5	36
42	–	–	1	20	3	28	5	34
45	–	–	–	–	2	26	4	32
48	–	–	–	–	2	23	3	30
52	–	–	–	–	1	20	3	27
56	–	–	–	–	–	–	2	24
62	–	–	–	–	–	–	1	20

Age	40 Years Rem. Life	40 Years % Good	45 Years Rem. Life	45 Years % Good	50 Years Rem. Life	50 Years % Good	55 Years Rem. Life	55 Years % Good
0	40	100	45	100	50	100	55	100
2	38	98	43	99	48	99	53	99
4	36	96	41	97	46	98	51	98
6	34	93	39	95	44	97	49	97
8	32	90	37	93	42	95	47	96
10	30	86	35	90	40	93	45	95
12	28	82	33	87	38	91	43	94
14	26	78	31	84	36	88	41	92
16	24	73	29	81	34	85	39	90
18	22	68	27	77	32	82	37	88
20	20	63	25	73	30	80	35	86
22	18	58	23	69	28	77	33	83
24	17	53	21	65	26	73	31	80
26	15	50	20	60	24	69	29	77
28	14	48	18	55	23	65	27	74
30	13	47	17	50	21	61	26	71
32	11	45	15	49	20	57	24	67
34	10	44	14	48	18	53	22	63
36	9	43	13	47	17	50	21	59
38	8	42	12	46	16	48	19	55
40	8	40	11	44	14	47	18	52
42	7	39	10	43	13	46	17	50
44	6	38	9	42	12	45	16	49
46	6	36	8	41	11	44	15	48
49	5	35	7	40	10	43	14	47
50	5	34	7	38	10	42	13	45
52	4	32	6	37	9	41	12	44
54	4	31	6	36	8	40	11	43
56	3	30	5	35	8	39	10	42
58	3	29	5	34	7	38	9	41
60	3	27	4	32	7	37	9	40
62	2	26	4	31	6	36	8	39
64	2	25	4	30	6	35	8	38
66	2	24	3	29	5	34	7	37
68	2	22	3	28	5	33	7	36
70	2	21	3	27	4	32	6	36
72	1	20	3	25	4	31	6	35
74	–	–	2	24	4	30	5	34
76	–	–	2	23	3	28	5	32
82	–	–	1	20	3	26	4	30
84	–	–	–	–	2	24	4	29
88	–	–	–	–	2	22	3	27
92	–	–	–	–	1	20	2	25
96	–	–	–	–	–	–	2	23
102	–	–	–	–	–	–	1	20

Age	60 Years Rem. Life	60 Years % Good	70 Years Rem. Life	70 Years % Good
0	60	100	70	100
2	58	99	68	99
4	56	99	66	99
6	54	98	64	99
8	52	97	62	98
10	50	96	60	98
12	48	95	58	97
14	46	94	56	96
16	44	93	54	96
18	42	92	52	95
20	40	89	50	94
22	38	87	48	93
24	36	85	46	92
26	34	83	45	91
28	32	81	42	89
30	30	78	40	87
32	29	75	39	85
34	27	72	37	83
36	25	69	35	81
38	24	66	33	79
40	22	63	31	76
42	21	60	30	73
44	20	56	29	70
46	18	52	27	67
48	17	49	26	64
50	16	48	25	61
52	15	47	23	58
54	14	46	22	56
56	13	46	21	54
58	12	45	20	52
60	11	44	19	50
64	10	42	17	48
68	9	40	15	46
72	8	38	13	44
76	7	36	12	43
80	6	35	11	41
86	5	32	9	39
92	4	29	8	36
100	3	25	6	33
108	2	22	4	29
112	1	20	3	27
122	–	–	2	24
130	–	–	1	20

Additional Costs for Commercial and Industrial Structures

Section Contents

Additional Structure Costs

Basements

Cost includes concrete floor and walls, open ceiling, minimum lighting, no plumbing, and no wall finish. Cost per square foot of floor at 12' wall height.

Area	500	1,000	1,500	2,000	3,000	4,000	5,000	7,500	10,000	15,000	20,000
Cost	36.97	33.08	28.86	26.10	25.38	22.17	21.35	20.63	18.15	17.31	16.29

Add or subtract the amount listed in the table below to the square foot of floor cost for each foot of wall height more or less than 12 feet.

Wall Height Adjustment
Square Foot Area

Area	500	1,000	1,500	2,000	3,000	4,000	5,000	7,500	10,000	15,000	20,000
Cost	2.46	1.82	1.61	1.30	1.07	.98	.92	.72	.58	.45	.43

Canopies, per S.F. of canopy area

Light frame, flat roof underside, plywood and paint or cheap stucco supported by wood or light steel posts, 4" to 6" wood fascia.	$15.24 to $15.30
Average frame, underside of good stucco, flat roof, cantilevered from building or supported by steel posts, 6" to 12" metal fascia.	$16.60 to $21.64
Same as above but with sloping shake or tile roof.	$18.39 to $24.06
Corrugated metal on steel frame.	$14.71 to $20.18

Canopy Lights, per S.F. based on one row of lights for 5' canopy

Recessed spots (1 each 6 linear feet)	$2.42
Single tube fluorescent	4.42
Double tube fluorescent	6.20

Public Address Systems, speakers attached to building. No conduit included.

Base cost, master control	$750 to $1,450
Per indoor speaker	155
Per outdoor speaker	310

Sound Systems, cost per unit

Voice only, per unit	$85 to $140
Music (add to above), small units	85 to 110
Music (add to above), large units	110 to 330

Larger installations cost the least per unit.

Docks for unloading trucks. Cost per S.F. of dock at 4' height

L x W	10'	20'	30'	50'	100'	200'
5'	27.21	24.16	22.06	19.96	18.49	17.13
10'	24.16	21.01	18.39	16.49	15.24	14.71
15'	21.23	17.86	15.76	13.66	12.61	11.66
20'	19.01	15.24	13.24	12.19	11.34	10.72

Cost includes compacted fill, three concrete walls, concrete floor, and rock base.

Intercommunication Systems

Master control, base cost	$1,430 to $4,300
Cost per station	110 to 165
Nurses call system, per station	165 to 300

Security Systems

Control panel	$710 to $1,148
Each door or window secured	71 to 136
Heat detectors, each	98 to 180
Smoke detectors, each	190 to 273
Motion detectors, each	273 to 327

Loading Ramps, cost per S.F. of ramp

Size	
Under 300 S.F.	$8.30
Over 300 S.F.	7.77

Dock Levelers and Lifts, cost each

Dock leveler, manual	$6150
Dock leveler, mechanical	3,000
Powered platform dock leveler	
6' x 6' recessed	2,670
6' x 8' recessed	3,025
Electro-hydraulic, pit recessed scissor lift	
5,000 lb. capacity, 6' x 8'	8,005
10,000 lb. capacity, 8' x 10'	14,170
20,000 lb. capacity, 8' x 12'	23,375

Additional Structure Costs

Doors, with hardware

Exterior, commercial, cost per door	
Glass in wood (3' x 7')	$790 to $1,230
1/4" plate in aluminum (3' x 7')	1,280 to 2,070
Automatic, tempered glass (3' x 7')	6,070 to 9,200
Residential type (3' x 7')	300 to 550
Interior, commercial and industrial, cost per S.F.	
Hollow core wood	$12.80 to $13.77
Solid wood	13.26 to 17.14
Hollow core metal	28.87 to 34.17
Fire, cost per S.F.	
Hollow metal, 1-3/4"	$45.59 to $60.28
Metal clad, rolling	38.15 to 61.00
Metal clad, swinging	53.45 to 76.19

Elevators, Freight, Electric, car and equipment, per shaft, car speed in feet per minute, 2 stop

Capacity	50 to 75	100 to 150	200
2,500 lbs	$50,300	—	—
3,000	53,050	$60,770	$70,250
3,500	63,860	63,660	71,490
4,000	59,540	66,230	76,632
5,000	63,654	72,615	81,990
6,000	70,150	79,516	88,480
8,000	79,520	88,480	100,430
10,000	91,270	100,425	115,160

For manual doors, **add** $4,560 for each stop. For power operated doors, **add** $6,900 for each additional stop. **Add** $6,900 per car for self-leveling cars. **Add** for double center opening doors, per stop $290. **Add** for deluxe cab (raised panel, interior, drop ceiling) $3,450.

Elevators, Freight, Hydraulic, 100 F.P.M.

Shaft, car, machinery	
2,500 lb. capacity	$44,780
6,000 lb. capacity	75,230
Cost per stop	
Manual doors	7,650
Automatic doors	16,660

Roll-Up Metal Warehouse Door with chain operator, cost each

10' x 10'	$1,760
12' x 12'	2,340
14' x 14'	2,660
Fusible link (add to above)	495
Motor controlled (add to above)	250

Draperies, cost per square yard of opening

	54" high	68" high	96" high
Minimum	$20.00	$20.00	$22.00
Good quality	45.00	46.00	54.00
Better quality	55.00	60.00	74.00

Escalators, cost per flight up or down

Total Rise	32" W	40" W	48" W
10' to 13'	$104,400	$116,060	$115,750
14'	107,650	112,000	120,420
15'	111,498	116,200	125,825
16'	114,800	123,487	126,150
17'	117,800	126,150	127,205
18'	120,425	129,550	127,730
19'	124,770	130,500	128,585
20'	129,540	134,750	133,150
21'	133,260	136,350	136,030

Add for glass side enclosure: $12,450- $14,650.

Dumbwaiters, includes door, traction type

	1st Two Stops	Add. Stops
Hand operated, 25 fpm (no doors)		
25 lb.	$1,875 to $3,400	$1,430
75 lb.	2,513 to 4,220	1,430
Electrical, with machinery above, floor loading		
100 lb., 50 fpm	$7,000 to $11,270	$2,690
300 lb., 50 fpm	7,272 to 11,260	2,680
500 lb., 50 fpm	7,660 to 12,130	2,680
500 lb., 100 fpm	11,260 to 18,140	(5 stop)

Elevators, Passenger, Electric, car and machinery cost, per shaft

Capacity	200 F.P.M., 10 Stops	350 F.P.M., 5 Stops	500 F.P.M., 5 Stops
2,000 lbs.	$80,000	$85,400	$184,700
2,500 lbs.	98,000	91,400	190,800
3,000 lbs.	99,000	96,600	193,000
3,500 lbs.	99,500	102,000	194,100
4,000 lbs.	100,100	108,500	195,600
4,000 lbs (Hospital)	101,500	110,700	201,300

Add for each additional stop: 200 or 350 F.P.M. units, $5,700; 500 F.P.M. units, $9,400. Deduct for multi-shaft applications, $2,700 to $5,700 per additional shaft. Add for rear-opening door: $8,100 to base cost, plus $5,850 per door.

Additional Structure Costs

Fire Extinguishers, cost each

Fire hose and cabinet	$300 to $600
Extinguisher cabinets	80 to 180
Extinguishers, chemical	65 to 160
Extinguishers, carbon dioxide	180 to 360

Fill, compacted under raised floor, includes perimeter retaining wall but not slab, per C.F.

Up to 10,000 S.F.	$.90 to $1.40
Over 10,000 to 50,000 S.F	.70 to 1.00

Fire Escapes

Type	Unit	Cost
Second story	Each	$3,470 to $4,740
Additional floors	Per story	2,040 to 3,060

Fire Sprinklers, cost per S.F. of area served

Area	Wet Pipe System Normal	Special*
to 2,000	$3.58	$4.42
2,001 to 4,000	2.48	3.58
4,001 to 10,000	2.20	3.04
Over 10,000	1.94	2.76

Area	Dry Pipe System Normal	Special*
to 2,000	$3.87	$4.69
2,001 to 4,000	2.48	3.37
4,001 to 10,000	2.38	3.30
Over 10,000	2.20	3.04

Costs include normal installation, service lines, permit and valves. *Special hazard systems are custom engineered to meet code or insurance requirements and are usually so identified by a metal plate attached to the riser.

Overhead Suspended Heaters, per unit

25 MBTU.	$ 890 to $1,070
50	1,035 to 1,140
75	1,143 to 1,267
100	1,308 to 1,483
150	1,600 to 1,720
200	1,850 to 1,930
250	2,060 to 2,070

Fireplace

	1 Story	2 Story
Freestanding wood burning heat circulating prefab fireplace, with interior flue, base and cap.	$1,350	—
Zero-clearance, insulated prefab metal fireplace, brick face.	1,930	$2,550
5' base, common brick, on interior face.	2,570	2,880
6' base, common brick, used brick, face brick or natural stone on interior face with average wood mantle.	4,000	4,300
8' base, common brick, used brick or natural stone on interior face, raised hearth.	5,600	6,280

Electric Heating Units

Baseboard, per linear foot	$61.71 to $74.97
Add for thermostat	86.29
Cable in ceiling, per S.F.	1.89 to 2.42
Wall heaters, per K.W.	181.60 to 209.00

Heating and Cooling Systems

Cost per S.F. of Floor Area

Type and Use	Heating Only	Heating & Cooling
Urban stores	$2.63 to $3.91	$5.20 to $10.20
Suburban stores	2.63 to 3.91	3.70 to 8.16
Small food stores	1.99 to 3.32	4.39 to 8.67
Supermarkets	1.88 to 3.26	4.08 to 10.13
Discount houses	1.88 to 3.26	4.28 to 7.50
Department stores	2.68 to 3.91	4.28 to 10.30
Reinforced concrete	2.68 to 4.02	5.90 to 12.09
General offices		
Forced air	3.70 to 5.25	8.84 to 13.06
Hot & chilled water	—	10.30 to 17.85
Bank and savings	4.56 to 6.80	5.41 to 14.84
Medical-Dental		
Forced air	4.08 to 6.63	10.20 to 16.83
Hot & chilled water	—	13.82 to 19.53
Convalescent hospitals		
Forced air	4.08 to 6.05	7.55 to 12.09
Hot & chilled water	—	10.18 to 17.34
Funeral homes	3.70 to 5.20	10.18 to 17.34
Theaters	3.70 to 5.20	7.29 to 11.99
Restaurants	3.06 to 4.66	10.08 to 17.95
Industrial buildings	2.19 to 4.56	5.41 to 9.18
Interior offices	3.70 to 6.59	5.91 to 11.22

Use the higher figures where more heating and cooling density is required.

Additional Structure Costs

Kitchen Equipment, cost per linear foot of stainless steel fixture

Work tables	$645 to $780
Serving fixtures	270 to 1,500

Mezzanines, cost per S.F. of floor

Unfinished (min. lighting and plumbing)	19.38 to 23.56
Store mezzanines	32.64 to 41.41
Office mezzanines (without partitions)	35.29 to 45.80
Office mezzanines (with partitions)	45.80 to 71.71

Costs include floor system, floor finish, stairways, lighting, and partitions where applicable.

Seating, cost per seat space

Theater, economy	$125.00
Theater, lodge	230.00
Pews, bench type	60.50
Pews, seat type	85.00

Partitions, cost per S.F. of surface

Gypsum on wood frame, (finished both sides) 2" x 4" wood studs, 24" on center with 1/2" gypsum board, taped, textures and painted.	$4.50
Plaster on wood frame (finished both sides) 2" x 4" wood studs, 24" on center with 2 coats plaster over gypsum lath, painted with primer and 1 coat enamel.	$8.25

Pneumatic Tube Systems

Twin tube, two station system	
2-1/4" round, 500 to 1,500 feet	$16,000 to $29,000
3" round, 500 to 1,500 feet	16,500 to 31,700
4" round, 500 to 1,500 feet	17,200 to 36,500
4" x 7" oval, 500 to 1,500 feet	27,300 to 45,300

Automatic System, twin tube, cost per station	
4" round, 500 to 1,500 feet	$20,300 to $27,000
4" x 7" oval, 500 to 1,500 feet	27,250 to 29,500

Skylights. Plastic Rectangular Domes, cost per unit

	Single Plastic Panel		Double Plastic Panel	
Size	Skylight Only	With 4" or 9" Insulated Curb	Skylight Only	With 4" or 9" Insulated Curb
16" x 16"	$143	$243	$174	$258
16" x 24"	164	275	196	270
16" x 48"	185	317	258	380
24" x 24"	185	295	275	290
24" x 32"	196	317	301	380
24" x 48"	228	359	317	433
28" x 92"	391	581	618	718
32" x 32"	196	317	275	380
32" x 48"	265	370	354	439
32" x 72"	338	523	575	665
39" x 39"	275	370	370	435
39" x 77"	412	581	718	771
40" x 61"	380	523	575	781
48" x 48"	295	433	433	533
48" x 64"	412	644	644	771
48" x 72"	476	739	739	871
48" x 92"	623	798	955	1,215
48" x 122"	882	1,056	1,203	1,404
58" x 58"	476	655	771	892
60" x 72"	597	798	929	1,066
60" x 92"	761	967	1,158	987
64" x 64"	607	771	841	1,333
77" x 77"	871	1,098	1,404	1,538
94" x 94"	1,547	1,805	2,532	2,809

Triple dome skylights cost about 30% more than double dome skylights.

Additional Structure Costs

Plastic Circular Dome Skylights, cost each

	Single Plastic Panel		Double Plastic Panel		Additives	
Size	4" Curb	9" Curb	4" Curb	9" Curb	Ceiling Dome	Wall Liner
30"	$571	$582	$675	$707	$273	$273
36"	582	609	707	762	349	293
48"	772	848	989	1,033	467	349
60"	979	1,022	1,315	1,327	663	359
72"	1,469	1,446	1,947	2,034	870	435
84"	2,012	2,142	2,817	2,968	1,252	512
96"	2,958	3,088	4,122	4,404	1,360	609

The above costs are for single skylights. For three or more, deduct 20%.

Plastic Pyramid Skylights, cost each

		Installed Cost	
Size	Height	2 or Less	3 or More
39" x 39"	34"	$ 929	$ 772
48" x 48"	42"	1,315	1,098
58" x 58"	49"	1,457	1,588

Plastic Continuous Vaulted Skylights, cost per L.F.

Width	Single Panel	Double Panel
16"	$ 98	$142
20"	104	147
24"	119	169
30"	147	202
36"	158	212
42"	169	222
48"	202	245
54"	212	267
60"	222	283
72	245	304
84"	293	397

Ventilators, Roof, Power Type, cost each

Throat Dia.	2 or Less	3 or More	Add for Insulated Curb
6"	$ 424	$ 413	$76
8"	686	620	109
10"	827	767	131
12"	925	892	136
18"	1,012	963	147
24"	1,143	1,077	179
30"	1,958	1,816	190
36"	2,099	1,849	196
48"	4,611	4,187	240

Above costs are for a single-speed motor installation. Dampers and bird screens are included. Add: Explosive-proof units, add $343 each. Two-speed motors, add $597 to $882 each. Plastic coating, depending on size of unit, add $121 to $185 each.

Plastic Ridge Type Skylights, cost per linear foot

Width*	Single Panel	Double Panel
18"	$218	$325
24"	242	393
30"	302	448
36"	381	521
42"	393	627
48"	414	807

*Width is from ridge to curb following slope of roof.

Wire Glass Skylights, Exterior Aluminum Frame, cost each

24" x 48"	$349
24" x 72"	419
24" x 96"	560
48" x 48"	566
48" x 72"	696
48" x 96"	843

Ventilators, Roof, Gravity Type, cost each

Throat Dia.	2 or Less	3 or More	Add for Insulated Curb
8"	$185	$174	$90
12"	202	185	102
18"	311	304	142
24"	370	354	152

Heat and Smoke Vents, cost each

Size	Plastic Dome Lid	Aluminum Covered Lid
32" x 32"	$ 886	$ 1,007
32" x 48"	1,055	1,082
50" x 50"	1,196	1,239
50" x 62"	1,360	1,555
50" x 74"	1,252	1,621
50" x 92"	1,642	1,805
50" x 98"	1,914	1,935
62" x 104"	2,435	2,491
74" x 104"	2,730	3,001

Additional Structure Costs

Walk-In Boxes, cost per S.F. of floor area

Temperature Range	50	100	200	300	400	500	600
Over 45°	129.00	87.00	68.00	58.00	55.00	49.00	48.00
25.50° to 45°	144.00	104.00	79.00	69.00	62.00	58.00	55.00
0° to 25°	172.00	127.00	93.00	79.00	69.00	64.00	61.00
-25.50° to 0°	188.00	152.00	121.00	103.00	87.00	80.00	75.00

Cost Includes: Painted wood exterior facing, insulation as required for temperature, interior plaster, one 4 x 7 door per 300 S.F. of floor area. Costs are based upon 8' exterior wall height. Costs do not include machinery and wiring. Figure refrigeration machinery at $1,600.00 per ton capacity.

Material Handling Systems

Belt type conveyors, 24" wide.	
Horizontal sections, per linear foot	$184.00
Elevating, descending sections, per flight	306.00
Mail conveyors, automatic, electronic.	
Horizontal, per linear foot	1,460.00
Vertical, per 12' floor	18,870.00
Mail chutes, cost per floor, 5" x 14", aluminum	920.00
Linen chutes, 18 gauge steel, 30" diameter, per 10' floor	1,510.00
Disinfecting and sanitizing unit, each	622.00

Display Fronts

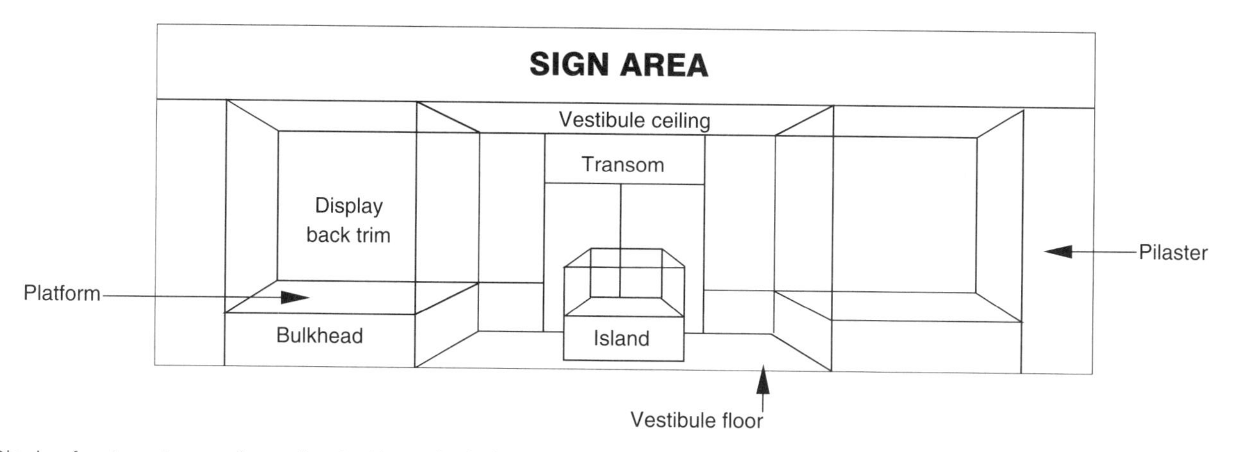

Display front costs may be estimated by calculating the in-place cost of each component or by estimating a cost per linear foot of bulkhead and multiplying by the bulkhead length. This section contains data for both methods. For most fronts, the cost per linear foot method is best suited for rapid preliminary estimates.

Bulkhead length is the distance from the inside of the pilaster and following along the bulkhead or glass to the inside of the opposite pilaster. This measurement includes the distance across entryways.

The cost per linear foot of bulkhead is estimated using the storefront specifications and costs in this section. This manual suggests linear foot costs for four quality types: low cost, average, good, and very good. Costs are related for each quality type in terms of flat or recessed type fronts.

Recessed type fronts include all components described in the specifications. Flat front costs do not include the following components: vestibule floor, vestibule and display area ceiling framing, back trim, display platform, lighting. The cost of automatic door openers is not included in front costs.

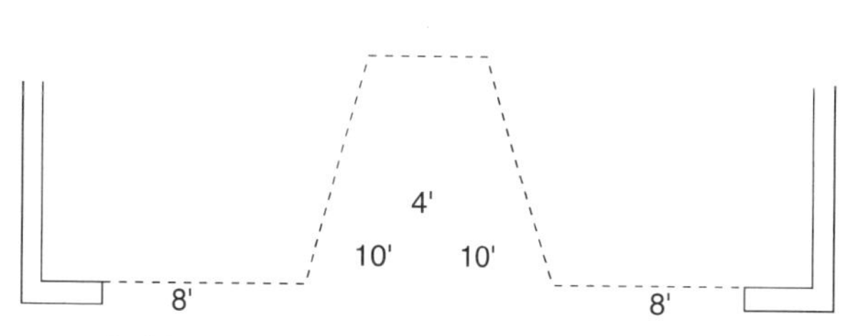

Bulkhead or storefront length 8'+10'+4'+10'+8' = 40'

Display Fronts

Display Front, Best

Display Front, Best

Display Front, Good

Display Front, Good

Display Front, Average

Display Front, Average

Display Front, Low Cost

Display Front, Low Cost

Display Fronts

All front costs are based upon a height of 10 feet from the floor level to the top of the display window. Variations from this standard should be adjusted by using the display window height adjustment costs shown with the front foot costs. These amounts are added or deducted for each foot of variation from the standard.

Bulkhead height variations will not require adjustment. Cost differentials, due to variations in bulkhead height, will be compensated for by equal variations in display window height if overall heights are equal.

Sign areas are based upon 4 foot heights. Cost adjustments are given for flat type and for recessed type fronts. A cost range is given for recessed type fronts because deeply recessed fronts will have lower linear foot costs for sign area components than will moderately recessed fronts because this cost is spread over a longer distance when recesses are deep.

Display island costs are estimated by applying 60 to 80 percent of the applicable linear foot cost to the island bulkhead length. Window height adjustments should be made, but sign height adjustments will not be necessary.

Components	Best	Good	Average	Low Cost
Bulkhead *(0 to 4' high)* (10% of total cost)	Vitrolite domestic marble or stainless steel.	Black Carrara flagstone, terrazzo or good ceramic tile.	Average ceramic tile, Roman brick or imitation flagstone.	Stucco, wood or common brick.
Window Area (30% of total cost)	Bronze or stainless steel. 1/4" float glass with mitered joints.	Heavy aluminum. 1/4" float glass, some mitered joints.	Aluminum. 1/4" float glass.	Light aluminum with wood stops. Crystal or 1/4" float glass.
Sign Area *(4' high)* (10% of total cost)	Vitrolite, domestic marble or stainless steel.	Black Carrara flagstone, terrazzo or good ceramic tile.	Average ceramic tile, Roman brick or imitation flagstone.	Stucco.
Pilasters (5% of total cost)	Vitrolite, domestic marble.	Black Carrara flagstone, terrazzo or good ceramic tile.	Average ceramic tile, Roman brick or imitation flagstone.	Stucco.
Vestibule Floor* (5% of total cost)	Decorative terrazzo.	Decorative terrazzo.	Plain terrazzo.	Concrete.
Vestibule and* Display Area Ceilings (10% of total cost)	Stucco or gypsum wallboard and texture.	Stucco or gypsum wallboard and texture.	Stucco or gypsum wallboard and texture.	Gypsum wallboard and texture.
Back Trim* (5% of total cost)	Hardwood veneer on average frame.	Gypsum wallboard and texture or light frame.	None.	None.
Display Platform Cover* (10% of total cost)	Excellent carpet.	Good carpet.	Average carpet.	Plywood with tile.
Lighting* (10% of total cost)	1 recessed spot per linear foot of bulkhead.	1 recessed spot per linear foot of bulkhead.	1 exposed spot per 2 linear feet of bulkhead.	1 exposed spot per 4 linear feet of bulkhead.
Doors (5% of total cost)	3/4" glass double doors.	Good aluminum and glass double door or single 3/4" glass door.	Average aluminum and glass double door.	Wood and glass.
Note: Use the percent of total cost to help identify the correct quality classification.				
Costs, Flat Fronts	$791.00/linear foot.	$530.00/linear foot.	$342.00/linear foot.	$298.00/linear foot.
Costs, Recessed Fronts	$838.00/linear foot.	$714.00/linear foot.	$406.00/linear foot.	$330.00/linear foot.
Display Window *Adjustment per Foot of Height*	$36.00/linear foot.	$33.00/linear foot.	$31.00/linear foot.	$31.00/linear foot.
Flat Front, Sign Area *Adjustment per Foot of Height*	$35.70/linear foot.	$22.45/linear foot.	$8.36/linear foot.	$3.05/linear foot.
Recessed Front, Sign Area *Adjustment per Foot of Height*	$12.65 to $18.40/ linear foot.	$7.42 to $10.80/ linear foot.	$3.40 to $5.10/ linear foot.	$1.70 to $1.95/ linear foot.

*Not included in flat front costs.

Display Fronts

Lighting, cost per fixture

Open incandescent	$46.40	to	$67.00
Recessed incandescent	67.00	to	135.00
Fluorescent exposed, 4' single	145.00	to	215.00
Fluorescent recessed, 4' single	154.50	to	275.00

Bulkhead Walls, cost per S.F. of wall

Up to 5' high, nominal 6" thick			
Concrete	$12.75	to	$18.87
Concrete block	8.98	to	13.26
Wood frame	5.61	to	9.95

Ceiling, cost per S.F. of floor

Dropped ceiling framing	$1.84	to	$2.60
Acoustical tile on wood strips	2.86	to	3.57
Acoustical plaster including lath	2.65	to	3.88
Gypsum board, texture and paint	2.06	to	3.06
Plaster and paint including lath	2.96	to	4.08

Entrances, cost per entrance

Aluminum and 1/4" float glass			
Single door, 3' x 7'	$1,360	to	$2,240
Double door, 6' x 7'	2,380	to	3,290
Stainless steel and 1/4" float glass			
Single door, 3' x 7'	2,650	to	4,160
Double door, 6' x 7'	3,650	to	5,960
3/4" tempered glass			
Single door, 3' x 7'	2,780	to	3,980
Double door, 6' x 7'	4,650	to	5,880

Includes door, glass, lock handles, hinges, sill and frame.

Exterior Wall Finish, cost per S.F. of wall

Aluminum sheet baked enamel finish	$3.77	to	$7.04
Brick veneer			
Common brick	$7.57	to	$9.87
Roman	9.72	to	14.34
Norman	9.74	to	14.34
Glazed	11.87	to	18.39
Carrara glass			
Black	18.92	to	29.17
Red	20.30	to	30.80
Flagstone veneer			
Imitation	12.85	to	16.52
Natural	19.89	to	31.62
Marble			
Plain colors	29.17	to	38.51
With color variations	53.24	to	62.68
Stucco	2.96	to	3.88
Terrazzo	19.38	to	26.62
Tile, ceramic	12.24	to	17.85

Display Platforms, cost per S.F. of platform area

Framing up to 5' high	$5.92	to	$9.18
Hardwood cover	3.57	to	5.81
Plywood cover	2.40	to	3.26

Glass and Window Frames, per S.F. of glass

Glass only installed			
1/4" float	$7.20	to	$10.00
1/4" float, tempered	9.50	to	12.70
1/4" float, colored	7.40	to	12.70
Store front, 1/4" glass in aluminum frame			
Anodized, 8' high	23.80	to	32.60
Anodized, 6' high	28.80	to	38.20
Anodized, 3' high	35.50	to	44.80
Satin bronze, 6' high	33.40	to	43.80
Satin bronze, 3' high	40.70	to	51.10

Satellite Receiver Systems

Satellite receiver systems are common in mountain and rural areas, where TV reception is limited. They are also often installed in residential or commercial areas, for homes, motels or hotels, restaurants and businesses. The cost of the unit varies with the size of the dish, amplifier, control unit, and type of mounting. A solid dish costs roughly 25% more than a mesh dish. Installed cost for an all-automatic, motorized system, including wiring to one interior outlet:

6' dish	$1,090	to	$1,850
8' dish	1,260	to	2,360
10' dish (mesh)	1,680	to	3,190
10' dish (solid)	2,100	to	4,200
12' dish	2,940	to	3,370
13-1/2' dish (mesh)	3,370	to	3,780

Deduct $420 for manual units.

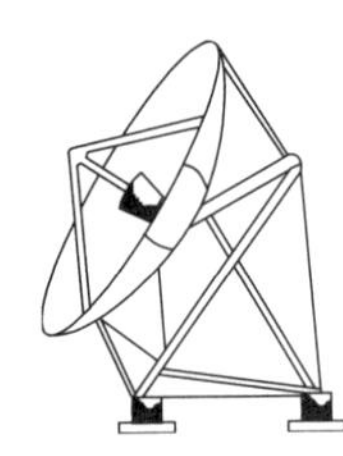

Signs

Lighted Display Signs, Cost per S.F. of sign area

Painted sheet metal with floodlights	$74.80 to $99.20
Porcelain enamel with floodlights	82.30 to 112.80
Plastic with interior lights	93.20 to 145.60
Simple rectangular neon with painted sheet metal faces and a moderate amount of plain letters	99.00 to 177.30
Round or irregular neon with porcelain enamel faces and more elaborate lettering	145.60 to 213.20
Channel letters - individual neon illuminated metal letters with translucent plastic faces, (per upright inch, per letter)	12.30 to 18.50

All of the above sign costs are for double-faced signs. Use 2/3 of those costs for single-faced signs. Sign costs include costs of installation and normal electrical hookup. They do not include the cost of a post. If signs are mounted on separate posts, post mounting costs must be added. These costs are for custom-built signs (one-at-a-time orders). Mass-produced signs will have lower costs.

Post Mounting Costs for Signs

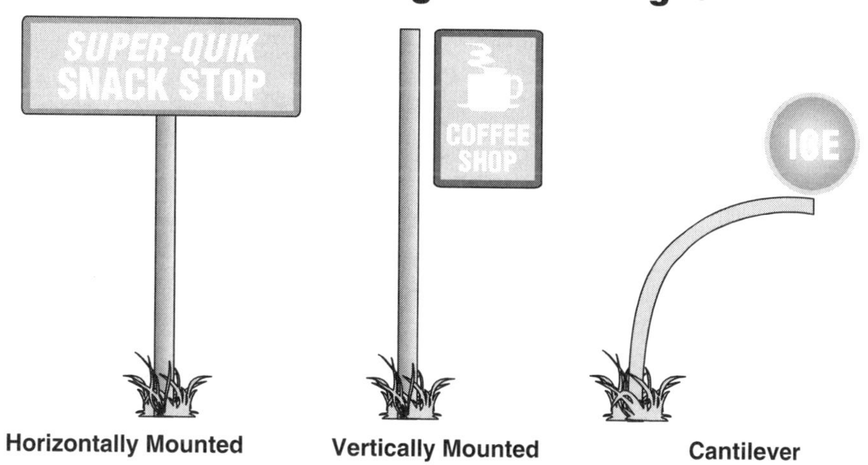

Horizontally Mounted | Vertically Mounted | Cantilever

Post Mounting Costs

	Pole Diameter at Base					
Post Height	4"	6"	8"	10"	12"	14"
15	$ 990	$ 1,160	$ 1,720	$2,330	$ 3,620	$3,620
20	1,160	1,400	2,000	2,420	4,300	4,850
25	1,260	1,580	2,070	2,870	4,800	5,290
30	1,460	1,880	2,140	3,140	5,070	5,870
35	—	2,050	2,420	3,420	5,740	6,260
40	—	2,160	2,940	3,700	5,960	7,020
45	—	—	3,450	4,210	6,510	7,270
50	—	—	3,700	4,800	6,930	7,970
55	—	—	—	5,070	7,270	8,670
60	—	—	—	5,360	7,780	9,070
65	—	—	—	—	8,350	9,640

If signs are mounted on separate posts, post mounting costs must be added. Post mounting costs include the installed cost of the post and foundation. On horizontally mounted signs, post height is the distance from the ground to the bottom of the sign. On vertically mounted signs, post height is the distance to the top of the post. For cantilevered posts, use one and one-half to two times the conventional post cost.

All of the above post costs are for single posts. Use 90% of the single post costs for each additional post.

Signs

If signs are mounted on buildings or canopies and if, because of the extra weight of the sign, extra heavy support posts or foundations are required, 125% of the post mounting cost should be used.

Sample calculations:

For example, the cost of a 4' x 25' plastic sign mounted on a 15' by 6" post shared by an adjacent canopy might be estimated as follows:

Sign cost 100 S.F. at $100.00	$10,000
Post cost $1,160 x 1/2	580
Total cost	**$10,580**

If this sign were mounted on a 8" post 20' high above the canopy with extra supports not needed, the cost might be estimated as follows:

Sign cost	$10,000
Post cost $1,960 x 1	2,000
Total cost	**$12,000**

Sign Rotator

Small signs	Less than 50 S.F.	$1,940 to $2,090
Medium signs	50 to 100 S.F.	2,040 to 4,130
Large signs	100 to 200 S.F.	4,080 to 7,500
Extra large signs	Over 200 S.F.	$42.30 per S.F. of sign area

Yard Improvements

Bumpers pre-cast concrete or good quality painted wood.

Typical 4-foot lengths $10.00 to $15.00 per linear foot.

Asphaltic Concrete Paving, cost per square foot

	Under 1,000	1,000 to 2,000	2,000 to 5,000	5,000 to 12,000	12,000 or more
2", no base	$2.41	$1.19	$1.02	$.89	$.81
3", no base	2.66	1.51	1.25	1.14	1.08
4", no base	2.92	1.84	1.58	1.51	1.25
2", 4" base	4.11	2.34	1.90	1.77	1.58
3", 6" base	4.50	2.72	2.34	2.16	1.96

Concrete Paving, small quantities, per S.F.

4" concrete (without wire mesh) no base	$2.32
4" concrete (without wire mesh) 4" base	2.65
5" concrete (with wire mesh) no base	2.94
6" concrete (with wire mesh) no base	3.20
Add for color	.33

Gates, coin or card operated

Single gate, 10' arm	$3,150.00 to $5,250.00
Lane spikes, 6' wide	1,265.00 to 1,580.00

Striping

Parking spaces	
Single line between spaces	$ 6.25 per space
Double line between spaces	9.82 per space
Pavement line marking, 4"	.32 per LF

Curbs, per linear foot

Asphalt 6" high berm	$4.99 to $6.04
Concrete 6" wide x 12" high	7.62 to 9.72
Concrete 6" wide x 18" high	9.46 to 11.82
Wood bumper rail 6" x 6"	8.94 to 10.24

Yard Improvements

Lighting

Fluorescent arm-type fixtures

	One arm unit, pole height			Two arm unit, pole height			
	12 ft.	16 ft.	30 ft.	12 ft.	16 ft.	30 ft.	Add for 3-fixture unit
2 tube, 60"	$1,310	$1,440	—	$1,680	$1,700	—	$680
4 tube, 48"	1,340	1,600	—	1,740	2,030	—	780
4 tube, 72"	1,530	1,700	$1,960	1,890	2,050	$2,340	900
6 tube, 72"	1,700	1,830	2,110	2,100	2,210	2,490	960
4 tube, 96"	1,760	1,900	2,200	2,170	2,340	2,690	1,050
6 tube, 96"	1,820	2,020	2,360	2,210	2,590	2,960	1,140

Mercury vapor

	1-fixture unit, pole height			2-fixture unit, pole height			4-fixture unit, pole height		
	12 ft.	16 ft.	30 ft.	12 ft.	16 ft.	30 ft.	12 ft.	16 ft.	30 ft.
400 watt	$2,660	$2,750	$3,720	$3,730	$4,000	$4,940	$6,710	$7,370	$8,380
1,000 watt	—	3,450	3,870	—	4,670	5,880	—	—	9,710

Incandescent scoop reflector

	1-fixture unit, pole height			2-fixture unit, pole height			
	12 ft.	16 ft.	30 ft.	12 ft.	16 ft.	30 ft.	Add for 3-fixture unit
500 watt	$790	$930	$1,190	$1,050	$1,120	$1,450	$270
1,000 watt	990	1,120	1,440	1,330	1,390	1,720	340

Mall and garden lighting

	1-fixture unit, pole height		
	3 ft.*	5 - 7 ft.	10 ft.
Incandescent 60-200 watt	$690	$ 1,000	$1,050
Mercury vapor 100-250 watt	780	1,280	2,700

*Area lights usually found in service stations.

Chain Link Fences, 9 gauge 2" mesh with top rail line posts 10' on center, per linear foot.

Height	Under 150 LF	150 to 1,000 LF	1,000 to 2,000 LF	Over 2,000 LF	3 strands barbed wire	Add for line posts 8' or less o.c.	Deduct For: No top rail	Deduct For: #11 gauge wire
4'	$10.54	$8.76	$8.34	$7.92	$.91	$1.23	$1.87	$1.06
5'	11.07	9.18	8.76	8.34	.91	1.38	1.87	1.11
6'	11.61	9.61	9.18	8.71	.91	1.49	1.87	1.14
8'	12.65	10.56	10.04	9.49	.91	1.62	1.87	1.23
10'	14.56	12.13	11.51	10.86	.91	1.73	1.87	1.44
12'	16.88	14.04	13.30	12.56	.91	1.87	1.87	1.67

Add for gates, per square foot of frame area $4.31 to $5.47

Add for gate hardware per walkway gate $21.85 to $27.54
per driveway gate $43.15 to $53.04

Drainage, per linear foot, runs to 50'

6" non-reinforced concrete pipe	$5.62	to	$6.51
8" non-reinforced concrete pipe	7.60	to	9.11
10" non-reinforced concrete pipe	8.37	to	15.71
12" non-reinforced concrete pipe	10.51	to	19.56

The cost varies with the depth of the pipe in the ground.

Drainage Items

Size and Type	Cost
Catch basin 4' x 4' x 4' deep	$1,930.00 ea.
For each additional foot in depth, add	104.00 ea.
Drop inlets	469.20 to 780.00 ea.
Manholes 4' diameter x 6' deep	1,410.00 ea.
For each additional foot in depth, add	130.00 per ft.

Wood Fences, per linear foot

4' high Redwood board	$16.01
6' high Redwood board	20.50
4' basketweave	18.97
6' basketweave	19.69
4' board and batt	10.40
6' board and batt	13.26
3' split rail	5.40
4' split rail	6.47
3' two rail Redwood picket	9.44
5' three rail Redwood picket	11.42
Typical 3' wide gate, 4' to 6' high	96.90

Agricultural Structures Section

Section Contents

General Purpose Barns

Quality Classification

General Purpose Barn, Class 1

General Purpose Barn, Class 3

Component	Class 1 Good Quality	Class 2 Average Quality	Class 3 Low Quality
Foundation (20% of total cost)	Continuous concrete.	Concrete or masonry piers.	Redwood or cedar mudsills.
Floor (5% of total cost)	Concrete.	Dirt, leveled & compacted.	Dirt, leveled & compacted.
Wall Structure (25% of total cost)	Good wood frame, 10' eave height.	Average wood frame, 10' eave height.	Light wood frame, 10' eave height.
Exterior Wall Cover (25% of total cost)	Good wood siding, painted.	Standard gauge corrugated iron, aluminum or average wood siding.	Light aluminum or low cost boards.
Roof Construction (9% of total cost)	Medium to high pitch, good wood trusses.	Medium to high pitch, average wood trusses.	Medium to high pitch, 2" x 4" rafters 24" to 36" o.c. or light wood trusses.
Roof Cover (5% of total cost)	Wood shingles.	Standard gauge corrugated iron or aluminum.	Light aluminum.
Electrical (8% of total cost)	Four outlets per 1,000 S.F.	Two outlets per 1,000 S.F.	None.
Plumbing (3% of total cost)	Two cold water outlets.	One cold water outlet.	None.

Note: Use the percent of total cost to help identify the correct quality classification.

Square Foot Area

Quality Class	1,000	2,000	3,000	4,000	5,000	6,000	7,000	8,000	9,000	10,000	11,000
1, Good	30.63	28.23	26.14	25.09	24.05	23.00	22.27	21.75	21.33	20.91	20.54
2, Average	22.73	20.51	19.00	18.25	17.54	16.91	16.43	16.11	15.83	15.58	15.34
3, Low	14.82	12.79	11.87	11.38	11.04	10.82	10.60	10.46	10.32	10.23	10.12

Hay Storage Barns

Quality Classification

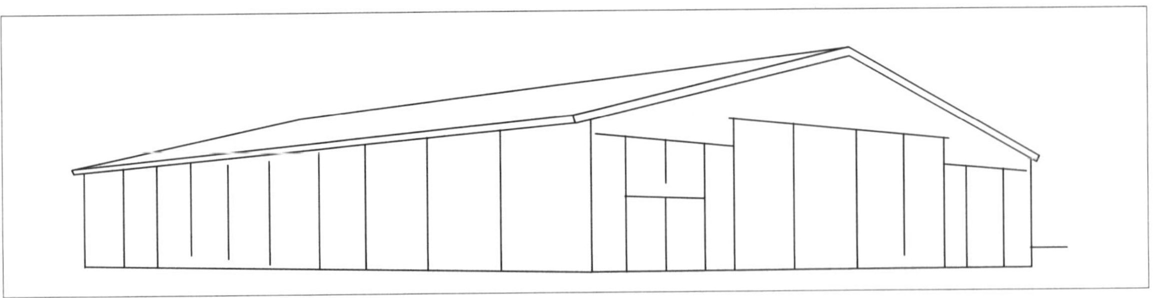

Hay Storage Barn, Class 2
Length between one and two times width.

Component	Class 1 Good Quality	Class 2 Average Quality	Class 3 Low Quality
Foundation (25% of total cost)	Continuous concrete.	Concrete or masonry piers.	Redwood or cedar mudsills.
Floor (5% of total cost)	Concrete.	Dirt, leveled & compacted.	Dirt, leveled & compacted.
Wall Structure (25% of total cost)	Good wood frame, 20' eave height.	Average wood frame, 20' eave height.	Light wood frame, 20' eave height.
Exterior Wall Cover (30% of total cost)	Good wood siding, painted.	Standard gauge corrugated iron or aluminum.	Light aluminum or low cost boards.
Roof Construction (10% of total cost)	Low to medium pitch, good wood trusses.	Low to medium pitch, average wood trusses.	Low to medium pitch, 2" x 4" rafters 24" to 36" o.c. or light wood trusses.
Roof Cover (5% of total cost)	Wood shingles.	Standard gauge corrugated iron or aluminum.	Light aluminum.
Electrical	None.	None.	None.
Plumbing	None.	None.	None.

Note: Use the percent of total cost to help identify the correct quality classification.

Square Foot Area

Quality Class	1,000	2,000	3,000	4,000	5,000	6,000	7,000	8,000	9,000	10,000	11,000
1, Good	23.63	21.22	19.65	18.45	17.73	17.14	16.53	16.17	15.81	15.44	15.18
2, Average	14.23	12.73	11.86	11.23	10.73	10.36	10.00	9.73	9.49	9.36	9.24
3, Low	12.54	11.26	10.38	9.74	9.36	9.00	8.73	8.48	8.35	8.23	8.11

Feed Barns

Quality Classification

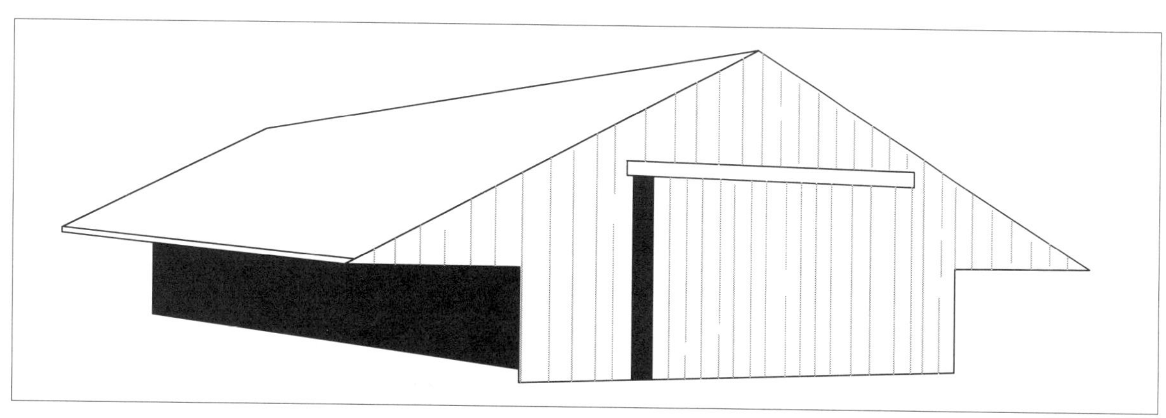

Feed Barn, Class 2

Component	Class 1 Good Quality	Class 2 Average Quality	Class 3 Low Quality
Foundation (20% of total cost)	Continuous concrete.	Concrete or masonry piers.	Redwood or cedar mudsills.
Floor (5% of total cost)	Concrete in center section.	Concrete in center section.	Dirt.
Wall Structure (25% of total cost)	Good wood frame, 8' eave height at drip line.	Average wood frame, 8' eave height at drip line.	Light wood frame, 8' eave eight at drip line.
Exterior Wall Cover (25% of total cost)	Open sides, good siding painted on ends.	Open sides, standard gauge corrugated iron, aluminum or average wood siding on ends.	Open sides and ends.
Roof Construction (9% of total cost)	Medium to low pitch, good wood trusses.	Medium to low pitch, average wood trusses.	Medium to low pitch, 2" x 4" rafters 24" to 36" o.c. or light wood trusses.
Roof Cover (5% of total cost)	Wood shingles.	Standard gauge corrugated iron or aluminum.	Light gauge corrugated iron or aluminum.
Electrical (7% of total cost)	Four outlets per 1,000 S.F.	Two outlets per 1,000 S.F.	None.
Plumbing (4% of total cost)	Two cold water outlets.	One cold water outlet.	None.

Note: Use the percent of total cost to help identify the correct quality classification.

Square Foot Area

Quality Class	1,000	2,000	3,000	4,000	5,000	6,000	7,000	8,000	9,000	10,000	11,000
1, Good	13.97	13.15	12.78	12.67	12.43	12.36	12.29	12.18	12.06	11.95	11.81
2, Average	12.36	11.68	11.35	11.18	11.05	10.98	10.91	10.86	10.79	10.73	10.67
3, Low	7.84	7.34	7.21	7.10	7.03	6.98	6.91	6.81	6.76	6.71	6.65

Shop Buildings

Quality Classification

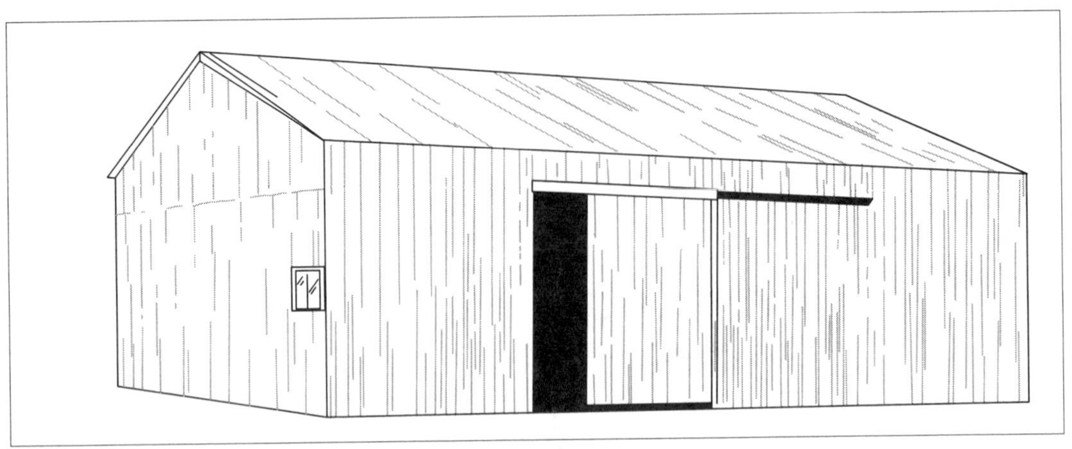

Shop, Class 2
Length between one and two times width.

Component	Class 1 Good Quality	Class 2 Average Quality	Class 3 Low Quality
Foundation (20% of total cost)	Continuous concrete.	Light concrete.	Light concrete.
Floor (5% of total cost)	Concrete.	Concrete.	Concrete.
Wall Structure (20% of total cost)	Good wood frame, 15' eave height.	Average wood frame, 15' eave height.	Light wood frame, 15' eave height.
Exterior Wall Cover (25% of total cost)	Good wood siding, painted.	Standard gauge corrugated iron, aluminum or average wood siding.	Light aluminum or low cost boards.
Roof Construction (5% of total cost)	Low to medium pitch, good wood trusses.	Low to medium pitch, average wood trusses.	Low to medium pitch, 2" x 4" rafters 24" to 36" o.c. or light wood trusses.
Roof Cover (5% of total cost)	Wood shingles.	Standard gauge corrugated iron or aluminum.	Light gauge corrugated iron or aluminum.
Electrical (5% of total cost)	Four outlets per 1,000 S.F.	Two outlets per 1,000 S.F.	Two outlets per 1,000 S.F.
Plumbing (7% of total cost)	Two cold water outlets.	One cold water outlet.	None.
Doors (5% of total cost)	One drive-thru door per 1,000 S.F. plus one walk-thru door.	One average sliding or swinging door per 2,000 S.F.	One light sliding or swinging door per 2,000 S.F.
Windows (3% of total cost)	Five percent of floor area.	None or few low cost.	None.

Note: Use the percent of total cost to help identify the correct quality classification.

Square Foot Area

Quality Class	1,000	1,500	2,000	2,500	3,000	4,000	5,000	6,000	8,000	10,000
1, Good	24.72	22.56	21.33	20.50	19.91	18.93	18.33	17.85	17.24	16.77
2, Average	20.10	18.34	17.36	16.61	16.09	15.37	14.86	14.47	13.97	13.62
3, Low	15.96	14.56	13.68	13.16	12.79	12.16	11.77	11.53	11.16	10.77

Machinery and Equipment Sheds

Quality Classification

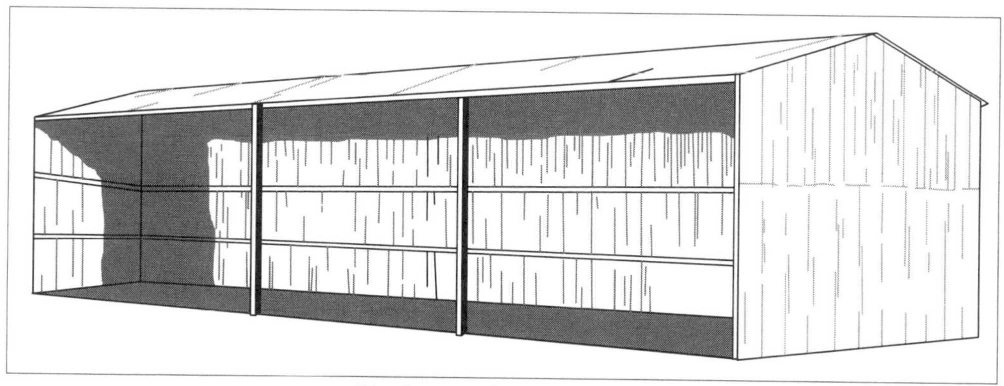

Equipment Shed, Class 3
Usually elongated, width between 15 and 30 feet, any length

Component	Class 1 Good Quality	Class 2 Average Quality	Class 3 Low Quality
Foundation (22% of total cost)	Continuous concrete.	Concrete or masonry piers.	Redwood or cedar mudsills.
Floor (5% of total cost)	Concrete.	Concrete.	Dirt, leveled & compacted
Wall Structure (25% of total cost)	Good wood frame, 10' eave height.	Average wood frame, 10' eave height.	Light wood frame, 10' eave height.
Exterior Wall Cover (30% of total cost)	Good wood siding, painted.	Standard gauge corrugated iron, aluminum or average wood siding.	Light aluminum or low cost boards.
Roof Construction (10% of total cost)	Low to medium pitch, gable or shed type, good wood framing.	Low to medium pitch, gable or shed type, average wood framing.	Low to medium pitch, shed type, light wood framing.
Roof Cover (5% of total cost)	Wood shingles.	Standard gauge corrugated iron or aluminum.	Light aluminum.
Electrical (3% of total cost)	Four outlets per 1,000 S.F.	Two outlets per 1,000 S.F.	None.

Note: Use the percent of total cost to help identify the correct quality classification.

All Sides Closed - Square Foot Area

Quality Class	500	1,000	1,500	2,000	2,500	3,000	3,500	4,000	4,500	5,000	6,000
1, Good	19.41	17.02	16.17	15.69	15.44	15.18	15.07	14.96	14.85	14.78	14.73
2, Average	14.97	13.11	12.36	11.99	11.86	11.68	11.54	11.48	11.41	11.35	11.23
3, Low	9.88	8.62	8.23	7.97	7.84	7.79	7.72	7.60	7.55	7.46	7.40

One Side Open - Square Foot Area

Quality Class	500	1,000	1,500	2,000	2,500	3,000	3,500	4,000	4,500	5,000	6,000
1 Good	17.85	15.07	14.23	13.76	13.38	13.15	13.02	12.89	12.78	12.67	12.56
2, Average	13.62	11.48	10.79	10.42	10.16	10.05	9.95	9.86	9.79	9.73	9.67
3, Low	8.73	7.40	6.98	6.71	6.58	6.47	6.40	6.32	6.27	6.20	6.15

Small Sheds

Quality Classification

Usually elongated, width between 6 and 12 feet, any length

Component	Class 1 Good Quality	Class 2 Average Quality	Class 3 Low Quality
Foundation (25% of total cost)	Continuous concrete.	Concrete or masonry piers.	Redwood or cedar mudsills.
Floor (5% of total cost)	Concrete.	Boards.	Dirt, leveled & compacted.
Wall Structure (25% of total cost)	Good wood frame, 8' eave height.	Average wood frame, 8' eave height.	Light wood frame, 8' eave height.
Exterior Wall Cover (30% of total cost)	Good wood siding, painted.	Standard gauge corrugated iron, aluminum or average wood siding.	Light aluminum or low cost boards.
Roof Construction (10% of total cost)	Low to medium pitch, gable or shed type, good wood framing.	Low to medium pitch, gable or shed type, average wood framing.	Low to medium pitch, shed type, light wood framing.
Roof Cover (5% of total cost)	Wood shingles.	Standard gauge corrugated iron or aluminum.	Light aluminum.
Electrical	None.	None.	None.

Note: Use the percent of total cost to help identify the correct quality classification.

All Sides Closed - Square Foot Area

Quality Class	50	60	80	100	120	150	200	250	300	400	500
1, Good	24.49	22.31	19.65	18.11	17.02	16.03	14.96	14.23	13.86	13.27	12.89
2, Average	18.12	16.61	14.62	13.48	12.60	11.86	11.18	10.61	10.29	9.86	9.61
3, Low	12.03	10.46	9.74	8.93	8.43	7.91	7.34	7.03	6.81	6.58	6.40

One Side Open - Square Foot Area

Quality Class	50	60	80	100	120	150	200	250	300	400	500
1, Good	18.33	17.24	15.69	14.73	13.86	13.02	12.01	11.34	10.79	10.14	9.59
2, Average	13.72	12.97	11.78	10.98	10.42	9.73	9.04	8.48	8.12	7.62	7.24
3, Low	8.87	8.35	7.60	7.10	6.71	6.27	5.83	5.45	5.25	4.88	4.62

Pole Barns

These prices are for pole barns with a low pitch corrugated iron or aluminum covered roof supported by light wood trusses and poles 15' to 20' o.c. The gable end is enclosed and the roof overhangs about 2' on two sides. Wall height is 18 feet. Where sides are enclosed, the wall consists of a light wood frame covered with corrugated metal.

All Sides Open - Side Length

End Width	34	51	68	85	102	119	136	153	170	187
20	7.12	6.85	6.71	6.66	6.61	6.54	6.53	6.52	6.51	6.50
25	6.68	6.42	6.31	6.24	6.20	6.15	6.14	6.13	6.12	6.11
30	6.38	6.13	6.04	5.96	5.93	5.90	5.88	5.86	5.85	5.85
35	6.14	5.93	5.83	5.74	5.70	5.68	5.67	5.66	5.65	5.65
40	6.06	5.83	5.71	5.65	5.62	5.60	5.58	5.53	5.53	5.52
45	5.94	5.71	5.62	5.53	5.50	5.47	5.46	5.45	5.43	5.43
50	5.80	5.59	5.46	5.41	5.37	5.36	5.33	5.32	5.32	5.31
60	5.77	5.54	5.43	5.37	5.34	5.32	5.31	5.30	5.29	5.29
70	5.71	5.48	5.39	5.34	5.31	5.29	5.28	5.25	5.24	5.24
80	5.68	5.46	5.36	5.31	5.28	5.25	5.21	5.21	5.20	5.18

Ends and One Side Closed, One Side Open - Side Length

End Width	34	51	68	85	102	119	136	153	170	187
20	12.83	11.31	10.54	10.13	9.87	9.67	9.53	9.40	9.33	9.28
25	11.73	10.35	9.67	9.28	9.00	8.84	8.74	8.62	8.53	8.47
30	11.02	9.70	9.06	8.72	8.45	8.29	8.18	8.09	8.01	7.95
35	10.46	9.23	8.62	8.26	8.04	7.87	7.78	7.71	7.61	7.55
40	10.08	8.90	8.32	7.97	7.77	7.60	7.49	7.41	7.33	7.28
45	9.81	8.65	8.06	7.76	7.51	7.39	7.27	7.20	7.13	7.08
50	9.53	8.37	7.83	7.51	7.29	7.17	7.08	6.98	6.91	6.88
60	9.28	8.15	7.62	7.29	7.13	6.98	6.88	6.79	6.72	6.69
70	9.06	7.97	7.44	7.16	6.98	6.84	6.70	6.65	6.60	6.53
80	8.85	7.73	7.27	7.00	6.78	6.66	6.55	6.50	6.44	6.38

All Sides Closed - Side Length

End Width	34	51	68	85	102	119	136	153	170	187
20	15.38	13.82	13.05	12.58	12.28	12.05	11.89	11.78	11.69	11.60
25	13.75	12.35	11.64	11.24	10.94	10.76	10.62	10.48	10.42	10.35
30	12.59	11.32	10.69	10.30	10.07	9.88	9.75	9.66	9.58	9.51
35	11.85	10.64	10.03	9.68	9.43	9.27	9.15	9.07	8.98	8.93
40	11.25	10.14	9.55	9.21	8.98	8.84	8.72	8.62	8.53	8.48
45	10.79	9.69	9.14	8.83	8.62	8.45	8.34	8.25	8.19	8.12
50	10.37	9.32	8.79	8.47	8.25	8.13	8.03	7.94	7.85	7.81
60	9.90	8.87	8.38	8.10	7.88	7.76	7.67	7.58	7.50	7.45
70	9.60	8.62	8.14	7.84	7.65	7.50	7.42	7.32	7.27	7.23
80	9.27	8.29	7.84	7.58	7.35	7.25	7.17	7.08	7.03	6.99

Side sheds tying into one side of a pole barn are priced as follows. The shed consists of one row of poles 14' to 16' high, spaced 15' to 20' o.c. A light wood truss covered with a low pitch sheet metal roof spans the distance between the poles and the barn side. If the sides are open, the cost will be between $5.57 and $7.42 per square foot of area covered. If all sides are enclosed with sheet metal and a light wood frame, the square foot cost will be $8.65 to $11.35.

Low Cost Dairy Barns

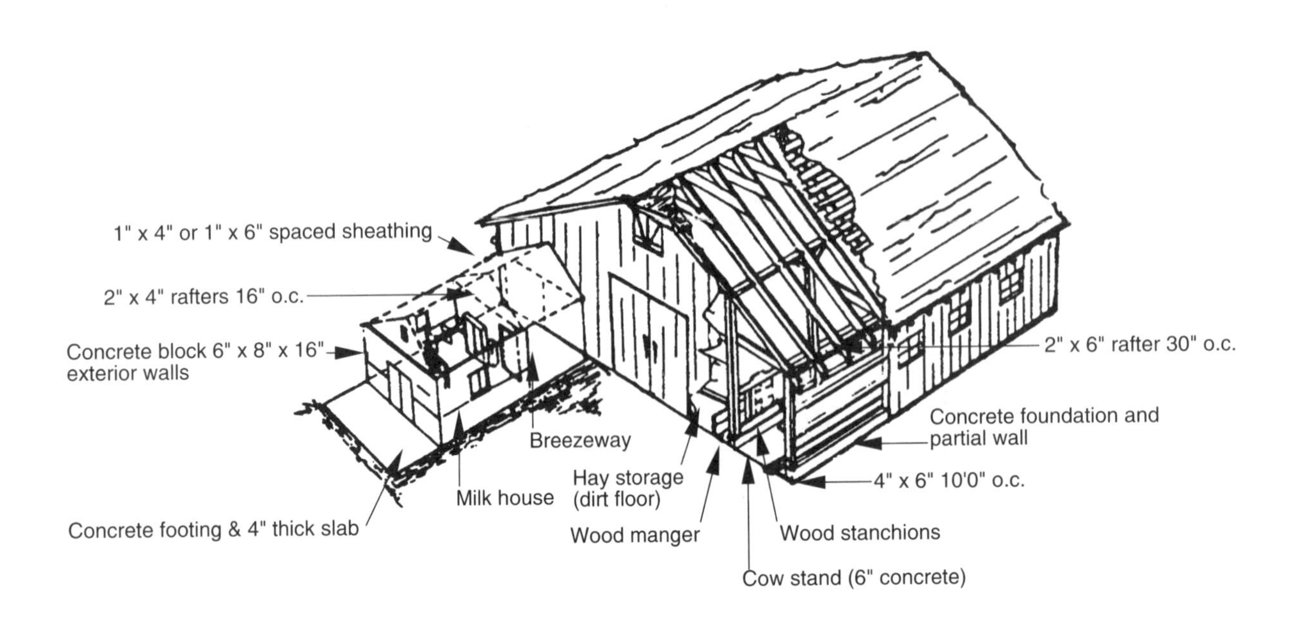

	Milk House	Dairy Barn
Foundation (20% of total cost)	Concrete.	Light concrete.
Floors (15% of total cost)	Concrete slab.	Concrete cow stands.
Walls (30% of total cost)	6" or 8" concrete block 36" high 2" x 4" @16" o.c. framing above.	Box frame, 4" x 6", 10' o.c.
Roof (15% of total cost)	Average wood frame, corrugated iron or aluminum cover.	Average wood frame, wood shingles, corrugated iron or aluminum cover.
Windows (5% of total cost)	Metal sash or metal louvers, on 5% of wall area.	Barn sash.
Interior (10% of total cost)	Smooth finish plaster.	Unfinished, wood stanchions.
Electrical (3% of total cost)	Minimum grade, fair fixtures.	None.
Plumbing (2% of total cost)	One wash basin.	None.
Square Foot Cost	$38.19 to $47.01 (including breezeway).	$14.71 to $17.53 (exclusive of milking equipment).

Stanchion Dairy Barns

Component parts of this dairy:
A Milking barn
B Feed room
C Milk, wash and equipment room

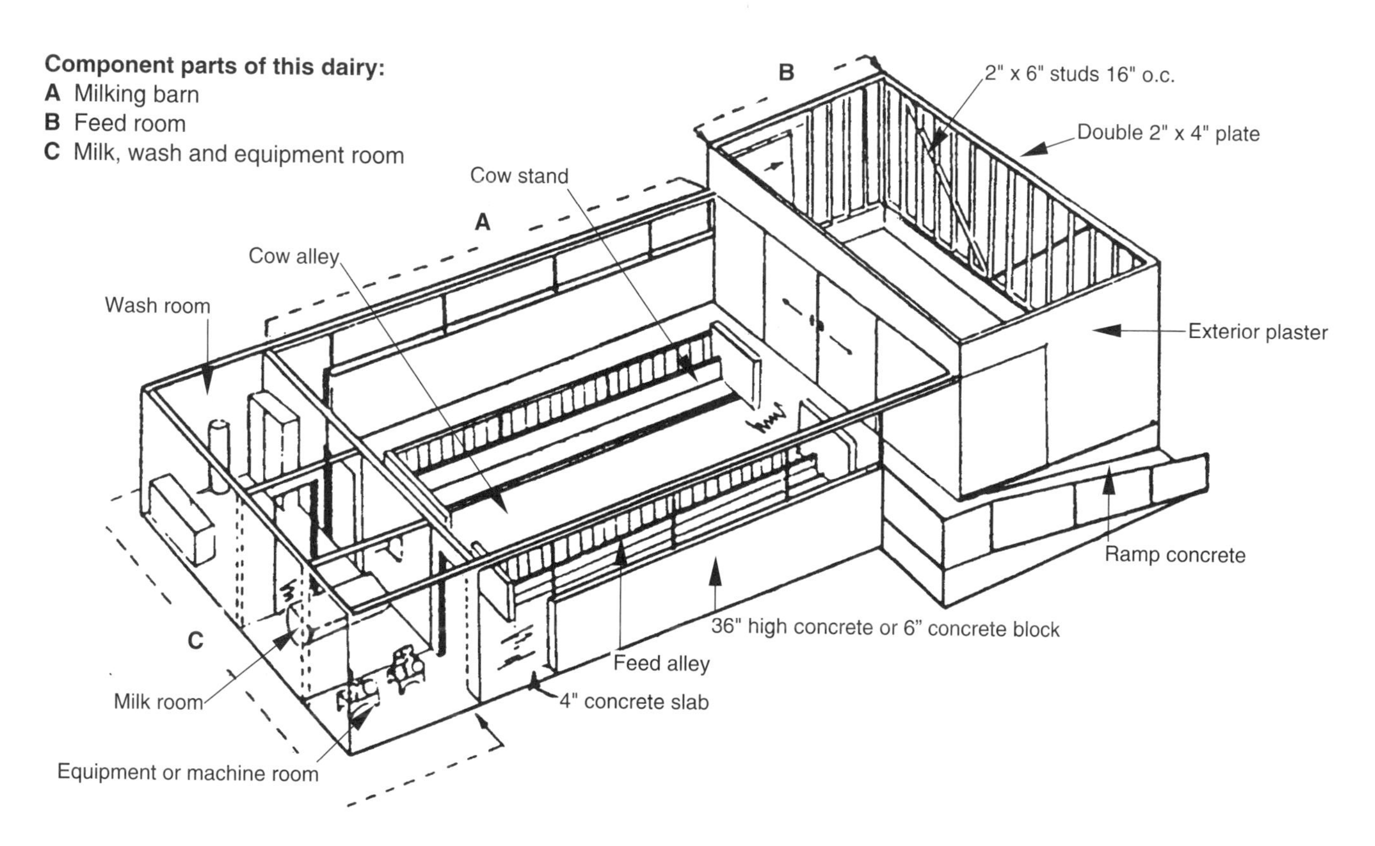

	Feed Room	Milk, Wash and Equipment Room	Milking Barn
Foundation (20% of total cost)	Reinforced concrete.	Reinforced concrete.	Reinforced concrete.
Floors (15% of total cost)	Concrete slab.	Concrete slab.	Concrete, well formed gutters and mangers.
Walls (30% of total cost)	2" x 4" or 2" x 6" @16" o.c. framing.	6" or 8" concrete block 36" high with 2" x 6" @16" o.c. framing above.	6" or 8" concrete block or reinforced concrete 36" high with 2" x 6" @ 16" o.c. framing above.
Roofs (15% of total cost)	Average wood frame, corrugated iron or aluminum cover.	Average wood frame, corrugated iron or aluminum cover.	Average wood frame, iron or aluminum cover.
Windows (5% of total cost)	None.	Metal sash or metal louvers on 10% of wall area.	Metal sash or metal louvers.
Interior (5% of total cost)	Unfinished.	Smooth finish plaster, cove base.	Smooth plaster 36" high, metal stanchions.
Electrical (4% of total cost)	Conduit, average fixtures.	Conduit, average fixtures.	Conduit, average fixtures.
Plumbing (6% of total cost)	None.	One wash basin with floor drains.	Floor drains and hose bibs.
Square Foot Costs	$15.37 to $25.43.	$53.40 to $68.66 (including breezeway).	$28.57 to $33.00 (exclusive of milking equipment).

Walk-Through Dairy Barns

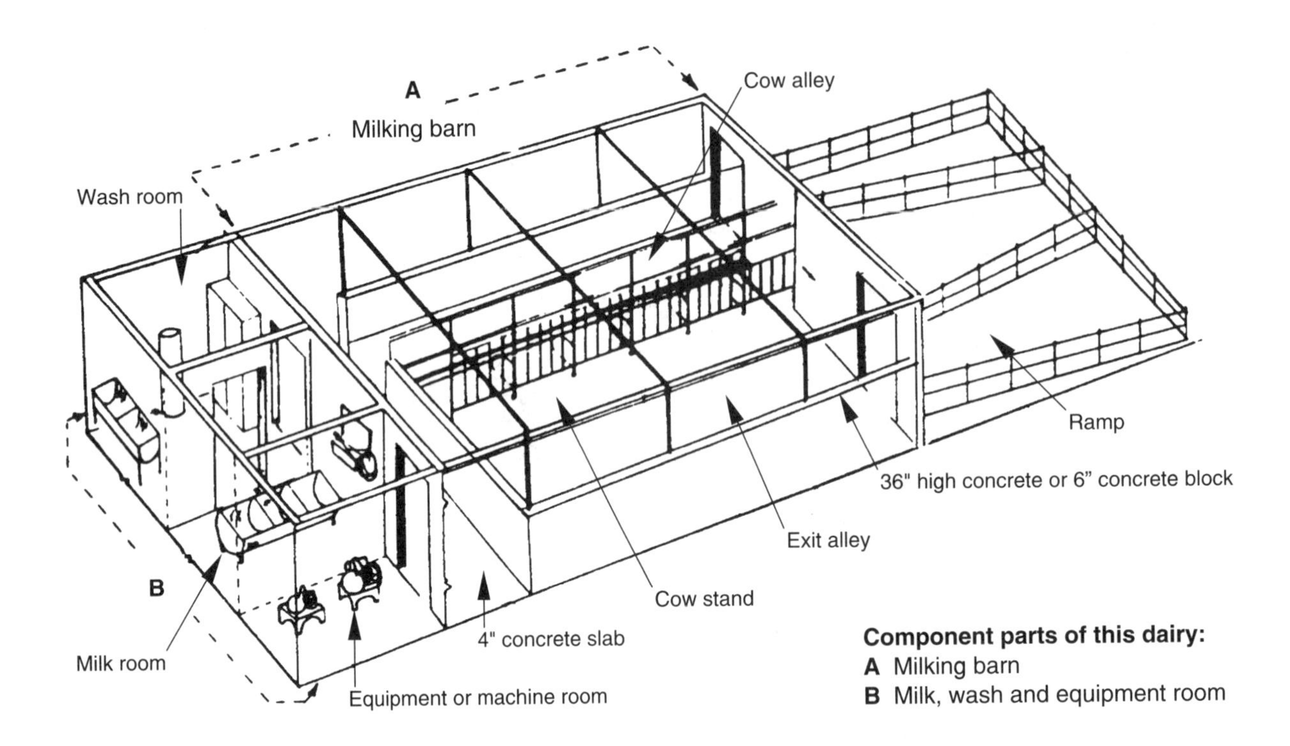

	Milking Barn	Milk, Wash, and Equipment Room
Foundation (20% of total cost)	Reinforced concrete.	Reinforced concrete.
Floors (15% of total cost)	Concrete, well-formed gutters and mangers.	Concrete slab
Walls (30% of total cost)	6" or 8" concrete block or reinforced concrete 36" high with 2" x 6" @16" o.c. framing above or all concrete block.	6" or 8" concrete block 36" high with 2" x 6" @16" o.c. framing above, or all concrete block.
Roofs (15% of total cost)	Average wood frame, corrugated iron or aluminum cover.	Average wood frame, corrugated iron or aluminum cover.
Windows (5% of total cost)	Metal sash or metal louvers.	Metal sash or metal louvers on 10% of wall area.
Interior (5% of total cost)	Smooth plaster 36" high, metal stanchions.	Smooth finish plaster, cove base.
Electrical (4% of total cost)	Conduit, average fixtures.	Conduit, average fixtures.
Plumbing (6% of total cost)	Floor drains and hose bibs.	One wash basin, floor drains.
Square Foot Costs	$33.11 (exclusive of milking equipment).	$34.40 (including breezeway).

Modern Herringbone Barns

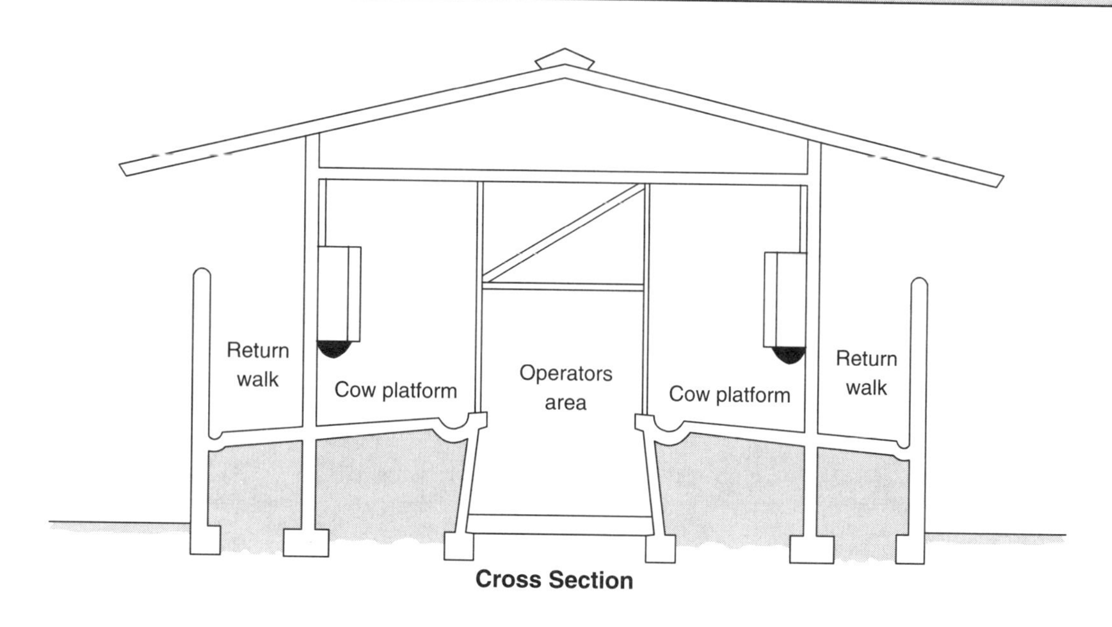

Cross Section

Milking Barn

Foundation (15% of total cost)	Reinforced concrete.
Floors (15% of total cost)	Concrete, wall-formed gutters and mangers.
Walls (30% of total cost)	6" or 8" concrete block or reinforced concrete 36" high with 2" x 6" @16" o.c. framing above or all concrete block.
Roof (15% of total cost)	Average wood frame, corrugated iron or aluminum cover.
Windows (5% of total cost)	Metal sash or metal louvers.
Interior (5% of total cost)	Smooth plaster 36" high.
Electrical (4% of total cost)	Conduit, average fixtures.
Plumbing (6% of total cost)	Floor drains and hose bibs.
Stanchions (5% of total cost)	Metal stanchions.
Cost per SF $35.00	Excluding stalls, feeding system and milking equipment.
Feed System & Stalls	Per double ten barn, $36,900

Milk, Wash, and Equipment Room

	Average Quality	**Good Quality**
Foundation (20% of total cost)	Reinforced concrete.	Reinforced concrete.
Floors (15% of total cost)	Concrete slab.	Concrete slab.
Walls (25% of total cost)	8" concrete block with 2" x 6" @16" o.c framing above or all concrete block.	8" concrete block with 2" x 6" @16" o.c. framing above or all concrete block.
Exterior (10% of total cost)	Stucco or concrete block.	Stucco and masonry veneer.
Roof (10% of total cost)	Corrugated iron or aluminum.	Wood shakes or mission tile, 3' to 4' overhang
Windows (5% of total cost)	Metal sash on 10% of wall area.	Metal sash on 10% of wall area.
Interior (5% of total cost)	Smooth finish plaster, cove base.	Smooth finish plaster, cove base.
Electrical (4% of total cost)	Conduit, average fixtures.	Conduit, average fixtures.
Plumbing (6% of total cost)	One wash basin, one water closet, one lavatory, floor drains.	One wash basin, one water closet, one lavatory, floor drains.
Square Foot Costs	$47.54 to $49.91.	$49.39 to $53.98.

Note: Use the percent of total cost to help identify the correct quality classification.

Miscellaneous Dairy Costs

Holding Corral and Wash Area Costs

Components	Cost
Floor or Ramp	Sloping concrete with abrasive finish, $3.16 to $3.67 per square foot.
Wall	5' to 6' high plastered interior, $42.02 to $52.53 per linear foot.
Metal rail fence	Welded pipe, posts 10' o.c. in concrete, top rail and 3 cables, $10.24 per linear foot.
Cable fence	Pipe posts 10' o.c. set in concrete, $8.40 + $.58 per linear foot per cable.
Gates	54" high, pipe with necessary bracing, $25.63 per linear foot.
Sprinklers	Hooded rainbird, $189.00 each, including plumbing and pump.
Roof	Pipe column supports, average wood frame, corrugated iron or aluminum cover, open sides, $4.47 to $6.57 per square foot.
Typical wash area (without sprinkler)	Sloped concrete floors, 5' concrete block exterior walls, welded pipe interior fences and gates, usual floor drains $23.87/S.F. with roof, $16.30/S.F. without roof.

Milk Line, 3" stainless steel

40-cow conventional barn	$18,800 total $470 per cow
Double ten herringbone barn	$11,340 total $567 per cow
Dual milk pump, add	$2,100
Incline filter, add	$2,300
Incline cold plate, add	$5,150 to $7,500

Refrigeration Compressors

7-1/2 HP	$6,180
10 HP	8,140
15 HP	10,250

Corral Costs

Components	Cost
4" concrete flatwork	$2.65 per square foot.
6" concrete flatwork	$3.20 per square foot.
12" curb	$10.00 per linear foot.
Cable fence	$10.00 per linear foot.
Water tank, concrete typical	$2,840 each.
Steel stanchions	$52.22 to $55.69 each. $20.81 to $27.85 per linear foot.
Steel lockable stanchions	$58.96 to $63.55 each. $23.26 to $27.85 per linear foot.
Light standards	$2,595 each.
Sump pumps, each	
3 HP	$2,783
4 HP	$3,572
10 HP	$5,988
Hay shelters	$5.41 to $6.50 per square foot.
Loafing sheds	$5.68 to $7.04 per square foot.
Typical 80-cow corral	$31,520 to $35,700 total. $394 to $446 per cow.

Refrigerated Holding Tanks

5,100 gallon	$51,700
4,080 gallon	48,700
3,570 gallon	46,700
3,060 gallon	41,500
2,550 gallon	38,300
2,040 gallon	35,700
1,530 gallon	31,600

Vacuum Pumps (air pumps)

Units	Cost
8	$6,180
12	10,100
20	13,970

Poultry Houses

Conventional Lay Cage Type

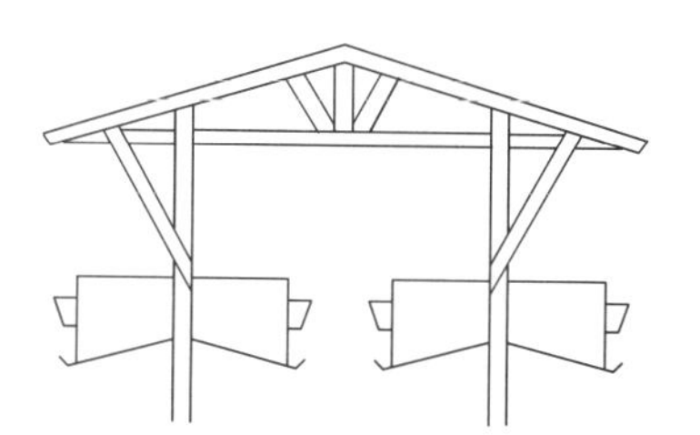

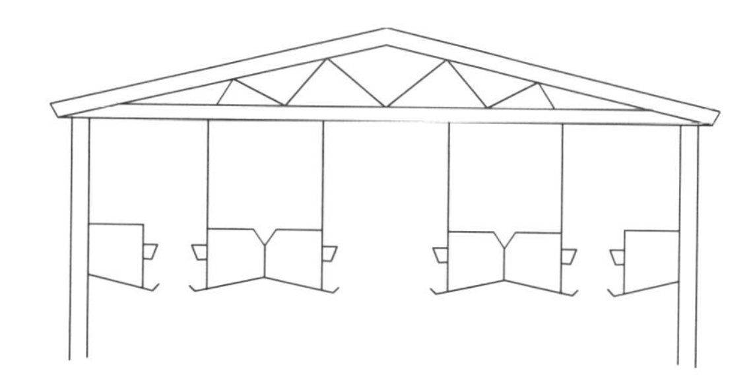

Basic Building

Component	Good Quality	Average Quality	Low Quality
Floors (20% of total cost)	2" concrete.	Dirt with 4' concrete walkways.	Dirt, leveled and compacted.
Foundations (15% of total cost)	Thickened slab.	Concrete piers.	Wood piers.
Frame (20% of total cost)	Light steel or average wood frame.	Average wood frame.	Light wood frame.
Roof Cover (5% of total cost)	Aluminum or corrugated iron.	Light aluminum or composition.	Light aluminum or composition.
Exterior (8% of total cost)	Plywood.	Vinyl curtains.	Wood lath.
Lighting (20% of total cost)	Good system, automatic controls.	Average system, automatic controls.	Minimum system, manual controls.
Plumbing (10% of total cost)	Good system.	Average system.	Fair system.
Insulation (2% of total cost)	Roof only.	None.	None.
Basic Building Cost Per S.F.	$5.84 to $7.04	$3.90 to $5.19	$3.36 to $3.90

Note: Use the percent of total cost to help identify the correct quality classification.

Equipment

Component	Best Quality	Good Quality	Average Quality	Low Quality
Cages (35% of total cost)	12" x 20" double deck.	12" x 20" single deck.	12" x 20" single deck.	12" x 12" single deck.
Water System (20% of total cost)	Automatic cup system.	Automatic cup system.	Simple "V" trough.	Simple "V" trough.
Feed System (30% of total cost)	Automatic system.	V trough.	V trough.	V trough.
Egg Gathering	Manual.	Manual.	Manual.	Manual.
Cooling (15% of total cost)	Pad and fan system.	Pad and fan system.	Simple fogging system.	Simple fogging system.
Cost Per S.F.	$15.76 to $17.54	$9.46 to $11.24	$7.14 to $8.94	$5.99 to $8.30

Note: Use the percent of total cost to help identify the correct quality classification.

Poultry Houses

Modern Controlled Environment Type

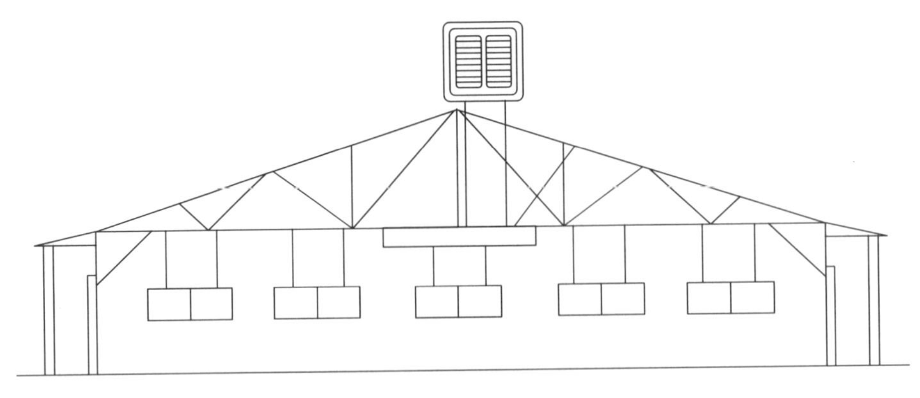

Basic Building	
Foundation (20% of total cost)	Concrete.
Floor (15% of total cost)	3-1/2 concrete slab.
Wall Frame (20% of total cost)	2" x 4" @ 24" o.c.
Roof & Cover (10% of total cost)	Wood trusses with 2" x 4" purlins @24" o.c., corrugated iron or aluminum cover
Exterior (18% of total cost)	Two rib aluminum or corrugated iron.
Interior (7% of total cost)	4" fiberglass roll with aluminum foil facing or 3/4" insulation.
Lighting (5% of total cost)	Fluorescent or good automatic incandescent system.
Plumbing (5% of total cost)	Good basic system.
Basic Building Cost Per S.F.	$9.64 to $11.68

	Equipment	
Component	**Single Deck**	**Stair Step**
Cages (45% of total cost)	12" x 20" single deck.	12" x 20" double deck.
Watering System (30% of total cost)	Automatic cup system.	Automatic cup system.
Feeding System	Manual.	Manual.
Egg Gathering System	Manual.	Manual.
Cooling (25% of total cost)	Evaporative coolers.	Evaporative coolers.
Heating	None.	None.
Building & Equipment Square Foot Cost	$21.23 to $21.84	$22.28 to $24.80

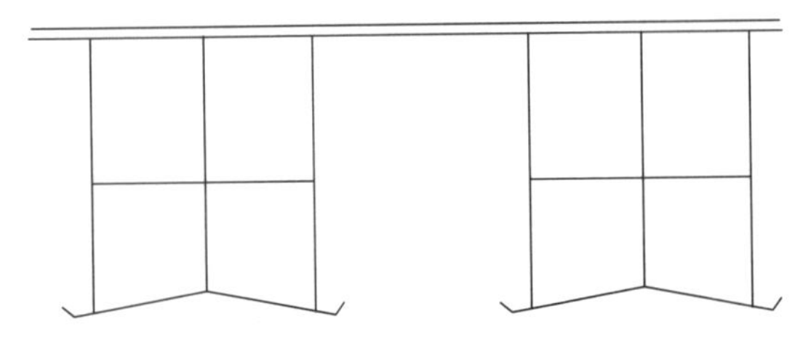

Single-deck cage system

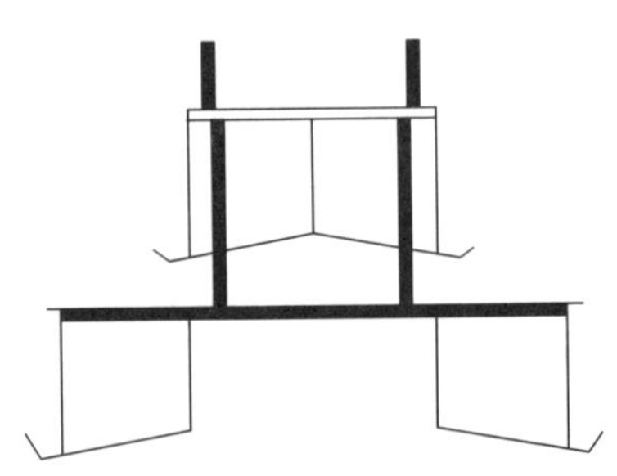

Poultry Houses

High Rise Type

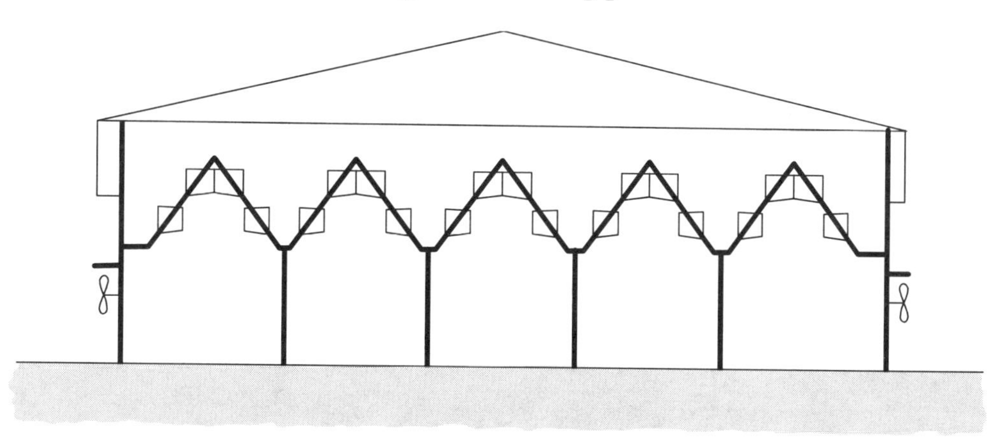

Basic Building	
Foundation (25% of total cost)	Concrete piers.
Floors (5% of total cost)	Dirt, leveled and compacted.
Wall Frame (20% of total cost)	2" x 4" @24" o.c.
Roof & Cover (15% of total cost)	Wood trusses with 2" x 4" purlins @24" o.c., corrugated iron or aluminum cover.
Exterior (18% of total cost)	Two rib aluminum or corrugated iron.
Interior (7% of total cost)	3/4" insulation.
Lighting (5% of total cost)	Fluorescent or good automatic incandescent system.
Plumbing (5% of total cost)	Good basic system.
Basic Building Cost Per S.F.	$10.39 to $12.83

Component	Equipment Flat Deck	Stair Step
Cages (33% of total cost)	12" x 20".	12" x 20".
Watering System (20% of total cost)	Automatic cup system.	Automatic cup system.
Feeding (25% of total cost)	Automatic system.	Automatic system.
Egg Gathering (15% of total cost)	Automatic system.	Automatic system.
Cooling (7% of total cost)	Negative pressure system.	Negative pressure system.
Heating	None.	None.
Building & Equipment Square Foot Cost	$19.54 to $21.43	$23.33 to $25.37

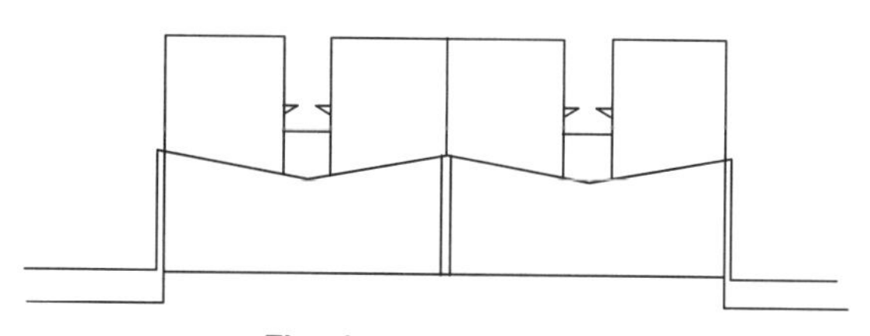

Flat-deck cage system

Poultry Houses

Deep Pit Type

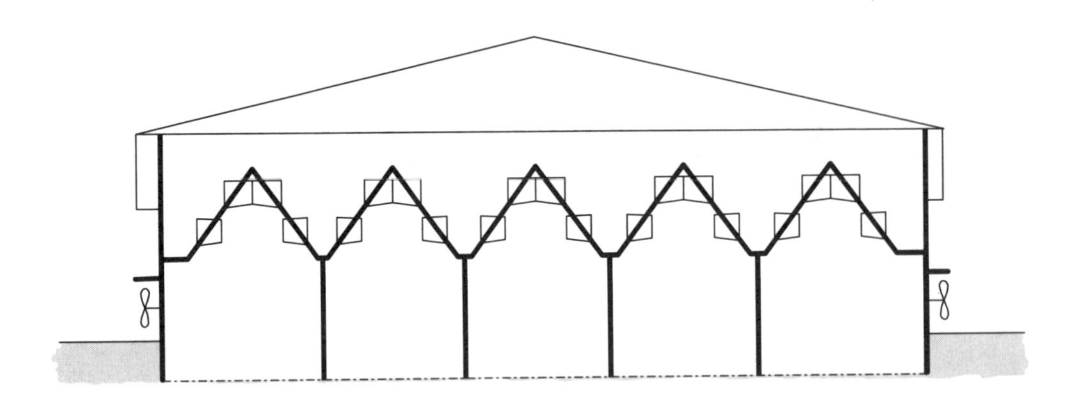

	Basic Building
Foundation (25% of total cost)	Concrete piers.
Floors (15% of total cost)	Concrete with waterproof membrane.
Wall Frame (15% of total cost)	2" x 4" @24" o.c.
Roof & Cover (15% of total cost)	Wood trusses with 2" x 4" purlins @24" o.c., corrugated iron aluminim cover.
Exterior (15% of total cost)	Two rib aluminum or corrugated iron.
Interior (5% of total cost)	3/4" insulation.
Lighting (5% of total cost)	Fluorescent or good incandescent.
Plumbing (5% of total cost)	Good basic system.
Basic Building Cost Per S.F.	$10.92 to $13.15

	Equipment	
Component	**Flat Deck**	**Stair Step**
Cages (33% of total cost)	12" x 20".	12" x 20".
Watering (20% of total cost)	Automatic cup system.	Automatic cup system.
Feeding (25% of total cost)	Automatic system.	Automatic system.
Egg Gathering (15% of total cost)	Automatic system.	Automatic system.
Cooling (7% of total cost)	Negative pressure system.	Negative pressure system.
Heating	None.	None.
Building & Equipment Square Foot Cost	$21.54 to $24.43	$24.43 to $28.95

Poultry Houses

Equipment Costs

Add these costs to the basic building cost

Component	Serving One Row of Cages	Serving Two Rows of Cages
Automatic feeders	$1.73 per bird	$.87 per bird
Automatic egg gathering	1.33per bird	.63 per bird
Automatic water cup system	1.31 per bird	.63 per bird
	4.36 per cup	2.94 per cup
"V" water trough	.30 per bird	.22 per bird
16" feed trough	.38 per bird	.27 per bird

Foggers

1/2" galvanized pipe	$1.89/linear foot
3/4" galvanized pipe	2.00/linear foot
1" galvanized pipe	2.10/linear foot

Roof sprinklers

$2.00 per linear foot

Evaporative coolers

$640 each. $1.58 per S.F. of building

Fans

30"	$595 each
36"	676.26 each
42"	784.38 each

Negative pressure air conditioning system

$1.20 to $1.58 per S.F. of building.

Cooling pads in walls

$.75 per S.F. of surface.

Heating systems

$1,840 per unit.

Cages, 12" x 20" or 18"

$6.52 each. $1.63 per bird.

Migrant Worker Housing

Quality Classification

	Class 1 Best Quality	Class 2 Prefabricated	Class 3 Good Quality	Class 4 Average Quality	Class 5 Low Quality
Slab Foundation (15% of total cost)	Spread footing around perimeter and thickened slab at partitions.	Spread footing around perimeter.	Thickened around perimeter.	Thickened around perimeter.	Thickened around perimeter.
Floor (10% of total cost)	4" concrete slab reinforced.	4" concrete slab reinforced.	4" concrete slab.	4" concrete slab.	4" concrete slab.
Walls (20% of total cost)	Masonry exterior walls, wood frame interior partitions and ceiling.	Metal building, prefabricated.	2" x 4" studs @16" o.c., 2" x 4" stud partitions.	Box construction 4" x 4" at 48" o.c.	Box construction 2" x 4" at 48" o.c.
Exterior Cover (15% of total cost)	Natural blocks.	Prefabricated metal building.	Average grade red-wood board and batt or horizontal siding or stucco finish.	Fair grade redwood or fir board and batt or horizontal board.	Poor grade of red-wood or fir, vertical or horizontal.
Interior Finish (15% of total cost)	Gypsum wallboard.	None.	Gypsum wallboard.	Plywood partitions.	None.
Roof Framing (10% of total cost)	Rafters, collar beams and ceiling joists.	Prefabricated metal building.	Rafters, collar beams and ceiling joists.	Very simple truss.	Rafters and tie at plate line.
Roofing (5% of total cost)	Composition shingles.	Metal.	Aluminum or wood.	Composition or sheet metal.	Composition or used sheet metal.
Doors (3% of total cost)	1 metal door and metal frame each room.	Metal doors.	1 average door each room.	3 or 4 average doors.	2 or 3 cheap doors.
Windows (2% of total cost)	1 steel sash or aluminum window in each room.	Approximately 20% of floor area.	1 steel or aluminum window in each room.	1 window each room.	Few and small.
Electrical (5% of total cost)	1 good light and plug in each room.	1 light, 1 plug for each 300 S.F.	1 light, 1 plug in each room.	1 pull chain light and plug for each 400 S.F.	1 pull chain light for each 500 S.F.

Note: Use the percent of total cost to help identify the correct quality classification.

Migrant Worker Housing, Class 1

Migrant Worker Housing, Class 4

Square Foot Area

Quality Class	400	600	800	1,000	1,200	1,500	2,000	2,500	3,000
1, Best	39.54	36.51	34.87	33.85	33.11	32.35	31.54	31.02	30.63
2, Prefabricated	36.22	33.47	31.98	31.01	30.37	29.63	28.91	28.43	28.07
3, Good	33.82	31.23	29.84	28.96	28.34	27.68	26.99	26.52	26.24
4, Average	28.48	26.30	25.12	24.40	23.87	23.33	22.73	22.32	22.07
5, Low	23.45	21.63	20.68	20.06	19.61	19.19	18.72	18.39	18.17

Costs do not include any plumbing. Add $654 to $784 per fixture.

Miscellaneous Agricultural Structures

Livestock Scales

Type	Size	Capacity	In-Place
Full-capacity beam	16' x 8'	5 ton	$10,330
Printing beam	16' x 8'	5 ton	12,850
Full-capacity beam	22' x 8'	10 ton	13,640
Printing beam	22' x 8'	10 ton	16,470

Additional Costs for Livestock Scales

Types and Size	Cost
Each foot arm is removed from scale	$119/L.F.
Angle iron stock rack for 16' x 8' scale	3,370 ea.
Angle iron stock rack (wood) for 16' x 8'	500 ea.
Angle iron stock rack for 22' x 8' scale	5,870 ea.

Scale pit has 4" concrete walls and slab poured in place. May be poured in or on top of the ground. If on top, compacted ramps and steps to the scale beam are included.

Motor Truck Scales

Five inch reinforced concrete platform. All-steel structure and scale mechanism. Reinforced concrete pit. Motor truck scales are of two general types, the beam type (either manual or type registering) and the full-automated dial type. The construction of both, insofar as the weight-carrying mechanism is concerned is very similar. The method of recording and weight capacity make the cost vary.

Capacity	Platform Size	Total Cost
20 tons	24' x 10'	$21,420
30 tons	34' x 10'	27,050
40 tons	40' x 10'	29,760
50 tons	45' x 10'	33,860
50 tons	50' x 10'	35,280
50 tons	60' x 10'	40,360
50 tons	70' x 10'	44,040

Above costs are for full-capacity beam scales.
Add $850 for the registering type beam.

Septic Tanks

2 bedroom home with 1,200 gallon tank	$1,840
3 bedroom home with 1,500 gallon tank	1,940
4 bedroom home with 2,000 gallon tank	2,060

Bulk Feed Tanks

Size and Type	Cost
3 to 4 ton	$1,331
5 to 6 ton	2,016
6 to 7 ton	2,122
7 to 8 ton	2,554
9 to 10 ton	2,660
10 to 12 ton	3,104
3 to 4 ton twin	1,520
4 to 5 ton twin	1,742
8 ton dairy feed	2,554
12 ton dairy feed	3,399
15 ton dairy feed	3,811

Tanks are equipped with a scissor-type opening chute

Domestic Water Systems Submersible pump, installed at 105' depth excluding well and casing

	Typical Installation		
	1/2 HP	3/4 HP	1 HP
Total cost	$1,953	$2,174	$2,554
Pressure tank size	82 gal.	82 gal.	120 gal.
Cost per ft. above or below 105' depth	$2.44	$2.65	$2.97
	1-1/2 HP	2 HP	3 HP
Total cost	$2,910	$3,707	$4,464
Pressure tank size	220 gal.	220 gal.	315 gal.
Cost per ft. above or below 105' depth	$3.14	$4.55	$6.22

6" wells average $31.52 per foot of depth
8" wells average $38.15 per foot or depth.

Pressure Tank Sizes and Installed Costs

42 gal. 16" dia. x 48" depth	$240 to $ 336
82 gal. 20" dia. x 60" depth	337 to 397
120 gal. 24" dia. x 60" depth	397 to 599
220 gal. 30" dia. x 72" depth	989 to 1,107
315 gal. 36" dia. x 72" depth	1,383 to 1,552
525 gal. 36" dia. x 120" depth	1,795 to 2,030

Typical Physical Lives in years for agricultural structures

Building Type	Good	Average	Low
Barns	40	30	20
Dairy barns	25	25	—
Dairy barns, low cost	—	20	20
Storage sheds	40	30	20
Poultry houses, modern	30	25	—
Poultry houses, conventional	—	25	20

To determine the useful life remaining, use the percent good table for residential structures on page 40.

Military Construction Costs

The Office of the Secretary of Defense has prepared the following square foot guidelines to reflect the cost of military construction for fiscal 2000 and 2001. The costs are based on construction of permanent facilities built on military bases worldwide. Use the "Construction Cost Indices" at the end of this section to adapt the square foot costs to any other area. The "Size Cost Adjustment Chart" should be used to determine the approximate cost of a structure larger or smaller than the typical size shown.

Included in these costs are all items of equipment which are permanently built-in or attached to the structure, including items with fixed utility connections.

The costs include items such as the following:

- Furniture, cabinets and shelving, built-in.
- Venetian blinds and shades.
- Window screens and screen doors.
- Elevators and escalators.
- Drinking water coolers.
- Telephone, fire alarm and intercom systems.
- Theater seats.
- Pneumatic tube systems.
- Heating, ventilating and air conditioning installations.
- Electrical generators and auxiliary gear.
- Waste disposers such as incinerators.
- Food preparation & serving equipment, built-in.
- Raised flooring.
- Hoods and vents.
- Chapel pews and pulpit.
- Refrigerators, built-in.
- Laboratory furniture, built-in.
- Cranes and hoists, built-in.
- Dishwashers.

The costs listed are the estimated contract award costs, excluding contingencies, supervision, and administration. They include construction to the five-foot line only, but do not include the cost of outside utilities or other site improvements.

Figures listed do not include the cost of piles or other special foundations which are considered as an additional supporting item. The cost of air conditioning is included to the extent authorized by the Construction Criteria Manual.

The costs of equipment such as furniture and furnishings which are loose, portable, or can be detached from the structure without tools are excluded from the unit costs. The cost of permanently attached equipment related directly to the operating function for which the structure is being provided, such as technical, scientific, production and processing equipment, is normally excluded from these costs. The following items are excluded from the costs on page 235.

- Furniture, loose.
- Furnishings, including rugs, loose.
- Filing cabinets and portable safes.
- Office machines, portable.
- Wall clocks, plug-in
- Food preparation and serving equipment, including appliances, portable.
- Training aids and equipment, including simulators.
- Shop equipment.
- Bowling lanes, including automatic pin spotting equipment, score table and players' seating.
- Automatic data processing equipment.
- Communications equipment.
- Photographic equipment, portable.
- Any operational equipment for which installation, mounting and connections are provided in building design and which are detachable without damage to the building or equipment.

Estimating Procedure

Determine the area relationship of the proposed building by dividing the gross area by the typical size as shown in the Square Foot Cost Table. Locate the quotient on the Area Relationship scale and trace vertically to the Factor Line, then trace horizontally to the Cost Relationship scale. This value is then multiplied by the unit cost in the Square Foot Cost Table and factored by the Construction Cost Index to determine the adjusted unit cost for the building.

Military Construction Costs

Facilities	Typical size Gross S.F.	Unit cost per S.F. FY 2000	FY 2001
Administrative office			
Multi-purpose	25,000	123.00	125.00
Data processing	21,000	149.00	151.00
Aircraft operations, no tower	20,000	150.00	152.00
Airfield control tower	3,000	215.00	215.00*
Applied instruction building			
General instruction	25,000	123.00	125.00
High Tech (Auto-Aid)	25,000	149.00	151.00
Barracks, dormitory (No Kitchenette equip)	100,000	135.00	137.00
Bowling alley			
8 lanes, pin setting equip. plus auto scoring	7,600	146.00	148.00
Chapel center	15,000	154.00	157.00
Commissary (sales store/equipment)	85,000	122.00	124.00
Enlisted service club	16,000	186.00	189.00
Exchange store, with cafeteria and snack bar	12,000	95.00	97.00
Family housing, U.S.	None	78.90	80.20
Family housing, outside U.S.	None	80.90	82.30
Family support			
Child development center	15,000	139.00	141.00
Education center	10,000	126.00	128.00
Youth center	15,000	120.00	122.00
Family service center	5,000	120.00	122.00
Fire station, community	7,500	135.00	137.00
Hangars maintenance			
Maintenance/Gen. purpose	38,000	130.00	132.00
High bay Maintenance	35,000	169.00	172.00
Medical facility			
Station hospital	None	174.00	177.00
Regional medical center	None	208.00	211.00
Medical clinic	30,000	148.00	150.00
Medical/Dental clinic	50,000	155.00	157.00
Dental clinic	15,000	182.00	185.00

Facilities	Typical size Gross S.F.	Unit cost per S.F. FY 2000	FY 2001
Mess hall, enlisted (with kitchen equipment)	16,000	208.00	211.00
Operations building			
General purpose	13,000	133.00	135.00
Headquarters	11,000	138.00	140.00
Squadron	12,000	145.00	147.00
Physical fitness training center	30,000	143.00	145.00
Recreation center	20,000	120.00	122.00
Reserve facility			
Center	20,000	117.00	119.00
Vehicle maintenance	6,000	121.00	123.00
Satellite Communications center	6,000	440.00	447.00
Library	12,000	126.00	128.00
School for dependents			
Elementary	None	97.30	99.60
Jr. high/middle	None	101.00	103.00
High school	None	105.00	107.00
Shops			
Vehicle maint. (wheeled)	30,000	122.00	124.00
Vehicle maint. (tracked)	25,000	129.00	131.00
Aircraft avionics	23,000	135.00	137.00
Installation maintenance	31,000	106.00	108.00
Parachute and dinghy	8,000	145.00	147.00
Aircraft machine shop	20,000	118.00	120.00
Storage facility			
Cold storage warehouse	6,000	145.00	147.00
Cold storage warehouse, w/processing	11,000	108.00	110.00
General purpose warehouse low bay	40,000	60.00	61.00
General purpose warehouse high bay	100,000	71.00	72.00
High Explosive magazine	5,000	177.00	180.00
Temporary lodging facility	15,000	127.00	129.00
Unmarried officers quarters	44,000	132.00	134.00

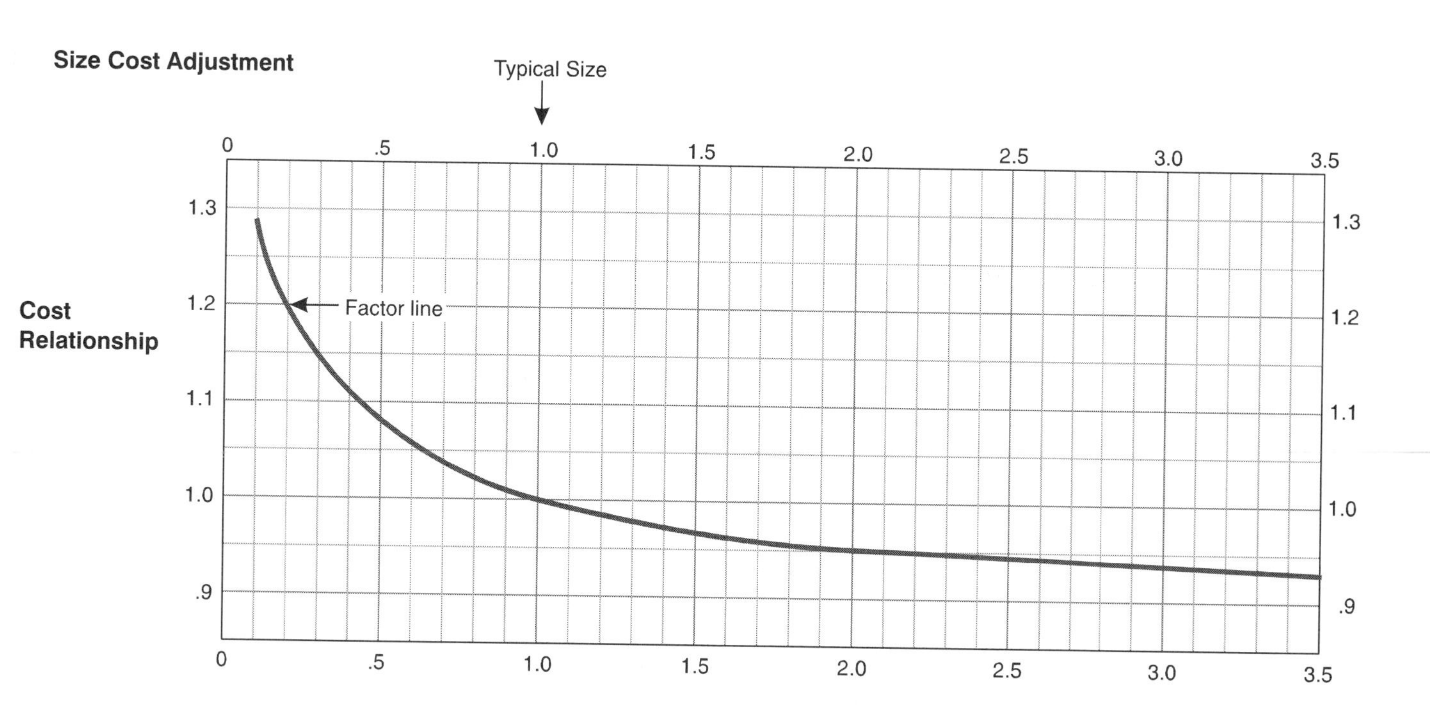

Military Construction Cost Indices

State	Index
Alabama	**0.85**
Mobile	0.84
Montgomery	0.86
Anniston AD	0.84
Fort McClellan	0.84
Fort Rucker	0.81
Maxwell AFB	0.86
Mobile Area	0.84
Redstone Arsenal	0.86
Alaska	**1.60**
Anchorage	1.50
Fairbanks	1.69
Adak NS	2.35
Eielson AFB	1.74
Elmendorf AFB	1.50
Fort Greeley	1.88
Fort Richardson	1.50
Ft Wainwright	1.69
Shemya AFB	2.42
Clear AFB	1.88
Arizona	**0.98**
Flagstaff	0.98
Tucson	0.98
Davis Monthan AFB	0.98
Fort Huachuca	1.00
Luke AFB	0.98
Navajo AD	0.96
Yuma MCAS	1.12
Yuma PG	1.12
Arkansas	**0.87**
Fort Smith	0.85
Pine Bluff	0.89
Fort Chaffee	0.84
Little Rock AFB	0.85
Pine Bluff Arsenal	0.89
California	**1.15**
San Diego	1.10
San Francisco	1.20
Beale AFB	1.25
Camp Pendleton MC	1.10
Centerville Beach	1.10
China Lake NWC	1.26
Edwards AFB	1.19
El Centro NAF	1.14
El Toro	1.05
Ft Hunter Liggett	1.25
Fort Irwin	1.26
Los Angeles area	1.12
March AFB	1.05
McClellan AFB	1.05
Monterey area	1.17
Oakland AB	1.20
Port Hueneme	1.07
Riverbank AAP	1.14
Sharpe AD	1.17
Sierra AD	1.27
Stockton area	1.15
Travis AFB	1.18
Vandenberg AFB	1.20
29 Palms MCB	1.29
Colorado	**1.03**
Colorado Springs	1.03
Denver	1.03
Air Force Academy	1.03
Cheyenne Mountain	1.08
Fitzsimons AMC	1.04
Fort Carson	1.06
Peterson AFB	1.03
Pueblo AD	0.95
Rock Mountain ARS	1.01
Schriever AFS	1.08
Connecticut	**1.05**
Bridgeport	1.05
New London	1.04
New London area	1.05
Stratford Eng. Plt	1.05
Delaware	**1.02**
Dover	1.01
Wilmington	1.03
Dover AFB	1.01
Florida	**0.87**
Miami	0.91
Panama City	0.82
Cape Canaveral	0.92
Eglin AFB	0.82
Homestead AFB	0.91
Jacksonville	0.91
Key West NAS	1.10
McDill AFB	0.88
Orlando	0.84
Panama City	0.82
Pensacola	0.84
Tyndall AFB	0.82
Georgia	**0.85**
Albany	0.79
Atlanta	0.91
Albany	0.79
Fort Benning	0.80
Fort Gillem	0.91
Fort Gordon	0.84
Fort McPherson	0.91
Fort Stewart	0.82
Robins AFB	0.79
Kings Bay	0.96
Moody AFB	0.83
Hawaii	**1.48**
Barbers Pt NAS	1.52
Barking Sands	1.61
Ford Island	1.59
Fort DeRussy	1.45
Fort Shafter	1.47
Hickam AFB	1.45
Honolulu	1.45
Kaneohe Bay NAS	1.52
Kaneohe MCAS	1.52
Pearl Harbor	1.47
Pohakuloa	1.80
Schofield Barracks	1.55
Tripler AMC	1.58
Wheeler AFB	1.55
Idaho	**1.06**
Boise	1.01
Mtn Home	1.11
Mtn Home AFB	1.11
Illinois	**1.15**
Belleville	1.09
Chicago	1.22
Great Lakes NTC	1.26
Rock Island Arsenal	1.06
Savannah AD	1.02
Scott AFB	1.16
Forest Park	1.22
Glenview	1.22
Indiana	**0.99**
Indianapolis	1.01
Logansport	0.96
Crane NWSC	1.01
Fort Ben Harrison	1.04
Grissom AFB	1.04
Jefferson PG	0.93
Iowa	**1.06**
Burlington	1.11
Des Moines	1.00
Iowa AAP	1.19
Kansas	**0.93**
Manhattan	0.96
Wichita	0.91
Fort Leavenworth	1.05
Fort Riley	1.09
Kansas AAP	0.89
McConnell AFB	0.99
Kentucky	**0.94**
Lexington	0.91
Louisville	0.97
Fort Campbell	1.06
Fort Knox	1.05
Lexington AD	0.98
Louisville NAS	0.97
Louisiana	**0.89**
New Orleans	0.95
Shreveport	0.84
Barksdale AFB	0.83
Fort Polk	0.89
Louisiana AAP	0.82
New Orleans AB	0.95
Maine	**1.06**
Bangor	1.07
Portland	1.05
Brunswick	0.95
Cutler Winter Hrbr	0.95
Maryland	**0.89**
Baltimore	0.89
Lexington Park	0.89
Aberdeen PG	0.90
Andrews AFB	0.95
Fort Detrick	0.88
Fort Meade	0.90
Fort Ritchie	0.88
Harry Diamond Lab	0.95
Indian Head	0.91
Patuxent River	0.89
Annapolis	0.90
Bethesda NMC	0.95
Cheltonham	0.95
Chesapeake Beach	0.90
Thurmont	0.88
Massachusetts	**1.12**
Boston	1.15
Fitchburg	1.08
Army Mat & Mech Lab	1.13
Fort Devens	1.15
Hanscomb AFB	1.12
Michigan	**1.17**
Detroit	1.19
Marquette	1.15
Detroit Arsenal	1.18
K. I. Sawyer AFB	1.16
Minnesota	**1.07**
Duluth	1.10
Minneapolis	1.04
Mississippi	**0.87**
Biloxi	0.92
Columbus	0.81
Columbus AFB	0.81
Gulfport	0.92
Keesler AFB	0.89
Meridian	0.95
Missouri	**0.96**
Kansas City	1.00
Sedalia	0.91
Fort Leonard Wood	1.06
Lake City AAP	0.98
St. Louis AAP	1.05
Whiteman AFB	1.01
Montana	**1.13**
Billings	1.15
Great Falls	1.12
Malmstrom AFB	1.12
Nebraska	**0.94**
Grand Island	0.91
Omaha	0.98
Cornhusker AAP	0.89
Offutt AFB	0.98
Nevada	**1.14**
Hawthorne	1.16
Las Vegas	1.12
Fallon	1.16
Hawthorne AAP	1.16
Nellis AFB	1.12
New Hampshire	**1.05**
Concord	1.06
Portsmouth	1.04
Portsmouth area	1.04
New Jersey	**1.18**
Newark	1.19
Trenton	1.17
Bayonne	1.18
Earle	1.21
Fort Dix	1.17
Fort Monmouth	1.17
McGuire AFB	1.17
Picatinny Arsenal	1.23
New Mexico	**1.01**
Alamagordo	1.02
Albuquerque	0.99
Cannon AFB	1.04
Holloman AFB	0.98
Kirtland AFB	0.99
White Sands MR	1.03
New York	**1.23**
Albany	1.05
New York City	1.40
Fort Drum	1.10
Seneca AD	1.09
Staten Island	1.35
USMA	1.28
Watervliet AFS	0.99
Niagara	1.18
Griffiss AFB	1.02
Plattsburg AFB	1.03
North Carolina	**0.85**
Fayetteville	0.85
Greensboro	0.85
Camp Lejeune	0.94
Cherry Point	0.94
Fort Bragg	0.88
New River	0.94
Pope AFB	0.88
Seymour AFB	0.82
Sunny Point	0.91
North Dakota	**1.04**
Grand Forks	1.01
Minot	1.08
Minot AFB	1.08
Ohio	**0.99**
Dayton	0.97
Youngstown	1.01
Ravenna AAP	1.00
Wright-Pttrsn AFB	0.97
Oklahoma	**0.87**
Lawton	0.87
Oklahoma City	0.86
Altus AFB	0.96
Fort Sill	0.91
McAlester AAP	0.84
Tinker AFB	0.86
Vance AFB	0.89
Oregon	**1.15**
Pendleton	1.18
Portland	1.12
Umatilla AD	1.25
Pennsylvania	**1.03**
Philadelphia	1.08
Pittsburgh	0.98
Carlisle Barracks	0.94
Indiantown Gap MR	1.00
Letterkenny AD	0.99
Mechanicsburg	0.94
New Cumbrlnd AD	0.94
Philadelphia Area	1.08
Tobyhanna AD	1.06
Warminster	1.06
Rhode Island	**1.05**
Newport	1.07
Providence	1.03
South Carolina	**0.87**
Charleston	0.89
Columbia	0.85
Beaufort	1.04
Charleston AFB	0.89
Fort Jackson	0.85
Shaw AFB	0.86
South Dakota	**0.95**
Rapid City	0.93
Sioux Falls	0.96
Ellsworth AFB	0.95
Tennessee	**0.87**
Chattanooga	0.83
Memphis	0.91
Memphis NAV	1.00
Volunteer AAP	0.82
Arnold AFB	0.87
Texas	**0.82**
San Angelo	0.81
San Antonio	0.82
Brooks AFB	0.82
Camp Bullis	0.82
Corpus Christi	0.90
Dyess AFB	0.86
Fort Bliss	0.91
Fort Hood	0.86
Fort Sam Houston	0.82
Goodfellow AFB	0.81
Kelly AFB	0.82
Kingville	0.91
Lackland AFB	0.82
Lone Star AAP	0.89
Longhorn AAP	0.80
Randolph AFB	0.82
Red River AD	0.89
Sheppard AFB	0.95
Dallas	0.89
Laughlin AFB	0.89
Reese AFB	0.90
Utah	**1.03**
Ogden	1.04
Salt Lake City	1.03
Dugway PG	1.06
Fort Douglas	1.03
Hill AFB	1.05
Tooele AD	1.06
Vermont	**0.92**
Burlington	0.90
Montpelier	0.93
Virginia	**0.92**
Norfolk	0.92
Richmond	0.92
Dahlgren	0.90
Fort A.P. Hill	0.86
Fort Belvoir	0.95
Fort Eustis	0.92
Fort Lee	0.94
Fort Monroe	0.92
Fort Myer	0.95
Fort Pickett	0.94
Fort Story	0.92
Quantico	0.92
Radford AAP	0.95
Vint Hill Farms	0.90
Langley	0.92
Washington	**1.07**
Spokane	1.06
Tacoma	1.08
Bremerton	1.16
Fairchild AFB	1.02
Fort Lewis	1.08
Indian Island	1.16
McChord AFB	1.08
Silverdale	1.11
Whidbey Island	1.15
Yakima Firing R.	1.08
Everett	1.11
West Virginia	**0.95**
Bluefield	0.93
Charleston	0.97
Sugar Grove	1.43
Wisconsin	**1.14**
Madison	1.15
Milwaukee	1.13
Badger AAP	1.22
Fort McCoy	1.19
Wyoming	**0.99**
Casper	0.98
Cheyenne	1.01
F.E.Warren AFB	1.01
Washington DC	**0.95**
Fort McNair	0.95
Walter Reed AMC	0.95
Bolling AFB	0.95

Index

M

N

O

P

R

S

T

U

V

W

XYZ

Practical References for Builders

Basic Engineering for Builders

If you've ever been stumped by an engineering problem on the job, yet wanted to avoid the expense of hiring a qualified engineer, you should have this book. Here you'll find engineering principles explained in non-technical language and practical methods for applying them on the job. With the help of this book you'll be able to understand engineering functions in the plans and how to meet the requirements, how to get permits issued without the help of an engineer, and anticipate requirements for concrete, steel, wood and masonry. See why you sometimes have to hire an engineer and what you can undertake yourself: surveying, concrete, lumber loads and stresses, steel, masonry, plumbing, and HVAC systems. This book is designed to help the builder save money by understanding engineering principles that you can incorporate into the jobs you bid. **400 pages, 8½ x 11, $36.50**

Construction Estimating Reference Data

Provides the 300 most useful manhour tables for practically every item of construction. Labor requirements are listed for sitework, concrete work, masonry, steel, carpentry, thermal and moisture protection, doors and windows, finishes, mechanical and electrical. Each section details the work being estimated and gives appropriate crew size and equipment needed. Includes a CD-ROM with an electronic version of the book with *National Estimator*, a stand-alone *Windows*™ estimating program, plus an interactive multimedia video that shows how to use the disk to compile construction cost estimates. **432 pages, 11 x 8½, $39.50**

Contractor's Guide to QuickBooks Pro 2001

This user-friendly manual walks you through QuickBooks Pro's detailed setup procedure and explains step-by-step how to create a first-rate accounting system. You'll learn in days, rather than weeks, how to use QuickBooks Pro to get your contracting business organized, with simple, fast accounting procedures. On the CD included with the book you'll find a QuickBooks Pro file preconfigured for a construction company (you drag it over onto your computer and plug in your own company's data). You'll also get a complete estimating program, including a database, and a job costing program that lets you export your estimates to QuickBooks Pro. It even includes many useful construction forms to use in your business.
328 pages, 8½ x 11, $45.25

National Building Assembly Costs

Complete assembly costs for all construction across the 16 CSI divisions. These prices include all the individual line items included in a typical construction assembly, added together to generate an assembly "unit" price. With unit price estimates from the assemblies section, and subcontractor unit prices from the line item cost information, you can put together an estimate for just about any project in minutes. It's far more accurate than a square foot cost estimate, yet takes a fraction of the time a stick-by-stick estimate takes. Includes a CD-ROM with estimating data you can use, and a demo version of WinEst LT, a powerful estimating program for light construction that includes the cost data from Craftsman's *National Construction Estimator*. **320 pages, 8½ x 11, $44.75**

Construction Forms & Contracts

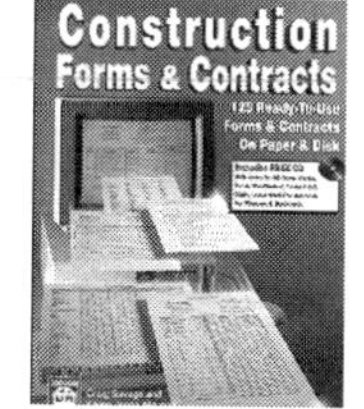

125 forms you can copy and use — or load into your computer (from the FREE disk enclosed). Then you can customize the forms to fit your company, fill them out, and print. Loads into *Word for Windows*™, *Lotus 1-2-3*, *WordPerfect*, *Works*, or *Excel* programs. You'll find forms covering accounting, estimating, fieldwork, contracts, and general office. Each form comes with complete instructions on when to use it and how to fill it out. These forms were designed, tested and used by contractors, and will help keep your business organized, profitable and out of legal, accounting and collection troubles. Includes a CD-ROM for *Windows*™ and Mac. **400 pages, 8½ x 11, $41.75**

National Construction Estimator

Current building costs for residential, commercial, and industrial construction. Estimated prices for every common building material. Provides manhours, recommended crew, and gives the labor cost for installation. Includes a CD-ROM with an electronic version of the book with *National Estimator*, a stand-alone *Windows*™ estimating program, plus an interactive multimedia video that shows how to use the disk to compile construction cost estimates. **616 pages, 8½ x 11, $47.50. Revised annually**

Estimating Tables for Home Building

Produce accurate estimates for nearly any residence in just minutes. This handy manual has tables you need to find the quantity of materials and labor for most residential construction. Includes overhead and profit, how to develop unit costs for labor and materials, and how to be sure you've considered every cost in the job. **336 pages, 8½ x 11, $21.50**

Moving to Commercial Construction

In commercial work, a single job can keep you and your crews busy for a year or more. The profit percentages are higher, but so is the risk involved. This book takes you step-by-step through the process of setting up a successful commercial business; finding work, estimating and bidding, value engineering, getting through the submittal and shop drawing process, keeping a stable work force, controlling costs, and promoting your business. Explains the design/build and partnering business concepts and their advantage over the competitive bid process. Includes sample letters, contracts, checklists and forms that you can use in your business, plus a CD-ROM with blank copies in several word-processing formats.
256 pages, 8½ x 11, $42.00

Steel-Frame House Construction

Framing with steel has obvious advantages over wood, yet building with steel requires new skills that can present challenges to the wood builder. This new book explains the secrets of steel framing techniques for building homes, whether pre-engineered or built stick by stick. It shows you the techniques, the tools, the materials, and how you can make it happen. Includes hundreds of photos and illustrations, plus a CD-ROM with steel framing details. **320 pages, 8½ x 11, $39.75**

Contractor's Guide to the Building Code Revised

This new edition was written in collaboration with the International Conference of Building Officials, writers of the code. It explains in plain English exactly what the latest edition of the *Uniform Building Code* requires. Based on the 1997 code, it explains the changes and what they mean for the builder. Also covers the *Uniform Mechanical Code* and the *Uniform Plumbing Code*. Shows how to design and construct residential and light commercial buildings that'll pass inspection the first time. Suggests how to work with an inspector to minimize construction costs, what common building shortcuts are likely to be cited, and where exceptions may be granted. **320 pages, 8½ x 11, $39.00**

National Painting Cost Estimator

A complete guide to estimating painting costs for just about any type of residential, commercial, or industrial painting, whether by brush, spray, or roller. Shows typical costs and bid prices for fast, medium, and slow work, including material costs per gallon; square feet covered per gallon; square feet covered per manhour; labor, material, overhead, and taxes per 100 square feet; and how much to add for profit. Includes a CD-ROM with an electronic version of the book with *National Estimator*, a stand-alone *Windows*™ estimating program, plus an interactive multimedia video that shows how to use the disk to compile construction cost estimates.
440 pages, 8½ x 11, $48.00. Revised annually